Multi-scale and Multifunctional Coatings and Interfaces for Tribological Contacts

This book covers developments in multi-scale and multifunctional coatings, including strategies in the preparation, characterization, and properties of both thin and thick multifunctional coatings along with their corresponding application. Various technologies for processing, characterization, and tribology effects of various coating surfaces and interfaces are discussed. It describes smart surfaces like piezoelectric materials, shape memory alloys, shape memory ceramics, magnetostrictive materials, electrostrictive materials, dielectric materials, and advanced ceramics.

- Explains multifunctional materials with respect to their tribology behavior at surface and interface.
- Covers analysis techniques for multifunctional surfaces and interfaces.
- Discusses emerging applications of multifunctional surfaces.
- Explores multifunctionality of thin films as well as thick coatings.

This book is aimed at graduate students and researchers in metallurgical engineering, materials science, and nanosciences.

Emerging Materials and Technologies

Series Editor: Boris I. Kharissov

The *Emerging Materials and Technologies* series is devoted to highlighting publications centered on emerging advanced materials and novel technologies. Attention is paid to those newly discovered or applied materials with potential to solve pressing societal problems and improve quality of life, corresponding to environmental protection, medicine, communications, energy, transportation, advanced manufacturing, and related areas.

The series takes into account that, under present strong demands for energy, material, and cost savings, as well as heavy contamination problems and worldwide pandemic conditions, the area of emerging materials and related scalable technologies is a highly interdisciplinary field, with the need for researchers, professionals, and academics across the spectrum of engineering and technological disciplines. The main objective of this book series is to attract more attention to these materials and technologies and invite conversation among the international R&D community.

Biodegradable Polymers, Blends and Biocomposites: Trends and Applications
Edited by A. Arun, Kunyu Zhang, Sudhakar Muniyasamy, and Rathinam Raja

Bioinspired Materials and Metamaterials: A New Look at the Materials Science
Edward Bormashenko

Computational Studies: From Molecules to Materials
Edited by Ambrish Kumar Srivastava

2D Semiconductors for Environmental Remediation
Edited by Honey John, Nisha T Padmanabhan, Sona Stanly, and Jith C Janardhanan

Materials from Natural Sources: Structure, Properties, and Applications
Edited by Ramesh Gardas, Neha Patni, and Amita Chaudhary

Dielectric Materials for Capacitive Energy Storage
Edited By Haibo Zhang and Hua Tan

Multifunctional Coordination Materials for Green Energy Technologies
Edited by Ghulam Yasin, Anuj Kumar, Sajjad Ali, Tuan Anh Nguyen, and Saira Ajmal

Advancements in Nanomaterials for Energy Conversion and Storage
Edited by Piyush Kumar Sonkar and Vellaichamy Ganesan

2D Materials-Based Sensors: Technology and Applications
Vinod Kumar Khanna

Metal Organic Framework Derived Materials: Design Strategies and Applications
Gomathi Nageswaran, Varsha M V, Arun Kumar Rajasekaran, and M Shashank Rao

Hydrogen Production, Storage, and Utilization: Technologies and Applications
Abbas Tcharkhtchi, Hamidreza Vanaei, Albert Lucas, and Sedigheh Farzaneh

MXenes for Energy Storage Applications: Emerging Characteristics, Compositions, and Synthesis Methods
Muhammad Rafique, M. Bilal Tahir, and Saira Anwar

Multi-scale and Multifunctional Coatings and Interfaces for Tribological Contacts
Ajit Behera, Kuldeep K Saxena, Dipen Kumar Rajak, and Shankar Sehgal

For more information about this series, please visit: www.routledge.com/Emerging-Materials-and-Technologies/book-series/CRCEMT

Multi-scale and Multifunctional Coatings and Interfaces for Tribological Contacts

Edited by
Ajit Behera, Kuldeep K Saxena,
Dipen Kumar Rajak, and Shankar Sehgal

CRC Press is an imprint of the
Taylor & Francis Group, an **informa** business

Designed cover image: Shutterstock

First edition published 2025
by CRC Press
2385 NW Executive Center Drive, Suite 320, Boca Raton FL 33431

and by CRC Press
4 Park Square, Milton Park, Abingdon, Oxon, OX14 4RN

CRC Press is an imprint of Taylor & Francis Group, LLC

ISBN: 978-1-032-55537-9 (hbk)
ISBN: 978-1-032-63533-0 (pbk)
ISBN: 978-1-032-63534-7 (ebk)

DOI: 10.1201/9781032635347

Typeset in Times
by KnowledgeWorks Global Ltd.

Contents

Introduction

With leading technology we are going to miniaturization of systems. With these small volumes, knowledge of surfaces and their connected interface in systems is most important. In the operational life of a system, it is exposed to wear and tear. Hence, the study of tribological contacts is essential. This book focuses on the emerging development and various strategies in the preparation, characterization, and properties of both thin and thick multifunctional coatings along with their corresponding applications. This book covers the introduction to multi-scale and multifunctional coatings and the various multifunctional activities with respect to tribology behavior for thin and thick surfaces. Smart surfaces like piezoelectric materials, shape memory alloys, shape memory ceramics, magnetostrictive materials, electrostrictive materials, dielectric materials, and advanced ceramics are discussed. The cutting-edge technology-based additive manufactured multifunctional thin films and thick films, high-temperature behavior of multifunctional thin films, self-cleaning multifunctional thin films, antiviral thin surfaces, multifunctional porous coatings, and abradable coatings are also discussed in this book along with their classifications.

Preface

This book focuses on the emerging development and various strategies in the preparation, characterization, and properties of both thin and thick multifunctional coating along with their corresponding applications. Section 1 of this book covers the introductory part of multi-scale and multifunctional coatings. Here various technologies for processing, characterization, and tribology effects on various coating surfaces/interfaces are discussed. Section 2 and Section 3 cover the various multifunctional activities with respect to tribology behavior for thin and thick surfaces, respectively. Smart surfaces like piezoelectric materials, shape memory alloys, shape memory ceramics, magnetostrictive materials, electrostrictive materials, dielectric materials, and advanced ceramics are discussed. The cutting-edge technology based additive manufactured multifunctional thin films and thick films, high-temperature behavior of multifunctional thin films, self-cleaning multifunctional thin films, antiviral thin surfaces, multifunctional porous coatings, and abradable coatings are also discussed in this book along with their classifications. Section 4 of this book covers various emerging applications and associated tribology effects for thin films as well as thick films. Finally, some future trends and the prospects in these research areas are also discussed to explain the coatings of tailored corrosion protection materials that are of the utmost relevance to ensure the reliability and long-term performance of coated parts as well as the product value of the coated materials.

About the Editors

Dr. Ajit Behera currently works as an assistant professor in the Metallurgical & Materials Department at the National Institute of Technology, Rourkela. He was born in 1987 in Odisha. He completed his PhD from IIT-Kharagpur in 2016. He received the National IEI Young Engineer Award in 2022, the Yuva Rattan Award in 2020, the C.V. Raman Award in 2019, and the young faculty award in 2017. He has published more than 170 publications, including books, book chapters, and journal articles.

Prof. Shankar Sehgal is a professor in mechanical engineering at UIET, Panjab University, Chandigarh, India. His area of research is materials, manufacturing, joining, and finite element model updating. He is a University Gold Medalist in B. E. (Mech. Eng.) from GNDEC Ludhiana. He completed his M. Tech. and PhD from the Indian Institute of Technology Delhi and Panjab University, respectively.

Prof. Sehgal has served as guest editor of many esteemed SCIe indexed journals. His Scopus h-index is more than 18. He served as coordinator of three international conferences, ICAMSE2022, ICAMSE2021, and ICAMSE2020, held at Panjab University, Chandigarh, India. He has also been the coordinator of two international faculty development programs in the field of materials and 3D printing. He served as zonal vice president of the Association for Machines and Mechanisms from 2014 to 2018.

Dr Dipen Kumar Rajak is Scientist of Alloys, Composites and Cellular Materials Division, CSIR-Advanced Materials and Processes Research Institute, Bhopal, Madhya Pradesh, India.

Dr Rajak has made significant contributions in the field of mechanical and materials science engineering. He did his doctor of philosophy work in the research area of crashworthiness analysis of empty/foam-filled structures at the Indian Institute of Technology (ISM) Dhanbad, Jharkhand, India.

Dr Rajak has 10 years of teaching and research experience in mechanical and materials engineering, and he is a life member in many professional bodies. He has presented many guest lectures at national and international gatherings. His field of interest is the development of metallic foams/composites for different applications where low density and high specific energy absorption are required in the automotive, aerospace, and thermal management sectors.

Dr Rajak has received many prestigious awards, and he has published more than 100 research articles in esteemed national and international journals. He has also authored two books, 16 book chapters, 18 patents, two designs, and one copyright.

Acknowledgement

We acknowledge the Advanced Materials Lab, NIT Rourkela, AMPRE Eng. Pvt. Ltd. for their extended support to carry out all the preparations for this book.

Contributors

Rashad Abaszade
Azerbaijan State Oil and Industry University
Baku, Azerbaijan

Younes Abrouki
Faculty of Science
Laboratory of Spectroscopy, Molecular Modeling, Materials, Nanomaterial, Water and Environment, CERNE
Mohammed V University in Rabat
Rabat, Morocco

Javeed Akhtar
Department of Chemistry
Mirpur University of Science and Technology (MUST)
Mirpur (AJK), Pakistan

Ftema W. Aldbea
Physics Department
Faculty of Science
Sebha University
Libya

Mayyadah S. Abed Al-Fatlawi
Materials Engineering
Department of Materials Engineering
University of Technology (UOT)
Baghdad, Iraq

Mohsen Khajeh Aminian
Department of Physics
Yazd University
Yazd, Iran

D.M.B.P. Ariyasinghe
Department of Engineering Technology
University of Jaffna
Kilinochchi, Sri Lanka

Sushmita Banerjee
Department of Environmental Sciences
Sharda University
Greater Noida, India

Ajit Behera
Department of Metallurgical & Materials Engineering
National Institute of Technology
Rourkela, India

Asit Behera
School of Mechanical Engineering
Kalinga Institute of Industrial Technology
Bhubaneswar, India

Ghita Amine Benabdallah
Laboratory of Spectroscopy, Molecular Modeling, Materials, Nanomaterials, Water and Environment, CERNE2D, ENSAM
Mohammed V University in Rabat
Rabat, Morocco

Mohammed Benchrifa
Faculty of Sciences
Laboratory of Solar Energy and Environment
Mohammed V University in Rabat
Rabat, Morocco

Meriem Bensemlal
Laboratory of Organic Bioorganic Chemistry and Environment
University Chouaib Doukkali
El Jadida, Morocco

Ravindra Bhardwaj
Mechanical Engineering
Birla Institute of Technology & Science
Pilani, Rajsthan, India
Dubai Campus, Dubai International Academic City
Dubai, U.A.E.

Ichraq Bouhouche
Laboratory of Spectroscopy, Molecular Modeling, Materials, Nanomaterials, Water and Environment, CERNE2D, ENSAM
Mohammed V University in Rabat
Rabat, Morocco

Khalid Bouiti
Laboratory of Spectroscopy, Molecular Modeling, Materials, Nanomaterials, Water and Environment, CERNE2D, ENSAM
Mohammed V University in Rabat
Rabat, Morocco

Gui-Bin Chen
Physics Department
Huaiyin Normal University
Jiangsu, P. R. China

Andrew Chun Yong Ngo
Institute of Materials Research and Engineering (IMRE)
A*STAR (Agency for Science, Technology and Research)
Fusionopolis Way, Singapore

Burak Dikici
Department of Metallurgical and Materials Engineering
Ataturk University
Erzurum, Turkiye

Abdelkader Djelloul
Science of Matter
Abbes Laghrour University
Khenchela, Algeria

Souad El Hajjaji
Faculty of Sciences
Laboratory of Spectroscopy, Molecular Modeling, Materials, Nanomaterials, Water and Environment, CERNE2D
Mohammed V University in Rabat
Rabat, Morocco

Khadija El-Moustaqim
Faculty of Sciences
Improvement and Valuation of Plant Resources
Ibn Tofaïl University—KENITRA-University Campus
Kenitra, Morocco

Mohamed Elouardi
Faculty of Science
Laboratory of Spectroscopy, Molecular Modeling, Materials, Nanomaterial, Water and Environment, CERNE2D
Mohammed V University in Rabat
Rabat, Morocco

Salar Karim Fatah
Department of Physics
Garmian University
Kalar-Kurdistan Region, Iraq

Praveen Gagrai
School of Mechanical Engineering
Kalinga Institute of Industrial Technology
Bhubaneswar, India

Driss Hmouni
Faculty of Sciences
Improvement and Valuation of Plant Resources
Ibn Tofaïl University-KENITRA-University Campus
Kenitra, Morocco

Che-Hua Yang
Graduate Institute of Manufacturing Technology
National Taipei University of Technology
Taipei, Taiwan

Xiaohu Huang
Institute of Materials Research and Engineering (IMRE)
A*STAR (Agency for Science, Technology and Research)
Fusionopolis Way, Singapore

Nurul Huda Abu Bakar
Faculty of Industrial Sciences and Technology
Universiti Malaysia Pahang
Gambang, Kuantan, Pahang

Meryem Idrissi
Technology School
Laboratory of Catalysis, Materials and Environment
Fez, Morocco

N. Jeyaprakash
School of Mechanical and Electrical Engineering
China University of Mining and Technology
Xuzhou City, China

Ajayi Oluwaseun K.
Department of Mechanical Engineering
Faculty of Technology
Obafemi Awolowo University
Ile-Ife, Nigeria

Siti Maznah Kabeb
Faculty of Industrial Sciences and Technology
University Malaysia Pahang
Gambang, Kuantan, Pahang

Sundara Subramanian Karuppasamy
Graduate Institute of Manufacturing Technology
National Taipei University of Technology
Taipei, Taiwan

Rahul Karyappa
Institute of Materials Research and Engineering (IMRE)
A*STAR (Agency for Science, Technology and Research)
Fusionopolis Way, Singapore

Elmira Khanmammadova
Azerbaijan State Oil and Industry University
Baku, Azerbaijan

Najoua Labjar
Laboratory of Spectroscopy, Molecular Modeling, Materials, Nanomaterials, Water and Environment, CERNE2D, ENSAM
Mohammed V University in Rabat
Rabat, Morocco

Nabil Lahrache
Laboratory of Spectroscopy, Molecular Modeling, Materials, Nanomaterials, Water and Environment, CERNE2D, ENSAM
Mohammed V University in Rabat
Rabat, Morocco

Yuhua Li
School of Mechanical Engineering
Xi'an University of Science and Technology
Xi'an, China

Hongfei Liu
Institute of Materials Research and Engineering (IMRE)
A*STAR (Agency for Science, Technology and Research)
2 Fusionopolis Way, Singapore

V. I. Loganina
Department of Quality Management and Construction Technology
Penza State University of Architecture and Construction
Penza, Russia

Fatima Ezzahra Maarouf
Laboratoire de chimieappliquée des matériaux (LCAM)
Faculty of Science
Mohammed V University in Rabat
Rabat, Morocco

Jamal Mabrouki
Faculty of Science
Laboratory of Spectroscopy, Molecular Modeling, Materials, Nanomaterial, Water and Environment, CERNE2D
Mohammed V University in Rabat
Rabat, Morocco

V. Madhushan
Department of Engineering Technology
University of Jaffna
Kilinochchi, Sri Lanka

Tzee Luai Meng
Institute of Materials Research and Engineering (IMRE)
A*STAR (Agency for Science, Technology and Research)
Fusionopolis Way, Singapore

Akash Mishra
Department of Metallurgical & Materials Engineering
National Institute of Technology
Rourkela, India

R. D. K. Misra
Center for Structural and Functional Materials Research and Innovation and Department of Metallurgical, Materials and Biomedical Engineering, University of Texas at El Paso, El Paso, TX, USA

Rasoul Moradi
Azerbaijan State Oil and Industry University
Baku, Azerbaijan,

Priyanshu Munda
Department of Metallurgical & Materials Engineering
National Institute of Technology
Rourkela, Odisha, India

Hamid Nasrellah
Higher School of Education and Training
Laboratory of Organic Bioorganic Chemistry and Environment
University Chouaib Doukkali
El Jadida, Morocco

Yee Ng
Institute of Materials Research and Engineering (IMRE)
A*STAR (Agency for Science, Technology and Research)
Fusionopolis Way, Singapore

Aditya Pandey
Department of Metallurgical & Materials Engineering
National Institute of Technology
Rourkela, Odisha, India

L.K. Pothal
School of Mechanical Engineering
Kalinga Institute of Industrial Technology
Bhubaneswar, India

Amlan Prabhujyoti Sahu
Department of Metallurgical & Materials Engineering
National Institute of Technology
Rourkela, India

Vyacheslav S. Protsenko
Ukrainian State University of Chemical Technology
Dnipro, Ukraine

Sabrina Roguai
LASPI2A Laboratory of Structures, Properties and Interatomic Interactions
Abbes Laghrour University
Khenchela, Algeria

Peter Rusinov
Kuban State Technological University
Krasnodar, Russia

Samson Rwahwire
Faculty of Engineering and Technology
Department of Polymer, Textile and Industrial Engineering
Busitema University
Tororo, Uganda

A. D. Ryzhov
Department of Information Systems
Penza State University of Architecture and Construction
Penza, Russia

José A. Sánchez-Fernández
Department of Polymerization Processes
Research Center of Applied Chemistry
Saltillo, Mexico

A. Sharma
CSIR-National Metallurgical Laboratory
Jamshedpur, India
Indian Institute of Technology Bombay (IITB)
Mumbai, India

Ivan Ssebagala
Department of Polymer, Textile and Industrial Engineering
Faculty of Engineering and Technology
Busitema University
Tororo, Uganda

Mehmet Topuz
Department of Mechanical Engineering
Van YuzuncuYil University
Van, Turkiye

Usman Lawal Usman
Department of Biology
Umaru Musa Yar'adua University
Katsina, Nigeria;
Department of Environmental Sciences
Sharda University
Greater Noida, India

Oktay Yigit
Department of Metallurgical and Materials Engineering
Firat University
Elazig, Turkiye

1 Introduction to Multifunctional Surfaces

Ajit Behera

1.1 INTRODUCTION

As the name suggests, multifunctional materials can perform multiple tasks and roles. They are supposed to be versatile and perform various tasks in order to meet various needs. In technology, biology, design, and other fields, these materials are being used more often. Multifunctional materials can be material, structures, or coatings that can be engineered to serve different useful functions. These materials can also be found in fields like material science, architecture, and engineering. Due to their versatility, these materials are applied in a wide range of fields and industries. Anticorrosion coatings and drag reduction surfaces can be used as multifunctional materials in aerospace and automobile sectors, respectively [1]. Anticorrosion coatings in metal surfaces increase lifespan in critical environments. So they can be very useful in aerospace and marine applications. Similarly, drag reduction surfaces reduce drag in aerospace applications, thus increasing performance and fuel efficiency of aerospace vehicles and automobiles. Self-healing surfaces, superhydrophobic surfaces, and anti-icing and antifogging surfaces are various surfaces that can be considered multifunctional surfaces [2]. Self-healing surfaces can repair minor damage in their surfaces, making them very useful for structural materials, coatings, and composites. Self-cleaning surfaces and superhydrophobic surfaces are used in car windshields, textiles, and building exteriors where self-cleaning is essential. To prevent the buildup of fog and ice, antifogging and anti-icing surfaces are used in transportation, optical devices, and aviation.

Multifunctional materials are also useful in the application of green building materials and air and water purification [3]. Multifunctional surfaces can remove contaminants and pollutants from water and air sources, thus proving to be environment friendly. Improvement of indoor air quality in buildings, temperature regulation, etc., can also be achieved by multifunctional materials, thus making them environmentally friendly and sustainable. Solar panels, antireflection coatings, and thermal management can also be done by multifunctional materials. Energy conversion efficiency can be increased, and heat dissipation can be enhanced in solar panels by use of multifunctional materials. Antireflection coatings made of multifunctional materials are very useful for optical lenses and solar cells. Multifunctional materials with enhanced thermal conductivity are used in electronic devices [4]. Smartphone and tablet screens, textiles, cookware, and kitchen appliances can be made of multifunctional materials. Anti-fingerprint, glare-reducing coatings and scratch-resistant features of multifunctional materials make them useful for smartphone applications. Easy-to-clean coatings and nonstick applications make them useful for kitchen appliances. Stain resistance, moisture wicking, and UV protection make them useful for the textile industry [5]. To maintain hygiene and prevent the spread of infections, multifunctional materials are used in health care. These materials can also be used in tissue engineering and drug delivery systems, and some materials compatible with biological tissues can be used as medical implants. To detect pathogens and molecules, these materials can also be used as biosensors [6]. Following are a few examples to illustrate the concept.

1.2 SMART SURFACES

Multifunctional materials can be embedded with actuators or sensors to form smart surfaces. The smart surfaces have greater efficiency and potency for various applications. Smart surfaces refer to materials or coatings that are designed to have enhanced functionalities beyond their traditional

DOI: 10.1201/9781032635347-1

roles. In building design and architecture, smart surfaces change their properties in response to external stimuli. For example, we can observe that multifunctional materials can alter opacity or color with respect to light or temperature. This property allows for dynamic control and automatic shading.

In smart surfaces, the surface material is often made up of shape memory alloy (SMA) [7]. As the name indicates, the shape memory alloy can remember its original shape. If one deforms the alloy, the alloy can return to its original state. That means it can show pseudoelasticity and shape memory. This property is due to a reversible phase transformation that occurs within the material's crystalline structure. Shape memory alloy surfaces can be prepared by depositing shape memory material on the substrate surface. This can be achieved by sputtering, physical vapor deposition, chemical vapor deposition, and other coating techniques. Shape memory materials are broadly used in medical devices, aerospace components, and actuators. Smart surfaces are used in a range of applications where controlled shape change is desirable [8]. For instance, smart surfaces may be used in stents, where the surface can be compressed during insertion and expanded when in position. Adaptive wings are used in aircraft that may incorporate smart surface SMAs that can change shape as climatic conditions and flight conditions change. Shape memory surfaces also face various challenges such as fatigue over repeated cycles and degradation of materials.

Energy harvesting means capturing one form of energy and converting it to another form of energy. A chromogenic surface is a surface that gives indication or changes its optical properties when a source of energy (light) is incident upon it or its voltage changes [9]. The combination of chromogenic surfaces with energy-harvesting techniques has various applications, such as adaptive camouflage and smart windows. A smart window uses a chromogenic surface that converts its optical properties when temperature changes or light is incident upon it. The incorporation of energy-harvesting techniques such as solar cells in windows makes them useful for harvesting electricity for homes. From this electricity various electrical devices can be powered. Again, solar cells may be used with a chromogenic surface that can absorb only certain spectra of light. This property can enhance heat-absorbing capacity as compared to traditional materials in which all spectra of light are absorbed. This can enhance the efficiency of solar cells. In some military appliances, adaptive camouflage is needed that may use chromogenic surfaces. These surfaces can change color to match their surroundings. By integrating energy-harvesting components, the energy generated from ambient light could power sensors or communication devices. Wearable devices incorporating chromogenic materials could change color based on the wearer's preference or environmental conditions. These wearables could also integrate energy-harvesting mechanisms to power themselves using ambient light or body heat. Nowadays smart buildings are being built that incorporate chromogenic surfaces. These surfaces change color depending upon the intensity of sunlight or temperature. The building also utilizes energy-harvesting equipment in which the whole building system is powered by sunlight. Chromogenic materials could be used in large outdoor advertising displays that change color or appearance based on the time of day, weather conditions, or even user interactions. Energy harvesting could help power these displays, reducing the need for external power sources. Chromogenic sensors that change color in response to specific environmental factors (e.g., pollution levels, temperature) could be integrated with energy harvesting to create self-powered monitoring systems [10].

1.3 ENERGY-HARVESTING SURFACE

An energy-harvesting surface is a surface that can absorb one source of energy and convert it to another form of energy. These surfaces have many applications. The main objective of the surface is to utilize the energy that is wasted. Also, these surfaces are used where there is a limited supply of traditional source of energy. Energy-harvesting surfaces have the potential to power various types of devices, including sensors, wearable electronics, remote monitoring systems, and low-power electronic gadgets. They are particularly useful in scenarios where frequent battery replacement or access to power sources is impractical or costly. Some of the generalized energy-harvesting surfaces are discussed below.

Solar energy-harvesting surfaces: These surfaces are equipped with photovoltaic cells that capture sunlight and convert it into electrical energy. This is a common and important application of energy harvesting from sunlight [11].

Thermal energy-harvesting surfaces: The energy is harvested as a result of temperature difference between the surface and the surrounding environment to generate electricity through thermoelectric effects [12].

Vibration and mechanical energy-harvesting surfaces: These surfaces often use piezoelectric materials. They convert mechanical vibrations into electrical power [13].

Radio frequency (RF) energy-harvesting surfaces: Through rectifying circuits and specialized antennas, electrical energy is harvested from radio frequency electromagnetic waves, such as those from cellular networks or Wi-Fi signals [14].

Wind energy-harvesting surfaces: Wind energy harvesting can be done by the use of wind turbines. These are commonly used in applications where wind energy is available, but traditional wind turbines are not feasible due to size or location constraints [15].

It is important to note that the efficiency of energy-harvesting surfaces can vary depending on factors such as the energy source, the technology used, and the specific application. While they may not generate large amounts of power, they can extend the operational lifetime and reliability of devices that require low levels of energy.

Piezoelectric surfaces: Piezoelectric surfaces work as an energy-harvesting source in certain circumstances. In a piezoelectric surface, material is used that can harvest energy [16]. When a piezoelectric material is subjected to mechanical stress or strain, electricity is generated in its surface. Due to the mechanical stress, there is an alteration in crystal properties, as a result of which voltage is generated. Similarly, when electric current is provided to the piezoelectric material, the material gets deformed. The electricity produced can be stored in a battery. In the context of energy harvesting, a piezoelectric material is used to create a surface that can capture and convert mechanical energy from sources like vibrations, impacts, or even ambient movements into usable electrical energy. This technology has gained significant attention due to its potential to provide power for small electronic devices, sensors, and even remote or low-power applications where traditional power sources are impractical. The piezoelectric material is integrated into a structure that allows it to deform in response to mechanical vibrations or forces. This structure could be a beam, membrane, or other flexible component that can convert the mechanical energy into strain on the piezoelectric material. The piezoelectric surface is exposed to mechanical vibrations or deformations. These can come from a variety of sources, such as ambient vibrations, footsteps, machinery vibrations, or even from vehicles passing on a road. Piezoelectric energy harvesting has found applications in various fields such as wearable electronics, smart infrastructure, wireless sensor networks, Internet of things (IoT) and structural health monitoring [17].

Dielectric surfaces: A dielectric surface is another type of energy-harvesting device that plays great role in solar energy storage. These materials don't conduct electricity, but they can store energy in the form of an electric charge. Dielectric materials are embedded with or coated in solar panels. In this process, generally the dielectric material is coated on photovoltaic cells. The coated material efficiently traps the sunlight, absorbs it, and reduces reflection. Dielectric metasurfaces, which are patterned structures made of dielectric materials, can also be designed to manipulate the incoming light's direction, polarization, and wavelength, further increasing the efficiency of solar panels. When a light source is incident on the panel, the electricity generated by the dielectric material can be stored. Dielectric materials can also be used in energy storage applications. Capacitors, which consist of two conductive plates separated by a dielectric material, can store and release energy quickly compared to batteries. The greater the dielectric constant of the material, the more energy it can store and the more preferable in industry [18].

Magnetostrictive surfaces: A magnetostrictive surface for energy harvesting refers to a technology that utilizes the magnetostrictive effect to convert mechanical vibrations or strains in a material into electrical energy. The magnetostrictive effect is a property exhibited by certain materials that causes them to change their magnetic properties when subjected to a mechanical stress or strain [19]. The change in magnetic property can be utilized to produce electricity. The electricity is produced through electromagnetic induction. The changes in the magnetic properties of the magnetostrictive material induce an electromagnetic voltage in nearby coils or conductors according to Faraday's law of electromagnetic induction. This voltage can then be harvested and used as electrical energy. The induced voltage is often alternating current (AC). To make it usable for most applications, it needs to be converted into direct current (DC) using a rectifier. The harvested energy can then be stored in batteries or capacitors for later use or directly supplied to power electronic devices.

Solar-responsive surfaces: Certain surfaces can harness solar energy by incorporating photovoltaic materials. These surfaces can generate electricity from sunlight and be integrated into various structures, such as windows, roofs, and facades. Solar-responsive surfaces, often referred to as "smart surfaces" or "smart coatings," are materials designed to interact with and respond to sunlight or solar radiation in various ways. These surfaces are engineered to have specific properties that can enhance energy efficiency, comfort, or other functionalities in buildings, vehicles, or other applications. Here are a few examples of solar-responsive surfaces and their functions [20].

- **Photovoltaic surfaces:** These surfaces are coated with solar cells that can convert sunlight directly into electricity. They are commonly used in solar panels to generate renewable energy for various applications, such as residential and commercial power generation [21].
- **Thermochromic surfaces:** Thermochromic materials change color in response to temperature changes. These surfaces can be used in buildings to modulate heat gain or loss. For example, windows coated with thermochromic materials can darken to reduce sunlight and heat transmission during hot days, and become transparent again when it's cooler [22].
- **Photochromic surfaces:** Photochromic materials change color when exposed to light, particularly UV light. These surfaces can be used in eyeglasses, windows, and other applications to automatically adjust their tint in response to sunlight, providing protection against glare and UV radiation [23].
- **Solar heat-reflective coatings:** These coatings are designed to reflect a significant portion of sunlight and solar heat, reducing the amount of heat absorbed by surfaces. This can help keep buildings cooler and reduce the need for air conditioning, leading to energy savings [24].
- **Solar thermal absorbers:** These surfaces are designed to absorb and retain solar heat, which can then be used for heating purposes. They are often used in solar water heaters and space heating systems [25].
- **Daylighting control surfaces:** These surfaces are engineered to optimize natural daylighting in buildings while minimizing glare and excessive heat gain. They can enhance indoor lighting quality and reduce the need for artificial lighting during the day.
- **Solar-powered sensors:** Some surfaces are integrated with sensors that are powered by solar energy. These sensors can be used for various purposes, such as environmental monitoring, security systems, and more [26].

Solar-powered ventilation: Certain surfaces can be designed to automatically open or close in response to sunlight, facilitating passive ventilation in buildings and improving indoor air quality [27].

Heat-reflective coatings: Heat-reflective coatings are also called reflective roof coatings or cool roof coatings. The coating reflects the sun's rays, including infrared and ultraviolet rays. Mainly these coatings are used on rooftops for cooling purposes. These materials have high reflectivity and low emissivity. They include pigmented substances that can inhibit the transfer of heat through conduction. They are typically available in liquid form and can be sprayed or rolled onto the surface. They form a protective layer that reflects sunlight and prevents excessive heat absorption. This can be used mostly in hotter regions or in applications where cooling is necessary. These coatings can be applied to various surfaces, including roofs, walls, and even pavements. B reflecting a larger portion of the sun's energy, the building remains cooler, which can lead to energy savings and increased indoor comfort. Heat-reflective coatings are often available in light colors, such as white or light gray, because these colors have higher reflectivity. However, advancements in technology have led to the development of coatings with improved reflectivity even in darker colors [28]. Benefits are:

Energy efficiency: By reducing the amount of heat absorbed by a building, heat-reflective coatings can lower the need for air conditioning and cooling systems, leading to energy savings.

Extended roof life: Excessive heat can cause thermal stress and degrade roofing materials over time. Reflective coatings can help prolong the life of roofing materials by minimizing temperature-related wear and tear.

Indoor comfort: Buildings with heat-reflective coatings often maintain a more comfortable indoor temperature, reducing the need for temperature control systems and enhancing occupant comfort.

Environmental impact: Reduced energy consumption can lead to lower greenhouse gas emissions, contributing to environmental sustainability.

1.4 SELF-CLEANING SURFACES

Some surfaces are engineered to repel dirt, water, and other substances, effectively keeping themselves clean. This can be useful in environments where regular cleaning is impractical or challenging, such as in outdoor installations or on buildings. Self-cleaning surfaces are materials that have the ability to remove dirt, dust, microorganisms, and other contaminants from their surface without the need for manual cleaning or external intervention. These surfaces can maintain their cleanliness and appearance over time through various mechanisms that prevent the accumulation of unwanted substances. There are two main types of self-cleaning surfaces: hydrophobic (water-repellent) and photocatalytic.

Hydrophobic self-cleaning surfaces: Hydrophobic means repellent of water. Hydrophobic self-cleaning surface means the surface can repel water, and in that process it can remove dirt and other contaminants. The lotus leaf is a well-known natural example of this phenomenon, where its surface structure and chemical composition repel water and maintain its cleanliness. Like the lotus leaf, other surfaces have been developed by researchers that can clean their own surfaces. These materials often feature micro- or nanostructured surfaces that reduce the contact area between water and the surface, creating a "lotus effect." Coatings for outdoor surfaces, self-cleaning glass windows, and even clothing that resists staining are some of the best examples of hydrophobic self-cleaning surfaces [29].

Hydrophilic self-cleaning surfaces: Hydrophilic means affinity for water. The hydrophilic effect in cleaning refers to the phenomenon whereby cleaning or dirt removal is done by the substance. The hydrophilic material also can remove hydrophobic substances from surfaces. When hydrophilic material is added to a surface, the surface tension of water decreases. Hence water can spread out over the surface. This increased wetting helps to loosen and lift dirt and grime, making it easier to remove them during the cleaning process. This effect is crucial in various cleaning processes, from household cleaning to industrial applications [30].

Photocatalytic self-cleaning surfaces: Photocatalytic self-cleaning surfaces use a photocatalyst to clean a surface. In this surface, photocatalytic reactions take place that degrade pollutants and organic matter. When ultraviolet (UV) light is incident on the surface either from the sun or from any artificial source, photocatalytic action takes place in the presence of the photocatalyst. In photocatalytic action, reactive oxygen species are produced which degrade organic compounds and kill microorganisms on the surface. Titanium dioxide (TiO_2) is a common photocatalyst used in self-cleaning materials. This type of self-cleaning surface is mostly used in applications like air purification systems, building exteriors, and outdoor signage. It can help reduce the need for regular cleaning and maintenance while also contributing to improved air quality [31].

Anti-icing and antifogging surfaces are designed to prevent the accumulation of ice and fog on various surfaces, respectively. These technologies are especially important for safety and efficiency in various applications, including transportation, infrastructure, and everyday items.

Anti-icing surfaces: Anti-icing surfaces are engineered to prevent the buildup of ice on surfaces, which is crucial in environments where ice accumulation can lead to hazardous conditions and reduced performance. There are several ways to create anti-icing surfaces [32].

Hydrophobic coatings: These coatings repel water and ice by reducing the surface's ability to form ice crystals. They prevent ice adhesion and make it easier for ice to slide off surfaces [33].

Surface with **De-icing chemicals:** Some surfaces are treated with de-icing chemicals that can be released when the temperature drops. These chemicals can melt ice on contact or prevent ice formation altogether [34].

Microtextures on surface: Surfaces with microtextures or nanostructures can inhibit ice formation. These structures trap air, reducing the contact area between the surface and water droplets, which prevents ice from adhering [35].

Surface with h**eating elements:** In some cases, surfaces are equipped with embedded heating elements that can warm the surface to prevent ice accumulation. This is commonly seen in aircraft wings and wind turbine blades [36].

Antifogging surfaces: Antifogging surfaces are designed to prevent the formation of fog on surfaces, ensuring clear visibility in various conditions. Fog forms when warm, moist air comes into contact with a cooler surface. To prevent this, surfaces can be treated with:

Hydrophilic coatings: These coatings encourage the even spread of water droplets, preventing the formation of foggy patches. They promote the formation of a thin, transparent water layer that doesn't distort vision [37].

Antireflective coatings: These coatings reduce the reflection of light on the surface, which can reduce glare and the perception of foggy conditions [38].

Superhydrophilic surfaces: Similar to hydrophilic coatings, these surfaces have a high affinity for water, causing it to spread out and form a uniform layer instead of forming droplets that cause fog [39].

Both anti-icing and antifogging technologies have applications in various industries. For example, in aviation, anti-icing surfaces are crucial for maintaining the aerodynamic performance of aircraft, while antifogging technologies are essential for ensuring pilots have clear visibility. In transportation, such as in automobiles and trains, antifogging surfaces can improve driver visibility during inclement weather. In everyday products like eyeglasses and camera lenses, antifogging coatings can enhance usability.

1.5 SELF-HEALING SURFACES

Self-healing surfaces refer to materials or coatings that have the ability to repair damage to themselves without the need for external intervention. These surfaces have the potential to revolutionize various industries, including electronics, automotive, aerospace, and construction, by extending the lifespan of products, reducing maintenance costs, and improving overall durability. The concept is inspired by biological systems that can heal themselves, such as human skin healing from cuts or bruises. There are a few different mechanisms that researchers have explored to achieve self-healing properties in materials [40].

Microcapsules: One approach involves embedding microcapsules filled with healing agents within the material. When damage occurs, such as a crack or scratch, these microcapsules rupture and release the healing agents into the damaged area, where they react and solidify to repair the material [41].

Vascular systems: Similar to the circulatory system in living organisms, some self-healing materials have vascular networks. When damage occurs, these vascular networks can deliver healing agents to the damaged area through channels or tubes [42].

Polymerization: Some materials are designed to have reversible polymerization properties. When the material is damaged, the polymer chains can rearrange and reconnect to restore the material's integrity [43].

Shape memory materials: Certain materials have the ability to "remember" their original shape and return to it after deformation. If a self-healing material with shape memory properties is damaged, it can revert to its original shape, effectively repairing the damage [44].

Chemical reactions: Some materials are engineered to have specific chemical reactions that can repair damage. For instance, a material might incorporate components that can react with each other to mend cracks or other types of damage [45].

Heat-induced healing: In this approach, materials are designed to heal themselves when exposed to heat. Damage causes the material to change its structure, and heating it can trigger a reversal of this change, effectively repairing the damage [46].

Light-activated healing: Materials that can be healed through exposure to specific wavelengths of light have also been explored. This involves incorporating light-sensitive compounds into the material that can initiate healing reactions upon exposure to light [47].

Electrochemical healing: Some self-healing materials rely on electrochemical reactions to repair damage. When damage occurs, the electrochemical reactions can be triggered to restore the material's structure [48].

Self-healing surfaces have applications in a wide range of industries. For example, in electronics, self-healing materials could lead to longer-lasting devices with reduced need for repairs. In the automotive and aerospace industries, self-healing coatings could help prevent corrosion and damage caused by environmental factors. In construction, self-healing materials could lead to more durable and longer-lasting structures.

1.6 NOISE-REDUCING SURFACES

Some surfaces are engineered to absorb or block sound waves, contributing to noise reduction in various environments. This can be particularly useful in urban areas or spaces where noise pollution is a concern. Noise-reducing surfaces, also known as acoustic surfaces or sound-absorbing materials, are designed to minimize or absorb sound waves in various environments. They are commonly used in architectural and interior design, industrial settings, and even in consumer products to improve acoustic comfort and reduce noise pollution. These surfaces work by converting sound energy into heat energy, thus decreasing the sound's intensity and preventing it from reflecting or echoing within a space.

Common types of noise-reducing surfaces are often made from materials like foam, fiberglass, or mineral wool. They are mounted on walls or ceilings to absorb sound and reduce echoes in a room. Acoustic panels are widely used in recording studios, theaters, offices, and other spaces where sound quality matters. These panels are installed in suspended ceilings to absorb sound and improve the acoustics of a room. They are commonly used in commercial spaces, schools, and offices. These panels consist of fabric stretched over a frame filled with sound-absorbing materials. They can be customized in terms of color and design to blend with the interior decor while also providing acoustic benefits. These suspended acoustic elements are often seen in large spaces like auditoriums and gymnasiums. They hang from the ceiling and help control sound reflections and noise levels. There are also wall coverings made from sound-absorbing materials. Similar to fabric panels, they can be textured or printed to match the aesthetic of a room. These are wallpapers designed with sound-absorbing properties. They can be used in residential and commercial spaces to reduce noise levels. Specialized flooring materials can absorb impact noise and footfall sounds, making them ideal for environments where noise from walking or movement is a concern.

Perforated panels: These are solid panels with small perforations that allow sound waves to pass through and be absorbed by sound-absorbing material behind them. They are often used in architectural design to integrate acoustic treatment with aesthetics.

Acoustic partitions: These are movable panels that can be used to create flexible spaces within a larger area. They also offer acoustic separation and noise reduction [49].

Green acoustic solutions: Some noise-reducing surfaces utilize natural materials like plants to absorb sound. Green walls or vertical gardens can act as both aesthetic features and acoustic treatments.

When considering noise-reducing surfaces, it's important to take into account factors such as the specific noise frequencies you want to address, the desired aesthetic, fire safety regulations, and the overall purpose of the space. Different materials and configurations are suitable for different applications, so consulting with an acoustic consultant or interior designer can help you choose the best solutions for your needs.

1.7 FLEXIBLE SURFACES

In materials science, there are surfaces that can change shape, texture, or flexibility in response to external factors like temperature, humidity, or pressure. These surfaces have applications in robotics, soft electronics, and adaptive structures. Flexible surfaces refer to materials or structures that can bend, deform, or change shape without breaking or losing their integrity. These surfaces are often designed to be adaptable, resilient, and capable of undergoing various forms of deformation while still maintaining their functionality. Flexible surfaces have a wide range of applications across different industries and technologies. Following are some examples and applications [50].

Flexible displays: Flexible OLED (organic light-emitting diode) displays are becoming more common in smartphones, tablets, and even TVs. These displays can be curved, rolled, or folded without damaging the screen or affecting the image quality.

Wearable electronics: Wearable devices like smartwatches and fitness trackers often incorporate flexible surfaces to provide a comfortable fit and allow movement while maintaining the functionality of the electronics.

Flexible sensors: Sensors made from flexible materials can conform to irregular surfaces, enabling applications in robotics, health care, and environmental monitoring. For example, flexible pressure sensors can be used to measure touch or pressure on curved surfaces.

Textile electronics: Flexible electronics can be integrated into fabrics to create smart textiles. These textiles can have embedded sensors, LEDs, and even conductive threads for various applications, including fashion, sports, and medical wearables.

Medical devices: Flexible surfaces are used in medical applications such as catheters and endoscopes. These devices can navigate through complex anatomies while minimizing trauma to tissues.

Foldable furniture: Flexible materials can be used to create foldable furniture and space-saving solutions. For example, foldable chairs and tables are designed to be easily stored and transported.

Aerospace: Flexible surfaces can be used in aircraft to provide morphing wings that can change shape during flight, improving aerodynamics and fuel efficiency.

Packaging: Flexible packaging materials, such as pouches and bags, are used to protect and store various products. They can be easily manipulated and resealed, making them convenient for consumers.

Solar panels: Flexible solar panels can be integrated into a variety of surfaces, including roofs and backpacks, to generate renewable energy in a more versatile manner.

Automotive industry: Flexible materials are used in car interiors and exteriors. For instance, car seats with flexible components provide better comfort, and flexible displays can be integrated into dashboards.

Robotics: Soft robotics often utilize flexible materials to create robots that can move in complex ways, adapt to their environment, and interact more safely with humans.

Entertainment and gaming: Flexible surfaces can be used in interactive installations, such as flexible touch screens or projection screens that respond to touch or movement.

The development of flexible surfaces involves advancements in materials science, engineering, and design. Researchers and engineers work on creating materials that can withstand repeated deformations while maintaining their performance characteristics. These flexible materials often involve polymers, elastomers, and other innovative compounds.

1.8 ANTIMICROBIAL SURFACES

Antimicrobial surfaces refer to materials that are designed to inhibit or kill the growth of microorganisms, such as bacteria, viruses, fungi, and other harmful pathogens. These surfaces are particularly important in settings where hygiene and cleanliness are essential, such as health care facilities, food processing areas, and public spaces [51]. The goal of antimicrobial surfaces is to reduce the spread of infections and improve overall hygiene. There are several ways in which antimicrobial surfaces can be created or enhanced.

Chemical treatments: Some surfaces are treated with chemical agents that have antimicrobial properties. These agents can be embedded in the material or applied as coatings. Common antimicrobial agents include silver ions, copper, zinc, and various types of organic compounds.

Nanostructures: Nanotechnology has enabled the creation of surfaces with nanostructures that have inherent antimicrobial properties. These structures can physically damage the microorganisms by puncturing their cell walls or interfering with their metabolic processes.

Photocatalytic coatings: Some antimicrobial surfaces utilize photocatalytic coatings, often based on titanium dioxide (TiO_2), which can break down organic matter (including microbes) when exposed to light, such as ultraviolet (UV) light.

Electrostatic interactions: Some surfaces can be modified to have an electric charge that attracts and kills microorganisms. This can be achieved through the use of materials with inherent electrical properties or by applying specialized coatings.

Natural materials: Some natural materials, such as certain types of wood, have inherent antimicrobial properties. These materials can be used as surfaces without the need for additional chemical treatments.

Biological agents: Some antimicrobial surfaces incorporate biological agents like enzymes or peptides that have the ability to break down the cell walls of microorganisms.

It's important to note that while antimicrobial surfaces can play a role in reducing the spread of pathogens, they are not a replacement for proper cleaning and hygiene practices. Regular cleaning and disinfection are still crucial to maintain a safe and healthy environment. Additionally, there are ongoing discussions about the potential risks of relying heavily on antimicrobial surfaces, such as the development of antimicrobial resistance.

1.9 SUMMARY

This chapter has presented a brief introduction to the novel behaviors of the surfaces of multifunctional materials along with their specific requirements, processing, and applications. A huge research gap still exists regarding multifunctional material surface phenomena, which indicates that there is a new horizon for futuristic applications.

REFERENCES

1. Faccini, M.; Bautista, L.; Soldi, L.; Escobar, A.M.; Altavilla, M.; Calvet, M.; Domènech, A.; Domínguez, E. Environmentally Friendly Anticorrosive Polymeric Coatings. Applied Science. 2021;11:3446. https://doi.org/10.3390/app11083446
2. Cao, M.; Tang, M.; Lin, W.; Ding, Z.; Cai, S.; Chen, H.; Zhang, X. Facile Fabrication of Fluorine-Free, Anti-Icing, and Multifunctional Superhydrophobic Surface on Wood Substrates. Polymers (Basel). 2022 May 11;14(10):1953. doi:10.3390/polym14101953
3. Wang, H.; Chiang, P.-C.; Cai, Y.; Li, C.; Wang, X.; Chen, T.-L.; Wei, S.; Huang, Q. Application of Wall and Insulation Materials on Green Building: A Review. Sustainability. 2018;10:3331. https://doi.org/10.3390/su10093331
4. Keshavarz Hedayati, M.; Elbahri, M. Antireflective Coatings: Conventional Stacking Layers and Ultrathin Plasmonic Metasurfaces, A Mini-Review. Materials (Basel). 2016 Jun 21;9(6):497. doi:10.3390/ma9060497
5. Tabata, E.; Ito, T.; Ushioda, Y.; Fujima, T. Fingerprint Blurring on a Hierarchical Nanoporous Layer Glass. Coatings. 2019;9:653. https://doi.org/10.3390/coatings9100653
6. Hasan, A.; Nurunnabi, M.; Morshed, M.; Paul, A.; Polini, A.; Kuila, T.; Al Hariri, M.; Lee, Y.K.; Jaffa, A.A. Recent Advances in Application of Biosensors in Tissue Engineering. BioMed Research International. 2014;2014:307519. doi:10.1155/2014/307519
7. Shreekrishna, S.; Nachimuthu, R.; Nair, V.S. A Review on Shape Memory Alloys and Their Prominence in Automotive Technology. Journal of Intelligent Material Systems and Structures. 2023;34(5):499–524. doi:10.1177/1045389X221111547
8. Dayyoub, T.; Maksimkin, A.V.; Filippova, O.V.; Tcherdyntsev, V.V.; Telyshev, D.V. Shape Memory Polymers as Smart Materials: A Review. Polymers. 2022;14:3511. https://doi.org/10.3390/polym14173511
9. Duinong, M.; Rasmidi, R.; Chee, F.P.; Moh, P.Y.; Salleh, S.; MohdSalleh, K.A.; Ibrahim, S. Effect of Gamma Radiation on Structural and Optical Properties of ZnO and Mg-Doped ZnO Films Paired with Monte Carlo Simulation. Coatings. 2022;12:1590. https://doi.org/10.3390/coatings12101590
10. Prauzek, M.; Konecny, J.; Borova, M.; Janosova, K.; Hlavica, J.; Musilek, P. Energy Harvesting Sources, Storage Devices and System Topologies for Environmental Wireless Sensor Networks: A Review. Sensors. 2018;18:2446. https://doi.org/10.3390/s18082446
11. Piorno, J.R.; Bergonzini, C.; Atienza, D.; Rosing, T.S. HOLLOWS: A Power-aware Task Scheduler for Energy Harvesting Sensor Nodes. Journal of Intelligent Material Systems and Structures. 2010; 21(13):1317–1335. doi:10.1177/1045389X10377033
12. Enescu, D. 2019. Thermoelectric Energy Harvesting: Basic Principles and Applications. IntechOpen. doi:10.5772/intechopen.83495
13. Aabid, A.; Raheman, M.A.; Ibrahim, Y.E.; Anjum, A.; Hrairi, M.; Parveez, B.; Parveen, N.; Mohammed Zayan, J. A Systematic Review of Piezoelectric Materials and Energy Harvesters for Industrial Applications. Sensors. 2021;21:4145. https://doi.org/10.3390/s21124145
14. Mouapi, A. Radiofrequency Energy Harvesting Systems for Internet of Things Applications: A Comprehensive Overview of Design Issues. Sensors. 2022;22:8088. https://doi.org/10.3390/s22218088

15. Perera, S.M.H.D.; Putrus, G.; Conlon, M.; Narayana, M.; Sunderland, K. Wind Energy Harvesting and Conversion Systems: A Technical Review. Energies. 2022;15:9299. https://doi.org/10.3390/en15249299
16. Chandra Sekhar, B.; Dhanalakshmi, B.; Srinivasa Rao, B.; Ramesh, S.; Venkata Prasad, K.; Subba Rao, P.S.V.; Parvatheeswara Rao, B. 2021. Piezoelectricity and Its Applications. IntechOpen. doi:10.5772/intechopen.96154
17. Mahapatra, S.D.; Mohapatra, P.C.; Aria, A.I.; Christie, G.; Mishra, Y.K.; Hofmann, S.; Thakur, V.K. Piezoelectric Materials for Energy Harvesting and Sensing Applications: Roadmap for Future Smart Materials. Advanced Science 2021;8: 2100864. https://doi.org/10.1002/advs.202100864
18. Ward, A. 2016. Dielectric materials for advanced applications. doi:10.13140/RG.2.1.3481.5600
19. Liu, H.; Liu, H.; Zhao, X.; Li, A.; Yu, X. Design and Characteristic Analysis of Magnetostrictive Vibration Harvester with Double-Stage Rhombus Amplification Mechanism. Machines. 2022;10:848. https://doi.org/10.3390/machines10100848
20. Bati, A.S.R.; Zhong, Y.L.; Burn, P.L. et al. Next-generation Applications for Integrated Perovskite Solar Cells. Communications Materials. 2023;4:2. https://doi.org/10.1038/s43246-022-00325-4
21. Ryu, D.; Meyers, F.N.; Loh, K.J. Inkjet-Printed, Flexible, and Photoactive Thin Film Strain Sensors. Journal of Intelligent Material Systems and Structures. 2015;26(13):1699–1710. doi:10.1177/1045389X14546653
22. Garshasbi, S.; Santamouris, M. Using Advanced Thermochromic Technologies in the Built Environment: Recent Development and Potential to Decrease the Energy Consumption and Fight Urban Overheating. Solar Energy Materials and Solar Cells. 2018;191:21–32. doi:10.1016/j.solmat.2018.10.023
23. Han, Y.; Yan, X.; Zhao, W. Effect of Thermochromic and Photochromic Microcapsules on the Surface Coating Properties for Metal Substrates. Coatings. 2022;12:1642. https://doi.org/10.3390/coatings12111642
24. Nayak, A.K. 2022. Bismuth series photocatalytic materials for the treatment of environmental pollutants. In Nanostructured Materials for Visible Light Photocatalysis (pp. 135–151). Elsevier.
25. Hayati Raad, S.; Atlasbaf, Z. 2023. Photovoltaic and Photothermal Solar Cell Design Principles: Efficiency/Bandwidth Enhancement and Material Selection. IntechOpen. doi:10.5772/intechopen.110093
26. Sivagami, A.; Angelo Kandavalli, M., & Yakkala, B. (2021). Design and Evaluation of an Automated Monitoring and Control System for Greenhouse Crop Production. IntechOpen. doi: 10.5772/intechopen.97316
27. Conceição, E.; Gomes, J.; Awbi, H. Influence of the Airflow in a Solar Passive Building on the Indoor Air Quality and Thermal Comfort Levels. Atmosphere. 2019;10:766. https://doi.org/10.3390/atmos10120766
28. Maharjan, S.; Liao, K.S.; Wang, A.; Curran, S.A. Highly Effective Hydrophobic Solar Reflective Coating for Building Materials: Increasing Total Solar Reflectance via Functionalized Anatase Immobilization in an Organosiloxane Matrix. Construction and Building Materials. 2020; 243:118189. https://doi.org/10.1016/j.conbuildmat.2020.118189
29. Xu, Q.; Zhang, W.; Dong, C.; Sreeprasad, T.S.; Xia, Z. Biomimetic Self-cleaning Surfaces: Synthesis, Mechanism and Applications. Journal of the Royal Society, Interface. 2016 Sep;13(122):20160300. doi:10.1098/rsif.2016.0300
30. Tański, T.; Zaborowska, M.; Jarka, P.; Woźniak, A. Hydrophilic ZnO Thin Films Doped with Ytterbium and Europium Oxide. Scientific Reports. 2022 Jul 5;12(1):11329. doi:10.1038/s41598-022-14899-z
31. Wei, Y.; Wu, Q.; Meng, H.; Zhang, Y.; Cao, C. Recent Advances in Photocatalytic Self-Cleaning Performances of TiO_2-based Building Materials. RSC Advances. 2023 Jul 11;13(30):20584–20597. doi:10.1039/d2ra07839b
32. Lin, Y.; Chen, H.; Wang, G.; Liu, A. Recent Progress in Preparation and Anti-Icing Applications of Superhydrophobic Coatings. Coatings 2018;8:208. https://doi.org/10.3390/coatings8060208
33. Tetteh, E.; Loth, E. Reducing Static and Impact Ice Adhesion with a Self-Lubricating Icephobic Coating (SLIC). Coatings. 2020;10:262. https://doi.org/10.3390/coatings10030262
34. Sasmal, A.; Nayak, A.K. Morphology-dependent Solvothermal Synthesis of Spinel NiCo2O4 Nanostructures for Enhanced Energy Storage Device Application. Journal of Energy Storage. 2023;58: 106342.
35. Volpe, A.; Gaudiuso, C.; Ancona, A. Laser Fabrication of Anti-Icing Surfaces: A Review. Materials (Basel). 2020 Dec 13;13(24):5692. doi:10.3390/ma13245692
36. Laakso, T.; Peltola, E. (2005). Review on blade heating technology and future prospects. BOREAS VII FMI Conference.
37. Adera, S.; Naworski, L.; Davitt, A. et al. Enhanced Condensation Heat Transfer Using Porous Silica Inverse Opal Coatings on Copper Tubes. Scientific Reports. 2021;11:10675. https://doi.org/10.1038/s41598-021-90015-x

38. Raut, H.; Venkatesan; A.G.; Nair; S.A.; Ramakrishna, S. Anti-Reflective Coatings: A Critical, In-Depth Review. Energy & Environmental Science. 2011;4:3779–3804. doi:10.1039/C1EE01297E
39. Drelich, J.; Chibowski, E.; Meng, D.; Terpilowski, K. Hydrophilic and Superhydrophilic Surfaces and Materials. Soft Matter. 2011;7:9804–9828. doi:10.1039/C1SM05849E
40. Paladugu, S.R.M.; Sreekanth, P.S.R.; Sahu, S.K.; Naresh, K.; Karthick, S.A.; Venkateshwaran, N.; Ramoni, M.; Mensah, R.A.; Das, O.; Shanmugam, R.A., Comprehensive Review of Self-Healing Polymer, Metal, and Ceramic Matrix Composites and Their Modeling Aspects for Aerospace Applications. Materials (Basel). 2022 Nov 29;15(23):8521. doi:10.3390/ma15238521
41. Chakraborty, R.; Vilya, K.; Pradhan, M.; Nayak, A.K. Recent Advancement of Biomass-Derived Porous Carbon Based Materials for Energy and Environmental Remediation Applications. Journal of Materials Chemistry A. 2022;10(13):6965–7005.
42. NikMdNoordinKahar, N.N.F.; Osman, A.F.; Alosime, E.; Arsat, N.; Mohammad Azman, N.A.; Syamsir, A.; Itam, Z.; Abdul Hamid, Z.A. The Versatility of Polymeric Materials as Self-Healing Agents for Various Types of Applications: A Review. Polymers. 2021;13:1194. https://doi.org/10.3390/polym13081194
43. Yang, Y.; Urban, M.W. Self-Healing of Polymers via Supramolecular Chemistry. Advanced Materials Interfaces. 2018;5:1800384. https://doi.org/10.1002/admi.201800384
44. Xin, X.; Liu, L.; Liu, Y. et al. Mechanical Models, Structures, and Applications of Shape-Memory Polymers and Their Composites. Acta Mechanica Solida Sinica. 2019;32:535–565. https://doi.org/10.1007/s10338-019-00103-9
45. Cooper, G.M. The cell: A molecular approach. DNA Repair. 2nd edition. Sunderland (MA): Sinauer Associates; 2000. Available from: https://www.ncbi.nlm.nih.gov/books/NBK9900/
46. Nayak, A.K.; Gopalakrishnan, T. Phase-and Crystal Structure-Controlled Synthesis of Bi2O3, Fe2O3, and BiFeO3 Nanomaterials for Energy Storage Devices. ACS Applied Nano Materials. 2022; 5(10):14663–14676.
47. Lee, H.P.; Gaharwar, A.K. Light-Responsive Inorganic Biomaterials for Biomedical Applications. Advanced Science. 2020;7:2000863. https://doi.org/10.1002/advs.202000863
48. Wan, X.; Mu, T.; Yin, G. Intrinsic Self-Healing Chemistry for Next-Generation Flexible Energy Storage Devices. Nano-Micro Letters. 2023;15:99. https://doi.org/10.1007/s40820-023-01075-9
49. Amran, M.; Fediuk, R.; Murali, G.; Vatin, N.; Al-Fakih, A. Sound-Absorbing Acoustic Concretes: A Review. Sustainability. 2021;13:10712. https://doi.org/10.3390/su131910712
50. El-Atab, N.; Mishra, R.B.; Al-Modaf, F.; Joharji, L.; Alsharif, A.A.; Alamoudi, H.; Diaz, M.; Qaiser, N.; Hussain, M.M. Soft Actuators for Soft Robotic Applications: A Review. Advanced Intelligent Systems. 2020;2:2000128. https://doi.org/10.1002/aisy.202000128
51. Serwecińska, L. Antimicrobials and Antibiotic-Resistant Bacteria: A Risk to the Environment and to Public Health. Water. 2020;12:3313. https://doi.org/10.3390/w12123313

2 Processing of Multifunctional Surfaces

Ftema W. Aldbea and A. Sharma

2.1 INTRODUCTION

Recently, there has been a growing trend toward the modification of multifunctional surfaces because it is considered a promising way to change the morphology and structure of surfaces, and it offers a wide range of potential changes to surface properties [1, 2]. The modification of multifunctional surfaces illustrates several important properties such as anticorrosion, self-cleaning, anti-icing, self-healing, bioadhesion, oil–water separation, water treatment, antireflectance, and many other interesting applications in medicine, science, technology, and the environment. The modification of surfaces can be performed by chemical, physical, or mechanical treatments [3]. In this chapter, the chemical and physical modification of multifunctional surfaces will be discussed in detail.

2.2 CLASSIFICATION OF SURFACE PROCESSING

The processing of multifunctional surfaces can be divided into two common categories: chemical and physical processing. For example, Chemical processing includes many types of processes such as sol-gel, chemical vapor deposition, atomic layer deposition, chemical etching, and biochemical treatment [4]. The physical modification methods for multifunctional surfaces include flame, corona discharge, ultraviolet (UV), gamma ray, electron beam, plasma, and femtosecond treatment [5]. Selected modification methods for multifunctional surfaces will be introduced and discussed in the next sections.

2.2.1 Chemical Processing

Many chemical processing methods are used to modify multifunctional surfaces, including chemical vapor deposition (CVD), atomic layer deposition (ALD), sol-gel, chemical etching, biochemical methods, and oxidation treatments. Among these methods, sol-gel, CVD, ALD, and chemical etching are explained in detail here.

2.2.1.1 Sol-gel Processing

Sol-gel processing is a common and effective way to prepare multifunctional surfaces directly. It is a wet chemical method that generally uses alkoxides to produce inorganic materials [6]. This method contains several steps to reach the desired product as shown in Figure 2.1. However, in the normal sol-gel method, inorganic sol particles undergo surface-preferred gelation to form an inorganic coat on the substrates, and the surface can be fabricated by using spin coating, dip coating or spray coating. Additionally, it allows for a way to modify and control the microstructure of materials at low temperatures [7, 8].

Many types of multifunctional materials are created by using the sol-gel method. For example, an anisotropic wetting surface is created by duplicating the surface structure of the outward wings of grasshoppers (Figure 2.2) [9]. The deposition of multilayer sol-gel coatings has been obtained by Souter et al. [10]. The prepared multilayers can selectively reflect radiation in the near-infrared region of the spectrum and at the same time allow high light transmission. Raquel et al. [11] have produced novel multifunctional coatings by sol-gel that present antireflection and self-cleaning capacities.

DOI: 10.1201/9781032635347-2

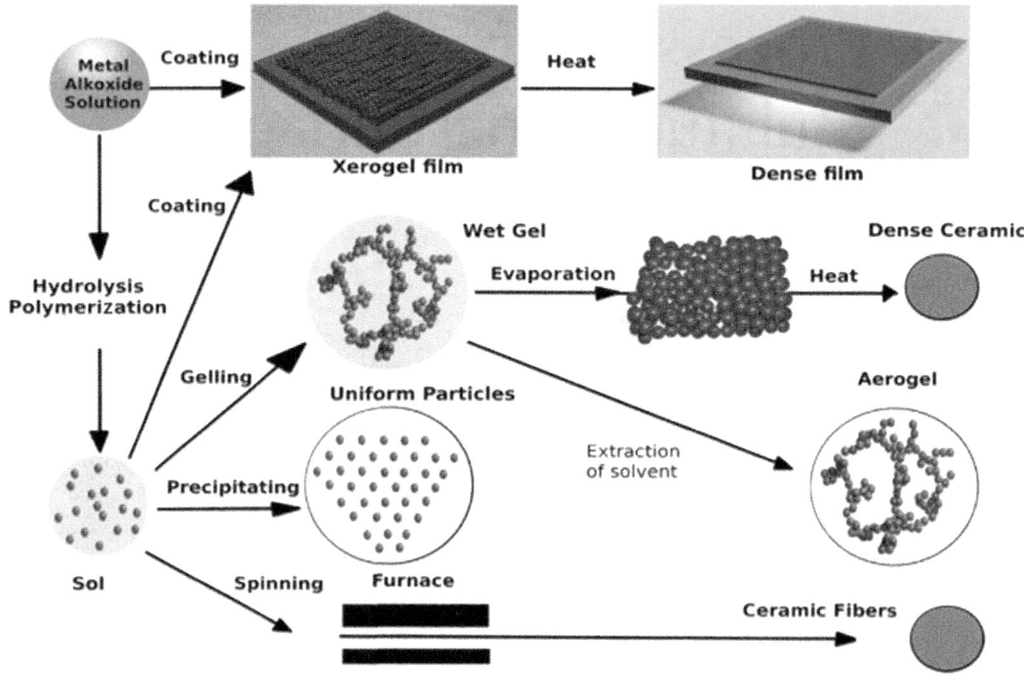

FIGURE 2.1 Sol-gel processing [7].

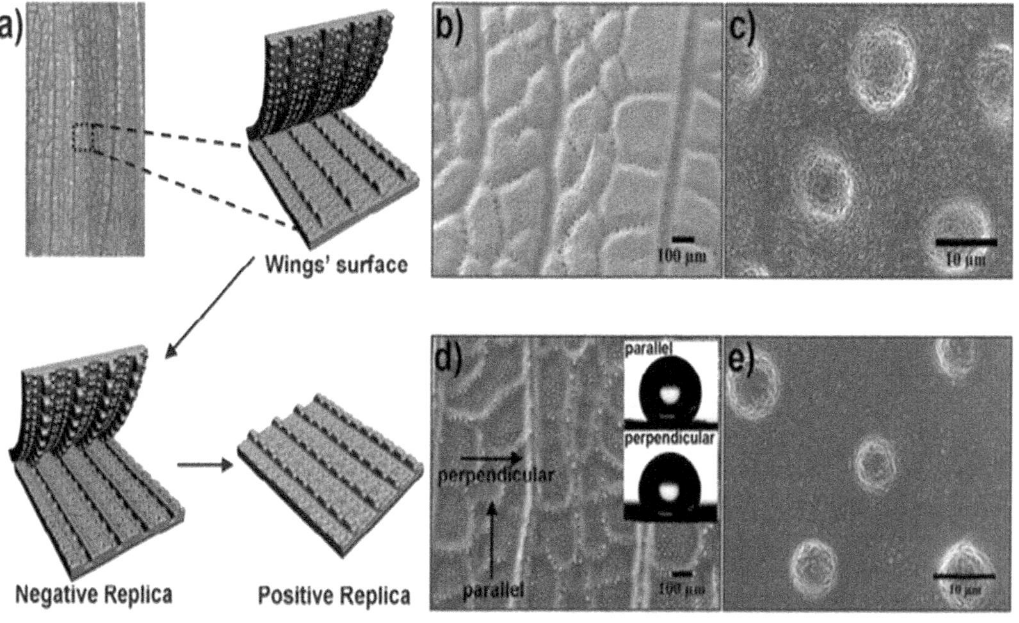

FIGURE 2.2 (a) The two-step replication process using sol-gel; (b, c) SEM images of first-step PDMS replica of the outward wings of grasshoppers; (d, e) SEM micrograph of second-step PDMS replica of the outward wings of grasshoppers. (d) Anisotropic high hydrophobicity of second-step PDMS replica [9].

2.2.1.2 Chemical Vapor Deposition (CVD)

Chemical vapor deposition is one of the most favorable methods in industry for surface treatment because it has a number of advantages, such as conformal coverage, uniform thickness, the ability to deposit a coating on one surface, and low deposition temperatures [12, 13]. Basically, the CVD process involves some steps to prepare the film. A substrate is exposed to a single or multicomponent volatile precursor in an inert atmosphere at a controlled high temperature and pressure. These volatile precursors will decompose onto the surface of the substrate, forming the desired thin-film material. The generated by-products are removed by flowing gas through the reactor. Figure 2.3 shows the CVD process [14, 15].

The CVD process is classified based on the energy used in the reactor into seven main types: temperature, pressure, wall/substrate, nature of precursor, depositing time, gas flow state, and activation/power source [15, 17]. These methods depend on the processing parameters that are represented above (Figure 2.4).

The CVD process is usually used to fabricate metal, nonmetallic, and multicomponent alloy films. It is also used to modify and control surfaces to enhance their physical and chemical properties [18]. The full range of wettability of a single-wall carbon nanotube (SWCNT) was achieved by modifying its surface using photo CVD [19]. The surface properties were enhanced from superhydrophilic with contact angle less than 5° (CA < 5°) to superhydrophobic with contact angle more than 150° (CA > 150°). Cristina et al. [20] have used PICVD to incorporate oxygen functionalities on the surface of boron nitrate nanotubes (BNNT). A highly oxidized surface, a decrease in water contact angle, and an increase in surface energy were observed for the treated BNNT surface. Tables 2.1 and 2.2 show contact angle and surface energy results.

Hot filament chemical vapor deposition (HFCVD) is a common process that is utilized to enhance the nucleation and growth of large area uniform multifunctional ultrananocrystalline diamond films on a W-coated 100 nm diameter Si substrate, followed by eliminating seeding of the substrate surface. In this process, the plasma generated by the electric field between positive bias filaments and negative substrates produces positively charged and natural Ar, CH_x (x = 1 2 3), C, and H in an Ar/

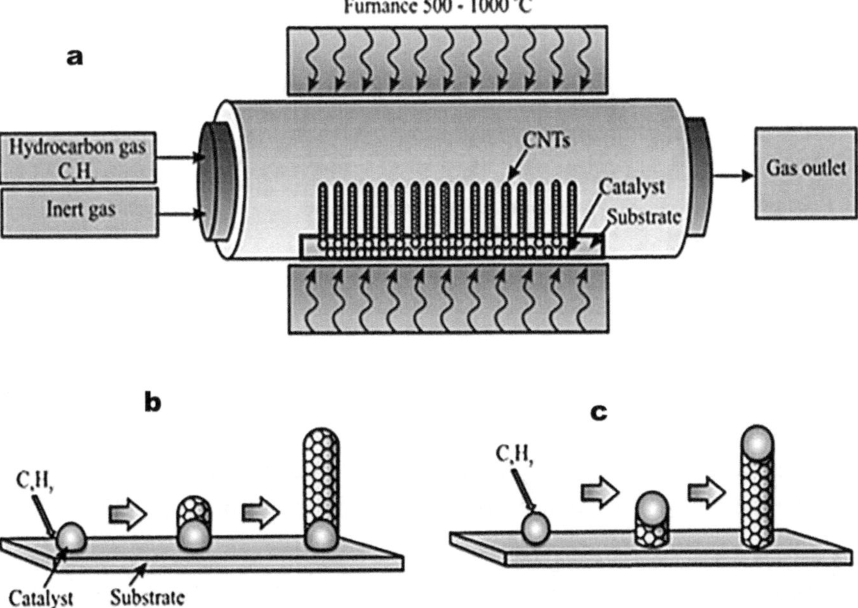

FIGURE 2.3 The steps of chemical vapor deposition (CVD) processing of carbon nanotubes (CNTs). (a) Simple CVD reactor, (b) base growth of CNT, and (c) tip growth model of CNT growth [16].

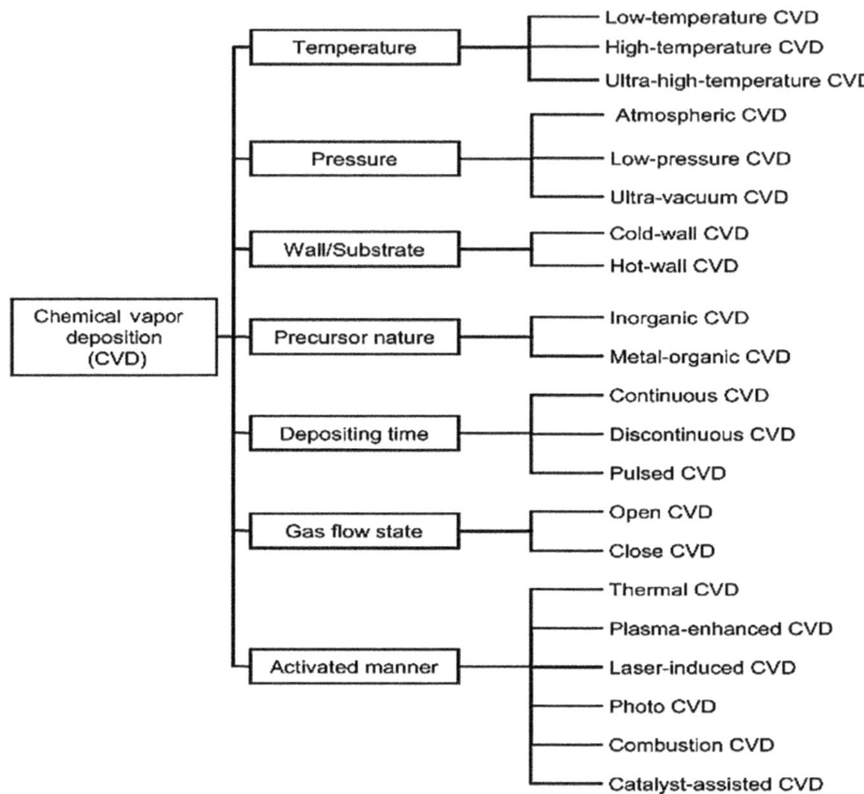

FIGURE 2.4 Types of chemical vapor deposition technique [17].

TABLE 2.1
Static Contact Angle and Surface Energy of Boron Nitride Nanotube (BNNT) Films [20]

	CA (°)			Surface Energy (mJ/m²)		
Sample	**Water**	**Formamide**	**Diiodomethane**	**Total**	**Dispersive**	**Polar**
Untreated BNNT	87.8 ± 7.1	33.0 ± 4.0	22.2 ± 3.8	55.43	55.01	0.41
Treated BNNT	16.8 ± 1.3	15.6 ± 3.1	12.9 ± 3.1	67.67	30.56	37.11

TABLE 2.2
Water Advancing, Receding Contact Angles (ARCA) and Surface Energy for Untreated and Treated Boron Nitride Nanotubes ($BNNT_S$) [20]

	Water ARCA (°)		
Sample	**Advancing**	**Receding**	**Total Surface Energy (mJ/m²)**
Untreated BNNT	65.9 ± 0.8	56.2 ± 3.4	50.15
Treated BNNT	21.2 ± 1.6	12.6 ± 1.7	71.55

CH_4-rich and electron/H_2 mixture. The C^+-based ions influencing the substrate surface nucleate a W-carbide layer, leading to the growth of a uniform ultrananocrystalline diamond film [21].

2.2.1.3 Atomic Layer Deposition (ALD)

The ALD process has become an essential technique for depositing thin films for a variety of applications because it has features such as simple and precise film thickness control, uniformity and excellent conformality, high-quality and dense films, low-temperature processing, multilayers, and a self-assembled nature [22, 23]. However, ALD is a chemical gas phase thin film deposition method based on saturated surface reaction. In this process, the precursors are pulsed one at a time alternately, and separated by inert gas purging to avoid gas phase reactions (Figure 2.5). One cycle of ALD consists of the following steps [22]:

1. Pulsing of the first gaseous precursor and its chemisorption on the substrate.
2. Purging of inactive gas from the excess of the precursor and reaction by-products formed.
3. Pulse of the second gaseous precursor and its surface reaction with the adsorbate formed from the first precursor.
4. Finally, repetition of the second step (purging of inactive gas from the excess of the precursor and reaction by-products formed.

Recently, ALD has become an important process for multifunctional surface processing [24], which is one of the main inspirations for the recent development of ALD. For example, two polymeric inverted membranes, polyacrylonitrile (PAN) and polyetherimide (PEI), have been treated by growing Al_2O_3 on their surface for oil antifouling. Here, the Al_2O_3 surface treatment reduces oil droplet surface coverage and enhances the membrane`s performance (Figures 2.6 and 2.7) [25].

Furthermore, coating one cycle of AL_2O_3 ALD exhibited a significant effect on the dye-sensitized solar cell SnO_2 nanoparticle electrode, where the short circuit current density improved more than that of the naked electrode (Figure 2.8). The open circuit photo voltage and fill factor were increased from 230 mV to 510 mV and from 35% to 65%, respectively [26].

Kemell et al. [27] have also used the ALD technique to deposit Al_2O_3 and TiO_2 on various polymer substrates such as polymethylmethacrylate (PMMA), polyetheretherketone (PEEK), polytetrafluoroethylene (PTFE), and ethylene-tetrafluoroethylene copolymer (ETFE). The treated films showed good properties and a lower water contact angle than uncoated surfaces (Figure 2.9).

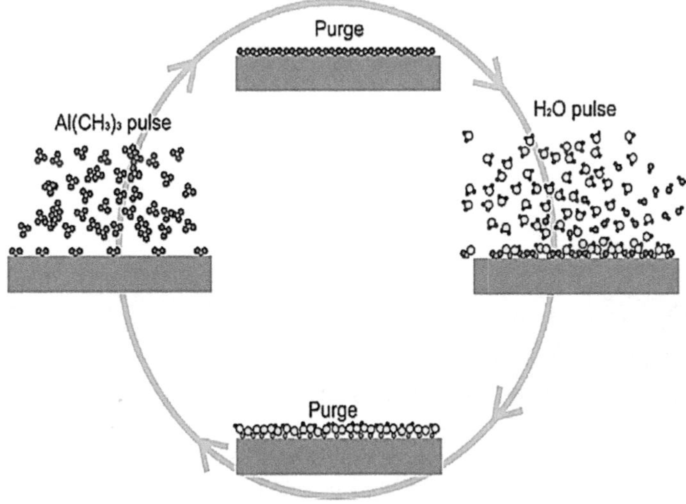

FIGURE 2.5 ALD processing [22].

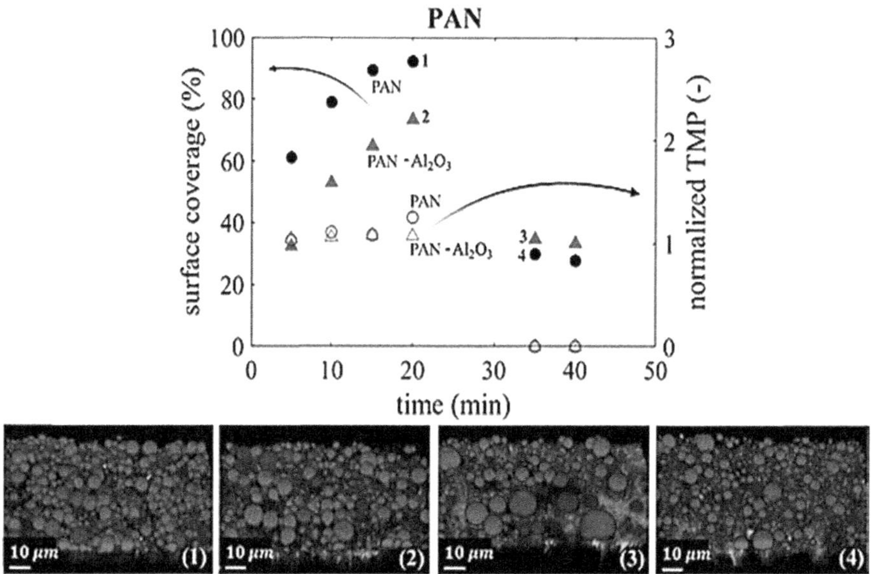

FIGURE 2.6 Top: membrane surface coverage and normalized transmembrane pressure (TMP) (TMP/TMP0); PAN vs. PAN-Al_2O_3 membrane performance at 300 LMH (PAN surface coverage, filled black dots; PAN-Al_2O_3 surface coverage, filled red triangles; PAN normalized TMP, empty black dots; and PAN-Al_2O_3 normalized TMP, empty red triangles). Bottom: confocal 3D reconstruction images of the membrane surface at representative times: red, oil droplets; green, membrane surface [25].

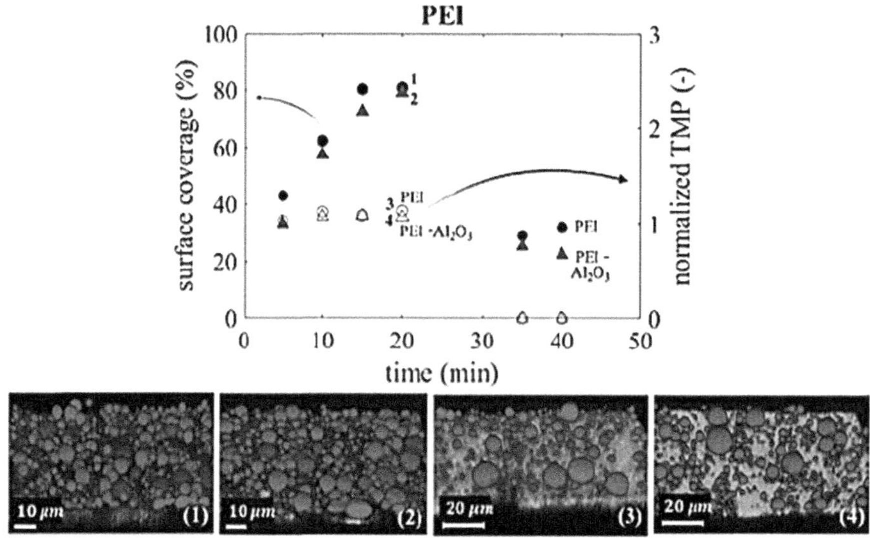

FIGURE 2.7 Top: droplet surface coverage and normalized TMP; PEI vs. PEI-Al_2O_3 membrane performance at 300 LMH (PEI surface coverage, filled black dots; PEI-Al_2O_3 surface coverage, filled blue triangles; PEI normalized TMP, empty black dots; and PEI-Al_2O_3 normalized TMP, empty blue triangle). Bottom: confocal microscopy of the membrane surface: red, oil droplets; green, membrane surface [25].

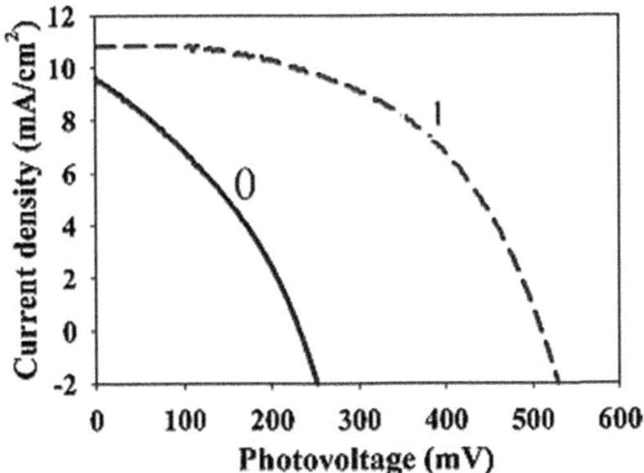

FIGURE 2.8 J-V curves for dye-sensitized solar cells (DSSCs) based on naked SnO_2 (solid red line) and on SnO_2 coated with one cycle of Al_2O_3 (dashed green line) [26].

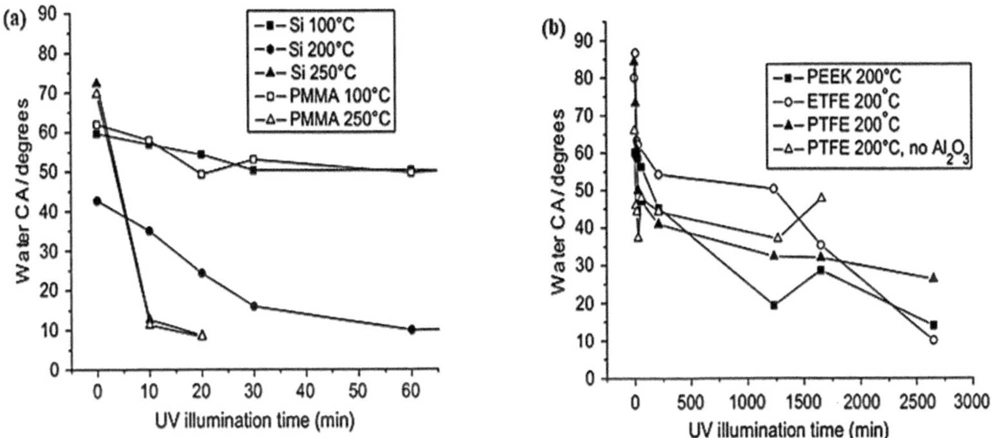

FIGURE 2.9 Average water contact angles of the TiO_2 films deposited on Si and different types of thermoplastics. (a) Si and PMMA, (b) PEEK, EFTE, and PTF as a function of UV illumination time [27].

2.2.1.4 Chemical Etching

Chemical etching is an effective method that is widely used to modify multifunctional surfaces. This process contains three main steps. Surfaces are firstly cleaned, mostly with deionized water/acetone/ethanol or hydrofluoric acid plus water/acetone/methanol. Second, the cleaning surface is immersed in the etching solution, which can be hydrochloric acid or hydrofluoric acid combined with hydrochloric acid to avoid the toxicity of HF [28]. Finally, there is a rinsing and drying process. The chemical etching processing was used to produce superhydrophobic surfaces. Here, the use of dislocation etchant dissolves the dislocation sites in the grains [29]. The surfaces were etched by using fluoroalkylsilane, and they exhibited showed superhydrophobic properties with a water contact angle larger than 150° (Figure 2.10).

Tian Shi et al. [30] employed chemical etching processing to modify a multifunctional surface (Al-Mg alloy) and promote its properties for anticorrosion behavior after being etched with fluoroalkylsilane. The processing of Al-Mg samples was executed by five steps represented as P, P-F, P-E-F, P-B-F, and P-E-B-F, where P, F, E, and B are polishing, fluoroalkyl-silane, etching, and boiling modification processing. They demonstrated that the isolation effect on seawater for modified surfaces was better than that for unmodified surfaces (Figure 2.11) [30, 31].

FIGURE 2.10 (a–c) SEM images of aluminum surfaces etched with a Beck`s dislocation etchant at room temperature for (a) 5 s, (b) 10 s, and (c) 15 s. (d) The photograph of 8 μL water drops [29].

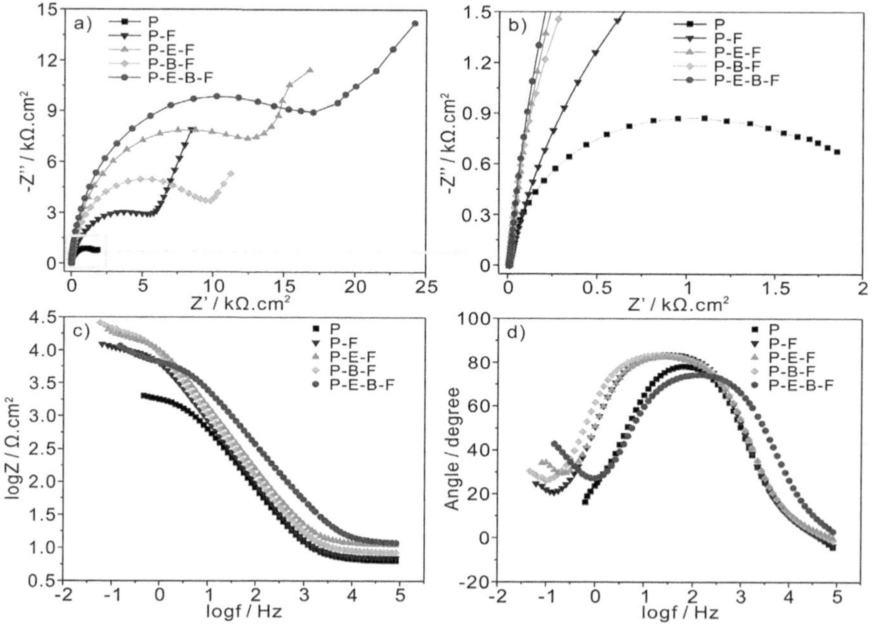

FIGURE 2.11 (a–b) Corresponding Nyquist plot and its partial enlarged details. (c–d) Bode plot from electrochemical impedance spectroscopy measurements of processed surfaces with 3.5% seawater [30].

2.3 PHYSICAL PROCESSING

Physical processing is the other approach used to modify multifunctional surfaces, and it offers more precise surface modification. The precise controllability enables the elimination of harsh surface roughening and surface damage. Physical surface modification is an environmentally safe and clean process [32]. This approach includes many methods, such as physical evaporation deposition (PVD), femtosecond laser surface processing, plasma, electron beam processing, imprinting, gamma-ray, and direct laser writing. In this section, the first four processing methods will be discussed in detail.

2.3.1 Physical Evaporation Deposition (PVD)

The physical evaporation deposition (PVD) method is used to produce or modify multifunctional surfaces for advanced applications [33]. PVD has many benefits: It can be used at low temperatures in a relatively inert environment [34], it improves the properties of surfaces, inorganic and some types of organic materials can be used, and it generates thin layers in nano to micro scale [35].

The deposition process steps using PVD begin with transferring the materials on an atomic level by evaporating the target material with a high-energy ion source under vacuum and inert gas to produce a vapor phase. After that, the vaporized material moves to the substrate surface and is placed in the chamber. Here, the reaction will occur only when the deposited materials are metal oxide, carbide, or nitrite. Finally, the vaporized atoms reach the substrate surface and are deposited as a thin layer [36]. The PVD process is shown in Figure 2.12.

The PVD method has been used to modify the surface of multifunctional materials because it covers and produces scalable surface areas with good properties for a variety of applications. Azizi et al. [37] have used the PVD process to modify an indium tin oxide (ITO) surface with gold for DNA biosensors. The modified surface morphology is changed, and the surface has homogenous layers on the DNA probe. Also, they observed a great improvement in electroactivity of the Au/ITO electrode, where the charge transfer resistance of ITO decreased to 89.01 Ω compared with that for the unmodified surface (381.01 Ω), and increasing in apparent charge transfer rate is more than four times. Furthermore, the modified surface Au/indium tin oxide (ITO) showed better results in detecting *Salmonella* DNA compared with other methods (Table 2.3).

The PVD process was used to study the electrochemical performance of a deposited C_{60} layer with different thicknesses on a carbon paste electrode (CPE) as reported by Lokesh et al. [43]. They found

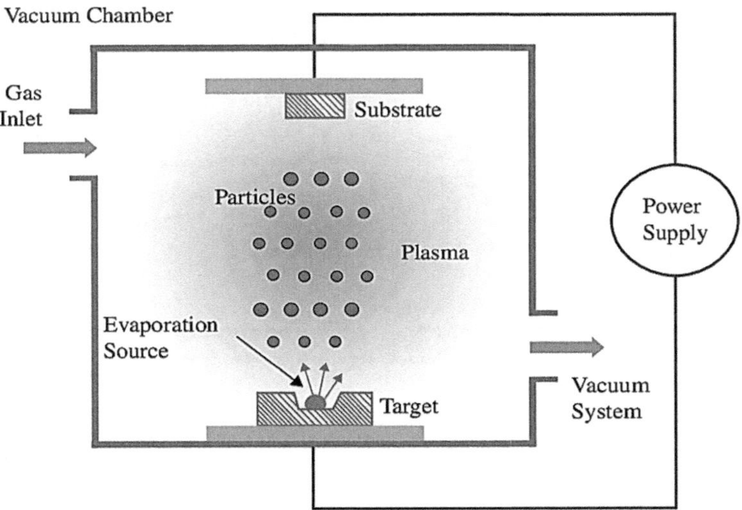

FIGURE 2.12 The steps of CVD processing [36].

TABLE 2.3
Results of *Salmonella* DNA Detection [38]

Method	Linear Range	Limit of Detection	References
SPR*	2–40 fM	2 fM	[39]
EIS	0.01–0.4 nM	0.084 nM	[40]
DPV	10–400 pM	2.1 pM	[41]
DPV	5–30 μM	0.208 μM	[42]
SPR	5–1000 nM	0.5 nM	[38]
DPV	20–240 fM	10 fM	[38]

* Surface plasmon resonance.

that the modified CPE was covered by a thin carbon layer, and the surface was much smoother than the unmodified CPE. Additionally, the C/CPE film exhibited high catalytic activity for dopamine and potassium ferricyanide (Figure 2.13) as well as high stability during the redox process.

X. Sun et al. [44] modified a graphitic carbon nitrate sheet (CNS) via the PVD process. They realized that the CNS sheet showed unchanged tri-s-triazine structure with existing rich surface

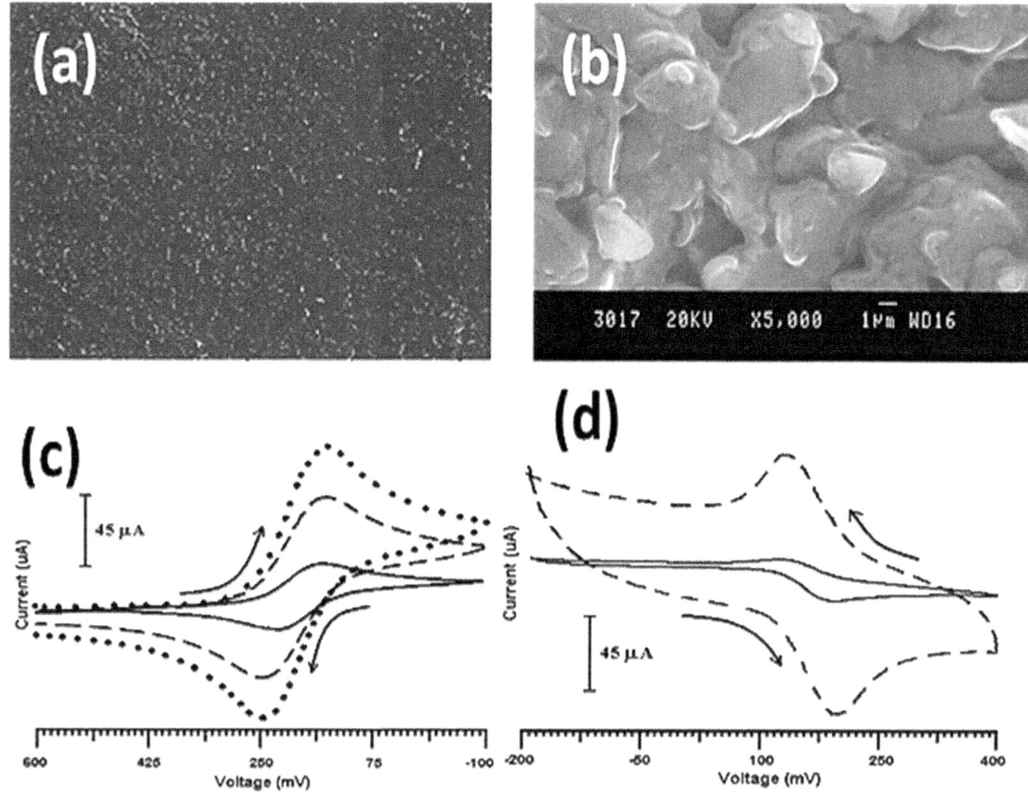

FIGURE 2.13 (a–b) SEM images of bar carbon paste electrode and C_{60} deposited carbon paste electrode. (c) CVs of ferricyanide at bare CPE (solid line), at C60/CPE (dotted line) where coating of C60 = 3 Å, and at C60/CPE (dashed line) where coating of C60 = 6 Å. Supp. electrolyte: KNO_3; scan rate: 100 mV s^{-1}. (d) CVs of dopamine at bare CPE (solid line) and at C60/CPE (dashed line). Supp. electrolyte: 0.2 mol L^{-1} phosphate buffer (pH 6.5); scan rate: 100 mV s^{-1} [43].

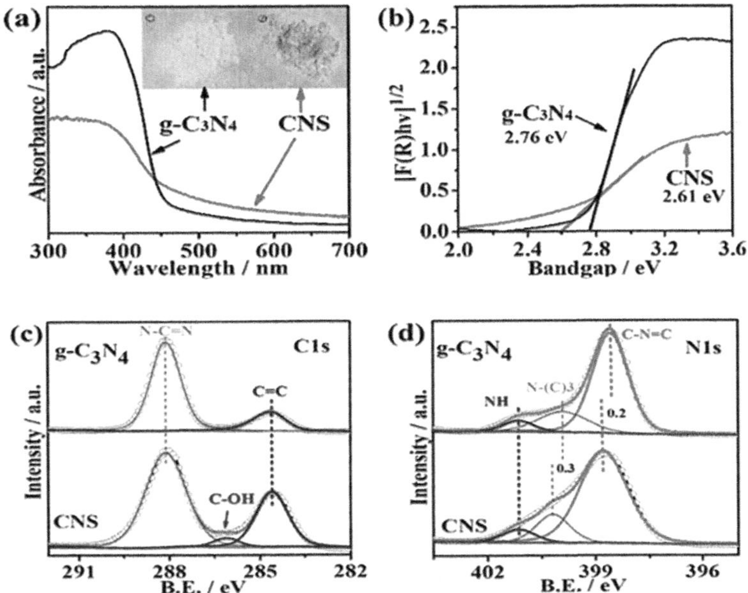

FIGURE 2.14 (a) UV/vis absorption spectra; (b) optical bandgap; the XPS spectra of (c) C1s and (d) N1s of CNS samples [44].

defects involving N-H, O-H, and C≡N groups. They also observed that the energy gap in the CNS was modified and reached the value of 2.76 eV (Figure 2.14).

2.3.2 Femtosecond Laser Surface Processing

Recently, the femtosecond laser (fs) process has become a favorable technique because it has shown promising applications in many aspects of science and technology [45]. It was selected as a good candidate technique to modify surfaces to enhance their properties, especially their wettability and optical properties [46–48]. This type of technique is able to form uniform periodic structure in nano/micro range with fewer defects [49]. The fundamental mechanisms of surface modification by fs are multiphoton phenomena and nonthermal processes [45]. Several materials are prepared and modified by using this technique to increase their performance. For example, the femtosecond laser is used to irradiate silicon surfaces. Such treated silicon surfaces showed a large range starting from low UV to a near infrared (200–2500 nm) [50]. Vorobyev et al. [51] used a femtosecond laser to treat some materials, such as platinum, titanium, and bare samples, to study their self-cleaning. They found that water droplets took away a large amount of dust particles from the modified surface and enhanced light absorption. In a solar application, femtosecond laser processing showed a remarkable effect on hafanium carbide and tantalum carbide surfaces, where the solar absorbance of a modified HfC surface was higher than that for an unmodified surface, as observed by Sciti et al. [52, 53]. In the field of optics, modified surfaces processed by femtosecond laser processing showed good antireflection and light transmittance properties [54]. Searchen et al. [55] produced a laser step-like structure in hydrogel using femtosecond laser processing, where this pattern appeared when laser cutting inside transparent materials. This chemical modification led to a decrease in the chemical depolymerization distance to reach the damage threshold in the original region.

Modification of antimicrobial surfaces by femtosecond laser processing introduces good results in medical, biological, clinical, and industrial applications. Chao Chen et al. [56] have worked on developing long-lasting and nonleaching bacterial surfaces. They combined micro scale patterning of borosilicate glass by ultrashort pulsed irradiation and a nonleaching layer-by-layer polyelectrolyte treatment of surfaces to obtain bactericidal surfaces. They found that the modified surfaces resulted in an improved antibacterial effect against two types of bacteria: *Staphylococcus aureus* (gram positive) and *Escherichia coli* (gram negative) (Figure 2.15).

2.3.3 Plasma Processing

The plasma technique is an effective process to enhance multifunctional surfaces for increasing their chemical and physical properties. The principle of plasma processing can be described in terms of the following steps: Firstly, ionizing states of materials electrically form charged particles (positive and negative ions). Then, the plasma will occur by generating electromagnetic radiation when the positive and negative charges reach the ground state [57, 58]. Generally, the plasma is classified according to different parameters such as temperature (low- or high-temperature plasma), pressure (low, medium, or atmospheric pressure plasma), mode (microwave radio frequency, corona and dielectric barrier discharge), gas (air, argon, helium, or oxygen), ionization (partially or completely ionized) and thermodynamic equilibrium (thermal or nonthermal plasma) [58]. However, plasma with all its types has played a fantastic role in the modification of surfaces in order to enhance their properties to be suitable for scientific, biomedical, and industrial applications, as well as food processing. Wang et al. [59] reported that increased antibacterial activity was achieved by coating Ag ions on TiO_2 surfaces using plasma processing. The modified surface showed a good antimicrobial effect against adherent and planktonic bacteria, as well as regulating the expression levels of biofilm-related genes toward inhibiting adhesion and biofilm formation. Nickel plasma modification of graphene surface was prepared by Wu et al. [60]. The modified surface exhibited excellent glucose sensitivity, low detection, a short time response, and long stability.

Cho et al. [61] used oxygen plasma to depose uniform nanostructure polyethylene terephthalate substrate to obtain antireflection and superhydrophobic films. They observed that the reflectance of the modified surface was about 4.2%, which was 56% lower than that of the unmodified surface. Khongnakorn et al. [62] have used plasma grafting polymerization processing to modify forward osmosis (FO) for protein recovery.

2.3.4 Electron Beam Processing

Electron beam processing is a common method used to modify a surface with excellent chemical, physical, and biomedical properties [3]. The main idea for the electron beam technique is that the transformation of electrons' kinetic energy to heat brings a specific heat distribution from the surface to the bulk. Here, the rates of heating and cooling are relatively high and lead to changes in the chemical composition, microstructure, and topography of the surface [63]. Compared with other surface processing methods, the modification of multifunctional surfaces using electron beam processing showed unique features such as low cost, a short process time, uniform energy electron beam distribution, and highly reproducible technological conditions [64]. Grenadyorov et al. [3] have used the electron beam to modify a titanium alloy surface to reduce blood hemolysis in cardiac assist devices. The result demonstrated that the destruction of red blood cells did not occur after the surface treatment compared with a similar pump without surface treatment. Valkov et al. [65] modified a Co-Cr-Mo alloy surface using electron beam processing and found that the phase composition of the modified alloy surface changed from $\varepsilon + \gamma$ to a single-phase γ structure. The surface roughness and the corrosion resistance of the modified Co-Cr-Mo alloy were significantly increased. Thite et al. [66] employed electron beam processing to synthesize silver particles in cotton fabric to produce a multifunctional cotton fabric surface. They reported that the silver particles were distributed uniformly in modified fabric samples. Also, they showed color changing from a light to a deeper shade, blocking reflectance and transmittance spectra in the UV region. The modified fabric surface exhibited better antibacterial behavior than the unmodified one, as shown in Figures 2.16 and 2.17.

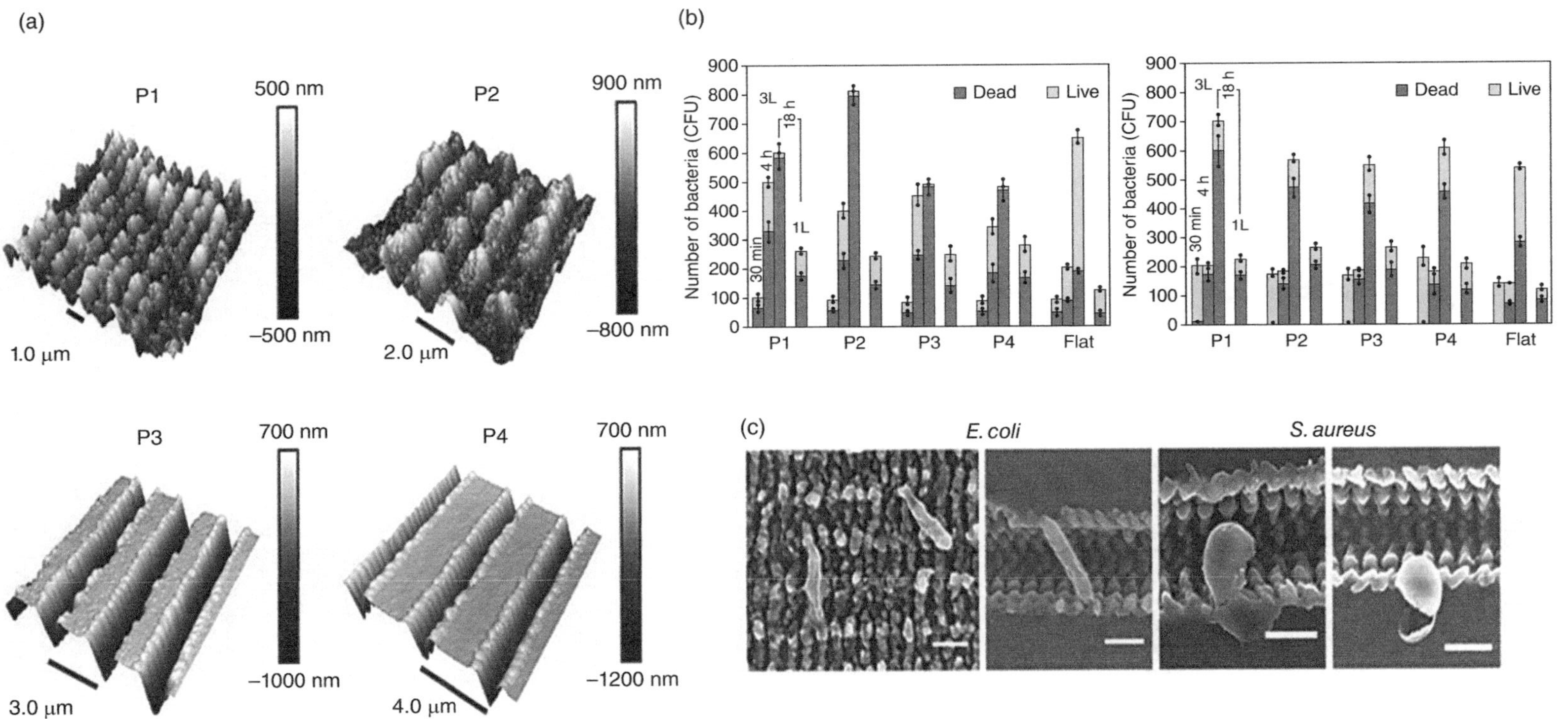

FIGURE 2.15 (a) Surface topography (3D rendering from AFM analysis) after laser patterning. (b) Adsorbed cell population and viability for *E. coli* and *S. aureus* on the treated surfaces. Each group of three columns represents the number of adsorbed bacteria on 3L-modified surfaces after 30 minutes, 4 hours, and 18 hours (from left to right). The data related to each bar correspond to the average number of dead (in red) and live (in green) colony-forming units (CFU) of bacteria incubated for each condition. (c) High-magnification SEM images of bacteria with ruptured cell envelopes. Scale bars, 1 μm [56].

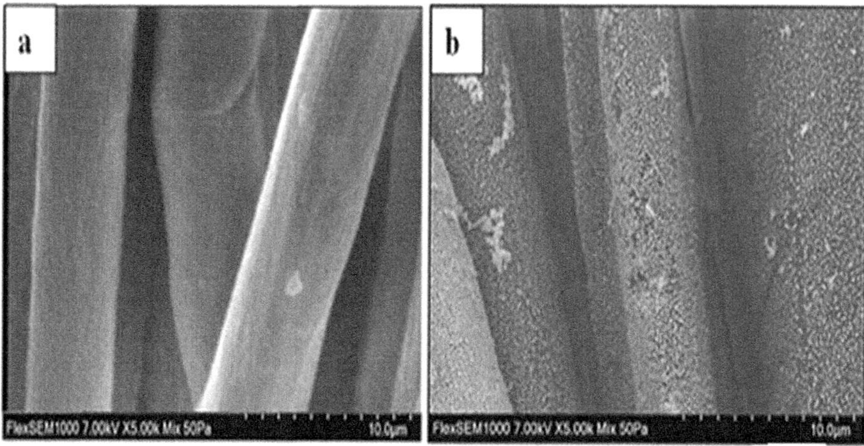

FIGURE 2.16 SEM images of cotton fabric (a) without treatment (control), (b) with electron beam radiation synthesized silver nanoparticles (1% wt/vol, 50 kGy) [66].

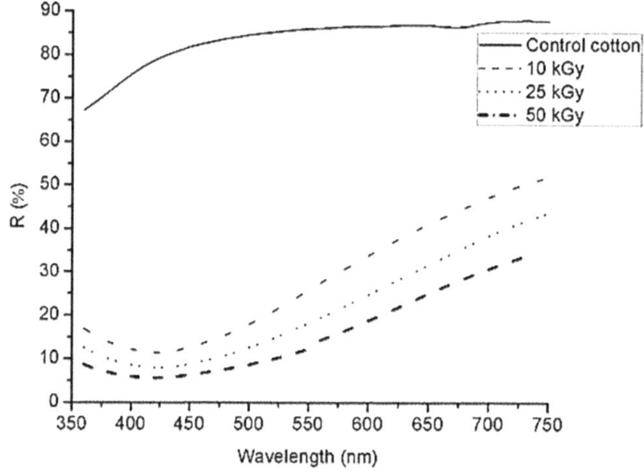

FIGURE 2.17 Reflectance spectra of fabric at 1% (wt/vol) concentration silver nitrate solution for padding [66].

2.4 OUTLOOK

This chapter focused on the most common chemical and physical processes that are used to modify multifunctional surfaces. These modified processes offer many ways to enhance the chemical and physical properties of multifunctional surfaces in order to promote their performance to be suitable for many scientific, technological, medical, biological, and industrial applications.

REFERENCES

1. V.P. Rotshtein, D.I. Proskurovsky, G.E. Ozur, Y.F. Ivanov, A.B. Markov, Surface modification and alloying of metallic materials with low-energy high-current electron beams. Surf. Coatings Technol. 180–181 (2004) 377–381. https://doi.org/10.1016/j.surfcoat.2003.10.085.
2. G.E. Ozur, D.I. Proskurovsky, V.P. Rotshtein, A.B. Markov, Production and application of low-energy, high-current electron beams. Laser Part. Beams. 21 (2003) 157–174. https://doi.org/10.1017/S0263034603212040.

3. A.S. Grenadyorov, A.A. Solovyev, K. V. Oskomov, S.A. Onischenko, A.M. Chernyavskiy, M.O. Zhulkov, V. V. Kaichev, Modifying the surface of a titanium alloy with an electron beam and a-C:H:SiOx coating deposition to reduce hemolysis in cardiac assist devices. Surf. Coatings Technol. 381 (2020) 125113. https://doi.org/10.1016/j.surfcoat.2019.125113.
4. M. Karaman, M. Gürsoy, M. Kuş, F. Özel, E. Yenel, Ö.G. Şahin, H.D. Kivrak, Chemical and Physical Modification of Surfaces, in: Surf. Treat. Biol. Chem. Phys. Appl., 2017: pp. 23–66. https://doi.org/10.1002/9783527698813.ch2.
5. M. Ozdemir, C.U. Yurteri, H. Sadikoglu, Physical polymer surface modification methods and applications in food packaging polymers. Crit. Rev. Food Sci. Nutr. 39 (1999) 457–477. https://doi.org/10.1080/10408699991279240.
6. T.-X. Fan, S.-K. Chow, D. Zhang, Biomorphic mineralization: From biology to materials. Prog. Mater. Sci. 54 (2009) 542–659. https://doi.org/10.1016/j.pmatsci.2009.02.001.
7. S. Mathew Simon, G. George, Sajna M S, Prakashan V P, T. Anna Jose, P. Vasudevan, A.C. Saritha, P.R. Biju, C. Joseph, N. V. Unnikrishnan, Recent advancements in multifunctional applications of sol-gel derived polymer incorporated TiO_2-ZrO_2 composite coatings: A comprehensive review. Appl. Surf. Sci. Adv. 6 (2021) 100173. https://doi.org/10.1016/j.apsadv.2021.100173.
8. R.B. Figueira, I.R. Fontinha, C.J.R. Silva, E.V. Pereira, Hybrid sol-gel coatings: Smart and green materials for corrosion mitigation. Coatings. 6 (2016). https://doi.org/10.3390/coatings6010012.
9. T. Zhang, M. Li, B. Su, C. Ye, K. Li, W. Shen, L. Chen, Z. Xue, S. Wang, L. Jiang, Bio-inspired anisotropic micro/nano-surface from a natural stamp: Grasshopper wings. Soft Matter. 7 (2011) 7973–7975. https://doi.org/10.1039/C1SM05366C.
10. A. Cannavale, F. Fiorito, M. Manca, G. Tortorici, R. Cingolani, G. Gigli, Multifunctional bioinspired sol-gel coatings for architectural glasses. Build. Environ. 45 (2010) 1233–1243. https://doi.org/10.1016/j.buildenv.2009.11.010.
11. R. Prado, G. Beobide, A. Marcaide, J. Goikoetxea, A. Aranzabe, Development of multifunctional sol-gel coatings: Anti-reflection coatings with enhanced self-cleaning capacity. Sol. Energy Mater. Sol. Cells. 94 (2010) 1081–1088. https://doi.org/10.1016/j.solmat.2010.02.031.
12. M.J. Hampden-Smith, T.T. Kodas, Chemical vapor deposition of metals: Part 1. An overview of CVD processes. Chem. Vap. Depos. 1 (1995) 8–23.
13. G. Li Puma, A. Bono, D. Krishnaiah, J.G. Collin, Preparation of titanium dioxide photocatalyst loaded onto activated carbon support using chemical vapor deposition: A review paper. J. Hazard. Mater. 157 (2008) 209–219. https://doi.org/10.1016/j.jhazmat.2008.01.040.
14. M. Fraga, R. Pessoa, Progresses in synthesis and application of sic films: From CVD to ALD and from MEMS to NEMS. Micromachines. 11 (2020). https://doi.org/10.3390/MI11090799.
15. M. Saeed, Y. Alshammari, S.A. Majeed, E. Al-Nasrallah, Chemical vapour deposition of graphene—Synthesis, characterisation, and applications: A review. Molecules. 25 (2020). https://doi.org/10.3390/molecules25173856.
16. O. Zaytseva, G. Neumann, Carbon nanomaterials: Production, impact on plant development, agricultural and environmental applications. Chem. Biol. Technol. Agric. 3 (2016) 1–26.
17. A.C. Ferrari, F. Bonaccorso, V. Fal'Ko, K.S. Novoselov, S. Roche, P. Bøggild, S. Borini, F.H.L. Koppens, V. Palermo, N. Pugno, Science and technology roadmap for graphene, related two-dimensional crystals, and hybrid systems. Nanoscale. 7 (2015) 4598–4810.
18. Z. Han, Z. Mu, W. Yin, W. Li, S. Niu, J. Zhang, L. Ren, Biomimetic multifunctional surfaces inspired from animals. Adv. Colloid Interface Sci. 234 (2016) 27–50. https://doi.org/10.1016/j.cis.2016.03.004.
19. S. Hosseininasab, N. Faucheux, G. Soucy, J.R. Tavares, Full range of wettability through surface modification of single-wall carbon nanotubes by photo-initiated chemical vapour deposition. Chem. Eng. J. 325 (2017) 101–113. https://doi.org/10.1016/j.cej.2017.05.034.
20. C.S. Torres-Castillo, J.R. Tavares, Covalent functionalization of boron nitride nanotubes through photo-initiated chemical vapour deposition. Can. J. Chem. Eng. (2022). https://doi.org/10.1002/cjce.24440.
21. J.J. Alcantar-Peña, E. de Obaldia, J. Montes-Gutierrez, K. Kang, M.J. Arellano-Jimenez, J.E. Ortega Aguilar, G.P. Suchy, D. Berman-Mendoza, R. Garcia, M.J. Yacaman, O. Auciello, Fundamentals towards large area synthesis of multifunctional ultrananocrystalline diamond films via large area hot filament chemical vapor deposition bias enhanced nucleation/bias enhanced growth for fabrication of broad range of multifunctional devices. Diam. Relat. Mater. 78 (2017) 1–11. https://doi.org/10.1016/j.diamond.2017.07.004.
22. M. Leskelä, J. Niinistö, M. Ritala, Atomic layer deposition. Compr. Mater. Process. 4 (2014) 101–123. https://doi.org/10.1016/B978-0-08-096532-1.00401-5.
23. P.O. Oviroh, R. Akbarzadeh, D. Pan, R.A.M. Coetzee, T.-C. Jen, New development of atomic layer deposition: Processes, methods and applications. Sci. Technol. Adv. Mater. 20 (2019) 465–496. https://doi.org/10.1080/14686996.2019.1599694.

24. S.M. George, Atomic layer deposition: An overview. Chem. Rev. 110 (2010) 111–131. https://doi.org/10.1021/cr900056b.
25. T. Itzhak, N. Segev-Mark, A. Simon, V. Abetz, G.Z. Ramon, T. Segal-Peretz, Atomic layer deposition for gradient surface modification and controlled hydrophilization of ultrafiltration polymer membranes. ACS Appl. Mater. Interfaces. 13 (2021) 15591–15600. https://doi.org/10.1021/acsami.0c23084.
26. C. Prasittichai, J.T. Hupp, Surface modification of SnO2 photoelectrodes in dye-sensitized solar cells: Significant improvements in photovoltage via Al2O3 atomic layer deposition. J. Phys. Chem. Lett. 1 (2010) 1611–1615. https://doi.org/10.1021/jz100361f.
27. M. Kemell, E. Färm, M. Ritala, M. Leskelä, Surface modification of thermoplastics by atomic layer deposition of Al2O3 and TiO2 thin films. Eur. Polym. J. 44 (2008) 3564–3570. https://doi.org/10.1016/j.eurpolymj.2008.09.005.
28. X. Li, Z. Huang, C. Zhi, Environmental stability of MXenes as energy storage materials. Front. Mater. 6 (2019) 2–10. https://doi.org/10.3389/fmats.2019.00312.
29. B. Qian, Z. Shen, Fabrication of superhydrophobic surfaces by dislocation-selective chemical etching on aluminum, copper, and zinc substrates. Langmuir. 21 (2005) 9007–9009. https://doi.org/10.1021/la051308c.
30. T. Shi, J. Kong, X. Wang, X. Li, Preparation of multifunctional Al-Mg Alloy surface with hierarchical micro/nanostructures by selective chemical etching processes. Appl. Surf. Sci. 389 (2016) 335–343. https://doi.org/10.1016/j.apsusc.2016.07.125.
31. D. Yu, J. Tian, J. Dai, X. Wang, Corrosion resistance of three-layer superhydrophobic composite coating on carbon steel in seawater. Electrochim. Acta. 97 (2013) 409–419.
32. F. Garbassi, M. Morra, E. Occhiello, F. Garbassi, Surface Analysis of Polymers, Polymer surfaces: from physics to technology, (1998) pp. 519–551.
33. X. Hou, H. Zhang, L. Jiang, Building bio-inspired artificial functional nanochannels: From symmetric to asymmetric modification. Angew. Chemie Int. Ed. 51 (2012) 5296–5307.
34. M.H. Staia, B. Lewis, J. Cawley, T. Hudson, Chemical vapour deposition of TiN on stainless steel. Surf. Coatings Technol. 76–77 (1995) 231–236. https://doi.org/10.1016/0257-8972(95)02527-8.
35. A.S.H. Makhlouf, 1 - Current and Advanced Coating Technologies for Industrial Applications, in: A.S.H. Makhlouf, I.B.T.-N. and U.-T.F. Tiginyanu (Eds.), Woodhead Publishing Series in Metals and Surface Engineering, Woodhead Publishing, 2011: pp. 3–23. https://doi.org/10.1533/9780857094902.1.3.
36. M.S. Rafique, M. Rafique, M.B. Tahir, S. Hajra, T. Nawaz, F. Shafiq, Synthesis methods of nanostructures. Nanotechnol. Photocatal. Environ. Appl. (2020) 45–56. https://doi.org/10.1016/B978-0-12-821192-2.00003-6.
37. S. Azizi, M.B. Gholivand, M. Amiri, I. Manouchehri, DNA biosensor based on surface modification of ITO by physical vapor deposition of gold and carbon quantum dots modified with neutral red as an electrochemical redox probe. Microchem. J. 159 (2020) 105523. https://doi.org/10.1016/j.microc.2020.105523.
38. D. Zhang, Y. Yan, Q. Li, T. Yu, W. Cheng, L. Wang, H. Ju, S. Ding, Label-free and high-sensitive detection of Salmonella using a surface plasmon resonance DNA-based biosensor. J. Biotechnol. 160 (2012) 123–128. https://doi.org/10.1016/j.jbiotec.2012.03.024.
39. A. Singh, H.N. Verma, K. Arora, Surface plasmon resonance based label-free detection of Salmonella using DNA self assembly. Appl. Biochem. Biotechnol. 175 (2015) 1330–1343.
40. N.T. Nguyet, L.T.H. Yen, V.Y. Doan, N.L. Hoang, V. Van Thu, H. lan, T. Trung, V.-H. Pham, P.D. Tam, A label-free and highly sensitive DNA biosensor based on the core-shell structured CeO2-NR@Ppy nanocomposite for Salmonella detection. Mater. Sci. Eng. C. 96 (2019) 790–797. https://doi.org/10.1016/j.msec.2018.11.059.
41. M. Amouzadeh Tabrizi, M. Shamsipur, A label-free electrochemical DNA biosensor based on covalent immobilization of salmonella DNA sequences on the nanoporous glassy carbon electrode. Biosens. Bioelectron. 69 (2015) 100–105. https://doi.org/10.1016/j.bios.2015.02.024.
42. T. García, M. Revenga-Parra, L. Añorga, S. Arana, F. Pariente, E. Lorenzo, Disposable DNA biosensor based on thin-film gold electrodes for selective Salmonella detection. Sensors Actuators B Chem. 161 (2012) 1030–1037. https://doi.org/10.1016/j.snb.2011.12.002.
43. S.V. Lokesh, B.S. Sherigara, Jayadev, H.M. Mahesh, R.J. Mascarenhas, Electrochemical reactivity of C60 modified carbon paste electrode by physical vapor deposition method. Int. J. Electrochem. Sci. 3 (2008) 578–587.
44. X. Sun, X. An, S. Zhang, Z. Li, J. Zhang, W. Wu, M. Wu, Physical vapor deposition (PVD): A method to fabricate modified g-C3N4 sheets. New J. Chem. 43 (2019) 6683–6687. https://doi.org/10.1039/C8NJ06509H.
45. E. Toyserkani, N. Rasti, Ultrashort pulsed laser surface texturing. Laser Surf. Eng. (2015) 441–453.
46. A.Y. Vorobyev, C. Guo, Direct femtosecond Laser surface nano/microstructuring and its applications. Laser, Photon. Rev. 7 (2013) 385–407.

47. K. Sugioka, Y. Cheng, Ultrafast lasers—Reliable tools for advanced materials processing. Light. Sci. Appl. 3 (2014) e149–e149.
48. K. Yin, H. Du, Z. Luo, X. Dong, J.-A. Duan, Multifunctional micro/nano-patterned PTFE near-superamphiphobic surfaces achieved by a femtosecond laser. Surf. Coatings Technol. 345 (2018) 53–60. https://doi.org/10.1016/j.surfcoat.2018.04.010.
49. T.R. Rublack, S. Krause, S. Schweizer, G.S. Seifert, Increasing solar-cell efficiency by femtosecond laser microstructuring. SPIE Newsroom. (2012). https://doi.org/10.1117/2.1201209.004466.
50. A.Y. Vorobyev, C. Guo, Antireflection effect of femtosecond laser-induced periodic surface structures on silicon. Opt. Express. 19 (2011) A1031–A1036.
51. A.Y. Vorobyev, C. Guo, Multifunctional surfaces produced by femtosecond laser pulses. J. Appl. Phys. 117 (2015) 33103.
52. D. Sciti, L. Silvestroni, D.M. Trucchi, E. Cappelli, S. Orlando, E. Sani, Femtosecond laser treatments to tailor the optical properties of hafnium carbide for solar applications. Sol. Energy Mater. Sol. Cells. 132 (2015) 460–466. https://doi.org/10.1016/j.solmat.2014.09.037.
53. D. Sciti, D.M. Trucchi, A. Bellucci, S. Orlando, L. Zoli, E. Sani, Effect of surface texturing by femtosecond laser on tantalum carbide ceramics for solar receiver applications. Sol. Energy Mater. Sol. Cells. 161 (2017) 1–6. https://doi.org/10.1016/j.solmat.2016.10.054.
54. X. Xie, Y. Li, G. Wang, Z. Bai, Y. Yu, Y. Wang, Y. Ding, Z. Lu, Femtosecond laser processing technology for anti-reflection surfaces of hard materials. Micromachines. 13 (2022). https://doi.org/10.3390/mi13071084.
55. E. Saerchen, S. Liedtke-Gruener, M. Kopp, A. Heisterkamp, H. Lubatschowski, T. Ripken, Femtosecond laser induced step-like structures inside transparent hydrogel due to laser induced threshold reduction. PLoS One. 14 (2019) 1–15. https://doi.org/10.1371/journal.pone.0222293.
56. C. Chen, A. Enrico, T. Pettersson, M. Ek, A. Herland, F. Niklaus, G. Stemme, L. Wågberg, Bactericidal surfaces prepared by femtosecond laser patterning and layer-by-layer polyelectrolyte coating. J. Colloid Interface Sci. 575 (2020) 286–297. https://doi.org/10.1016/j.jcis.2020.04.107.
57. H. Frey, Basic Principle of Plasma Physics BT - Handbook of Thin-Film Technology, in: H. Frey, H.R. Khan (Eds.), Springer Berlin Heidelberg, Berlin, Heidelberg, 2015: pp. 73–115. https://doi.org/10.1007/978-3-642-05430-3_4.
58. S. Birania, A.K. Attkan, S. Kumar, N. Kumar, V.K. Singh, Cold plasma in food processing and preservation: A review. J. Food Process Eng. 45 (2022) e14110. https://doi.org/10.1111/jfpe.14110.
59. J. Wang, J. Li, G. Guo, Q. Wang, J. Tang, Y. Zhao, H. Qin, T. Wahafu, H. Shen, X. Liu, X. Zhang, Silver-nanoparticles-modified biomaterial surface resistant to staphylococcus: New insight into the antimicrobial action of silver. Sci. Rep. 6 (2016) 32699. https://doi.org/10.1038/srep32699.
60. H. Wu, Y. Yu, W. Gao, A. Gao, A.M. Qasim, F. Zhang, J. Wang, K. Ding, G. Wu, P.K. Chu, Nickel plasma modification of graphene for high-performance non-enzymatic glucose sensing. Sensors Actuators B Chem. 251 (2017) 842–850. https://doi.org/10.1016/j.snb.2017.05.128.
61. E. Cho, M. Kim, J.-S. Park, S.-J. Lee, Plasma-polymer-fluorocarbon thin film coated nanostructured-polyethylene terephthalate surface with highly durable superhydrophobic and antireflective properties. Polymers (Basel). 12 (2020). https://doi.org/10.3390/polym12051026.
62. W. Khongnakorn, W. Bootluck, P. Jutaporn, Surface modification of FO membrane by plasma-grafting polymerization to minimize protein fouling. J. Water Process Eng. 38 (2020) 101633. https://doi.org/10.1016/j.jwpe.2020.101633.
63. G.P. Itkin, S.J. Shemakin, E.G. Shokhina, V.I. Burcev, P.V. Avramov, E.A. Volkova, D.V. Evljukhin, N.P. Shmerko, A. Mal'gichev, The first domestic implantable axial flow pump: Results of experimental studies in calves. Russ. J. Transplantology Artif. Organs. 15 (2014) 49. https://doi.org/10.15825/1995-1191-2013-3-49-58.
64. S. Valkov, M. Ormanova, P. Petrov, Electron-beam surface treatment of metals and alloys: Techniques and trends. Metals (Basel). 10 (2020). https://doi.org/10.3390/met10091219.
65. S. Valkov, S. Parshorov, A. Andreeva, M. Nikolova, P. Petrov, Surface modification of Co-Cr-Mo alloys by electron-beam treatment. IOP Conf. Ser. Mater. Sci. Eng. 1056 (2021) 12008. https://doi.org/10.1088/1757-899X/1056/1/012008.
66. A.G. Thite, K. Krishnanand, D.K. Sharma, A.K. Mukhopadhyay, Multifunctional finishing of cotton fabric by electron beam radiation synthesized silver nanoparticles. Radiat. Phys. Chem. 153 (2018) 173–179. https://doi.org/10.1016/j.radphyschem.2018.09.023.

3 Effect of Tribology on Multifunctional Coating Surface/Interface

Nurul Huda Abu Bakar and Siti Maznah Kabeb

3.1 FUNDAMENTALS OF TRIBOLOGY

Tribology is a study that deals with the design, friction, wear, and lubrication of interacting surfaces and encompasses how these surfaces and other tribo-elements behave in relative motion in natural and artificial systems. It is crucial in many industrial applications since friction is one of the main factors that cause corrosion or damage to the integrity of materials as the formation of microcracks is facilitated. Tribology is present in a wide variety of industries and engineering applications, such as products or assemblies (e.g., engines, curling stones), agriculture, exploration and construction (e.g., mine pumps, oil rigs), military activities, and others. Since every engineering material has a surface, any tribological interaction between a solid area and the environment or other materials may lead to the loss of the material from the surface of a solid body. The loss of material is known as wear.

Repetitive friction between two surfaces in sliding motion with respect to each other while exposed to a corrosive environment will initiate fretting corrosion that may lead to more rapid failure, such as abrasion corrosion through the formation of abrasive particles. Friction leads to energy dissipation as heat when the rubbing of surfaces results in the loss of resources [1]. The major cause of friction appears to be the forces of attraction known as adhesion and deformation [2]. Adhesive friction typically occurs in poorly lubricated sliding movements, such as between an automobile tire and the road, and the interacting surfaces are held together by the molecular binding force. Depending on the hardness of the material, adhesion and deformation friction might alternate randomly. Deformation friction occurs when a harder surface gets in between the asperities of softer surfaces, which leads to the formation of a characteristic microrelief on rubbing surfaces (Figure 3.1).

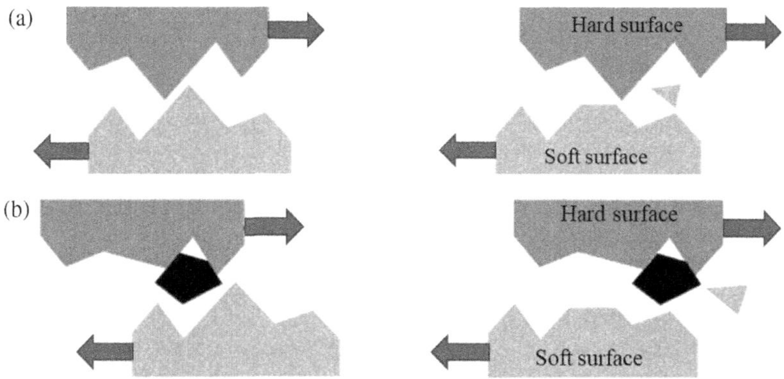

FIGURE 3.1 Abrasive wear mechanisms: Asperities on the harder material cut into the asperities of the softer material. The new particles become third bodies. (a) Two-body abrasive wear and (b) three-body abrasive wear. (Courtesy of the author.)

DOI: 10.1201/9781032635347-3

"Two body" wear is the wear that occurs without the interaction of debris. The formation of debris particles brings a "third body" into the sliding action, which can then significantly affect the wear process. This debris may be kept within the interface area, where it may become an "active" participant [3] in the wear process or may be ejected instantly after it forms, in which case it is referred to as "passive" debris. Active debris is typically small and may contain both metallic and oxide particles. Passive debris particles, on the other hand, tend to be much larger and may still retain a lot of their original shape and structure because they are immediately removed from the wear contact upon production. Passive debris is more likely to be metallic in metallic wear.

The primary effect of friction on objects is wear and tear, which shortens their lifespan. Kennedy et al. [4] reported that the service life of knee prosthetic tibial bearings made of ultra-high molecular weight polyethylene (UHMWPE) is estimated at less than 10 years owing to tribological failure. Patients suffer from aseptic loosening after joint replacement surgery, mainly caused by wear debris generated by the friction of artificial joint materials [5,6]. Excessive wear eventually leads to material failure, and wastage of materials also takes place. In buildings, wear and tear may result in water ingress and flooding, an increased risk of fire, and instability of structural and nonstructural components, and many buildings may present hazards from potential collapse, falling debris, damaged services, and unsanitary conditions. Numerous industrial situations involve impacting bodies in which failures due to wear can be costly, and safety, reliability, and quality can be significantly reduced.

There are many mechanisms by which wear occurs, and surfaces can wear due to one or more of these mechanisms simultaneously, including abrasive wear, adhesive wear, corrosive wear, erosive wear, and fatigue wear [7]. The process of wear might evolve over time or in response to operational conditions. Frictional heating typically speeds up wear through chemical and mechanical reactions. Abrasive wear results in the disintegration of the material on the surface when a hard surface or particles interact or slide on a soft surface and cause material loss. During abrasive wear, impurities like hard particles, dust, and sand in between the interacting surfaces are removed from the softer surfaces by hard particles (Figure 3.1). Moreover, adhesive wear occurs when two metals rub against each other vigorously enough to remove material from the less wear-resistant surface (Figure 3.2). Adhesive wear is dependent on physical and chemical factors such as material properties, the presence of a corrosive atmosphere or chemicals, and dynamics such as the velocity and applied load. The surfaces in motion are attracted to each other due to the molecular force of attraction, resulting in sticking.

Corrosive wear is an indirect wear mechanism in which the material surface gets disintegrated due to the chemical reaction between a corrosive medium (which can be either a chemical agent, lubricant, or atmospheric medium) and the material surface, which eliminates the corrosion prevention product. Corrosion wear can be considered an accelerated process of corrosion since it eliminates corrosive products and the passive protective layer more quickly than surfaces without any relative motion. In contrast, erosion wear is similar to adhesive wear—that is, the progressive material removal from a target surface due to the impact of particles traveling with significant velocity. Fatigue wear is damage that occurs as a result of strain that has been applied on the surface for a particular number of cycles to a certain critical limit (Figure 3.3). The cyclic loading creates stress

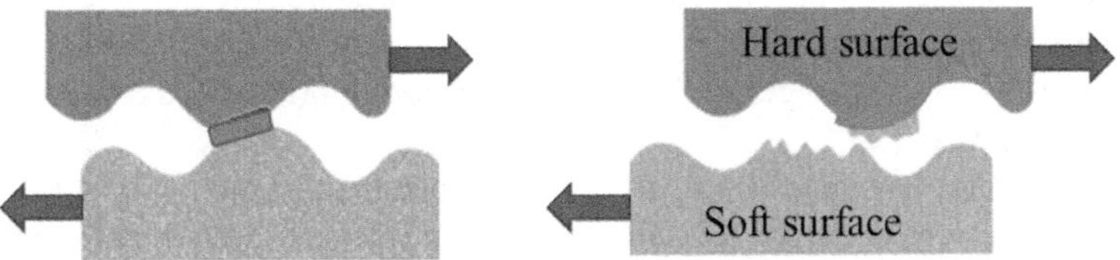

FIGURE 3.2 Adhesive wear mechanisms: Opposing asperities bond to each other and shear off as one surface slides over the other. (Courtesy of the author.)

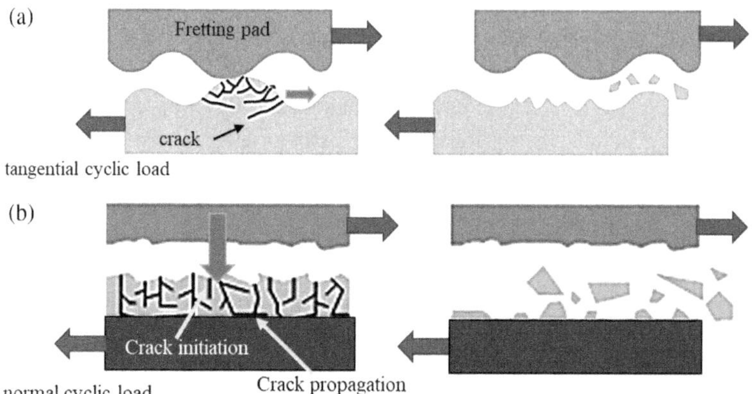

FIGURE 3.3 Fatigue wear mechanisms: Cyclical loading causes accumulation of micro-damage that breaks off as wear particles. (a) Fretting fatigue wear tangential cycling load and (b) fatigue wear of an overlay normal cyclic load. (Courtesy of the author.)

concentrations, causing defects to grow due to the multiple reversals. The new surface is thereby vulnerable to additional corrosive degradation.

The friction and wear can be prevented or minimized by introducing a friction-reducing film between moving surfaces in contact in order to separate the two sliding surfaces or by developing new wear-resistant materials [8]. This lubricant can be a solid, fluid, or plastic substance, with oil and grease being the most commonly used. In addition to reducing friction, lubricants also reduce wear, shield equipment from corrosion, regulate temperature and contaminant levels, transmit power, and act as fluid seals. Generally, lubricants can be classified into (a) hydrodynamic lubrication, (b) hydrostatic lubrication, (c) elastohydrodynamic lubrication (EHL), and (d) boundary layer lubrication [9]. The hydrodynamic lubrication is formed when there is no contact of asperities between the surfaces at the sliding interface and there is a gap of separation at the interface (completely separated) due to thick lubricant film formation developed by wedge action of the oil. In the case of bearings, hydrodynamic lubrication primarily takes place at high rotational speeds and low bearing loads.

Hydrostatic lubrication is a form of hydrodynamic lubrication in which the metal surfaces are separated by a complete film of oil, but instead of being self-generated, the separating pressure is supplied by an external oil pump. In contrast to hydrodynamic lubrication, which is dependent on the relative speed between the surfaces, oil viscosity, load on the surfaces, and clearance between the moving surfaces, hydrostatic lubrication depends on the inlet pressure of lube oil and clearance between the metal surfaces. Meanwhile, EHL is a type of hydrodynamic lubrication (HL) in which the surfaces experience significant elastic deformation, which significantly changes the thickness and shape of the lubricant layer in the contact. The term underlies how crucial the elastic deflection of the bodies in contact is to the formation of the overall lubricating layer. The lubricant is pulled in between the surfaces as a result of this elastic deformation, increasing the pressure, which maintains the load. Boundary lubrication occurs when the solid surfaces are sufficiently close to one another that the contact is dominated by the surface interactions between monomolecular or multimolecular films of lubricants (liquids or gases). Failure in boundary lubrication is caused by adhesive and chemical (corrosive) wear.

Advanced techniques have been implemented to reduce friction and wear [10,11]. One of these techniques is to apply deposits in the formation of coating layers on friction pairs. The surface must be coated with a material that produces low friction and wear in tribologically demanding applications. The following section evaluates the multifunctional coatings designed with various types of materials and structures for the tribological applications.

3.2 MULTIFUNCTIONAL COATING SURFACE/INTERFACE

Multifunctional coatings are modern coatings that respond well to conditions in the external environment, such as active ions, high temperatures, humidity, or the pH of the electrolyte. The surface/interface refers to the layers and the interphases around the layers deposited on top of a substrate (Figure 3.4). Therefore, multifunctional coating at the surface/interface must consist of a topcoat that is able to intervene in the interaction between the coating and the environment, as well as protect the underlying layers and the substrate from harsh environments [12].

Earlier studies demonstrated that multifunctional coatings could be achieved by microencapsulation technologies or by introducing porous materials such as zeolites of a few nanometers in size. Meanwhile, recent studies reported that multifunctional coatings containing nanoparticles such as zirconia could improve scratch and abrasion resistance properties [13]. The improvement was said to be due to the reaction of the hydroxyl groups on the ZrO_2 layer (with the addition of silane-functionalized organic coupling agents) and the formation of new cross-links. In another study, carbon nanotubes (CNTs) were proposed as a novel multifunctional coating material that could replace any environmentally hazardous antifouling materials [14]. Due to their dispersibility, the CNTs can also be functionalized and deposited by large-area deposition techniques. Another promising trend to produce multifunctional coatings is by combining different elements that each provide a specific function. In this regard, nanomaterials such as multiwall CNT (MWCNT) and TiO_2 have been proven to offer the opportunity because of their smaller sizes and synergistic protection effect of the surface underneath.

In recent years, the embedment of micro/nanocapsules containing a healing agent for the multifunctional coatings has been one of the effective methods widely employed in the bulk polymer matrix to generate a self-healing action [10,15]. The microcapsules that contain healing agents need to be sufficiently strong, have a long shelf life, and have excellent host-material bonding to demonstrate decent self-healing effects in response to externally controlled mechanical damage. Examples of self-healing agents are graphene oxide polyurea micro/nanocapsules and poly-urea-formaldehyde (PUF) microcapsules, as reported by Madelatparvar [16] and Mamat [17].

Generally, multifunctional coatings aim to modify the coating properties in order to achieve additional functions while retaining their original characteristics and performance. Many factors, such as coating adhesion and constituents of the multifunctional coatings, need to be investigated thoroughly, especially for tribological protection. In this context, the multifunctional coatings should encompass both passive components that are inherited from "classical" coatings (barrier

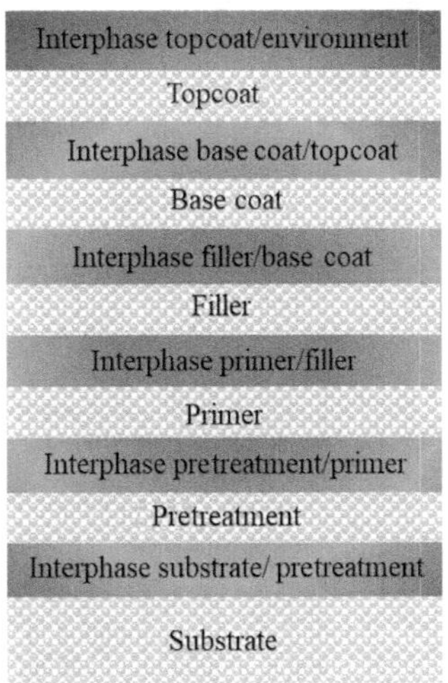

FIGURE 3.4 Multilayers deposited on a substrate for a multifunctional coating system. (Courtesy of the author.)

layers) and active components that enable quick responses of the coating properties to changes occurring either in the local environment surrounding the coating (e.g., temperature, humidity) or in the passive matrix of multifunctional coatings (e.g., cracks, local pH change).

3.3 ADVANCEMENT IN MULTIFUNCTIONAL COATINGS FOR TRIBOLOGICAL APPLICATIONS

Multifunctional coatings for tribological applications are widely found in construction equipment such as oil rigs and mine slurry pumps, tools such as knives or milling cutters, machine elements such as gears and bearings, and fluid machines such as blades, as well as biomedical implants, electronic devices, or even objects for daily use. The multifunctional coatings are intended to reduce the wear and tear that could be caused by the action of individual-specific wear mechanisms or by a combination of several basic mechanisms.

One of the important concepts of multifunctional coating for tribological characteristics is stress relaxation and crack deflection (Figure 3.5). The wear resistance and resistance to fracture (higher critical load) and lubrication performances can be enhanced through the multilayer design. This is because the stress distribution of multilayer coatings is influenced by the layer thickness. Additionally, from an engineering perspective, a multilayer coating with thin, soft layers can minimize the maximum bending stress when a normal force is applied to the coating's surface.

Numerous ways have been implemented on multifunctional coatings to fulfill the following requirements: (a) good adhesion to the substrate, (b) adequate hardness to withstand abrasion, (c) sufficient toughness and shear strength to prevent peeling, (d) compatibility to the substrate, (e) good chemical stability, (f) appropriate shear strength to resist contact pressures, (g) activity in forming a tribological film of lubricant or oxide, (h) adaptability to the substrate [18], and (i) improved tribological protection behaviors such as friction and lubrication. A tribological coating might be subjected to a variety of specifications, but the ideal coating should meet the abovementioned set of specifications [19].

He et al. [20] reported that multifunctional coatings that combined an $Al_{50}Ti_{50}N$ underlayer and an $Al_{60}Ti_{40}N$ upper layer deposited by physical vapor deposition (PVD) managed to improve the hardness, adhesion, and tribological characteristics. The primary factor relies on the stability of the cubic AlTiN phase due to the different quantities of atoms of Al and Ti which are able to distribute the residual stress. In addition, the bilayer of the Al/Ti coating is able to withstand a greater critical load due to the microstructure and the bonding force generated by the subsequent deposition. The bilayer coating possibly required a greater amount of force to overcome the bonds present between the two coating layers and the sublayer to the substrate. In contrast, the monolayer coatings could

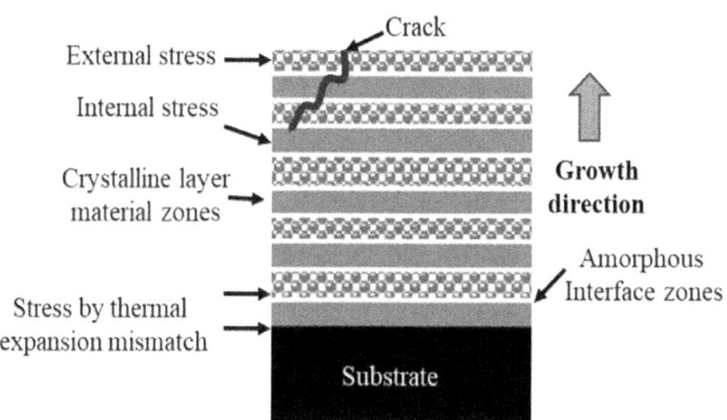

FIGURE 3.5 Mechanisms of the multilayer designing of coating. (Courtesy of the author.)

not withstand high critical loads as the thickness and adhesion directly relate to the degree of bonding between the coating and substrate [20].

Apparently, the thickness of the multilayer coating affects the tribological characteristics, and therefore optimizing the thickness is crucial. The thickness of the multilayered coatings is often connected to the stacking sequence, such as alternating hard/soft layers. The process offers a significant shear zone to prevent the fracture of hard (and brittle) layers upon deflection induced by the applied load. In this regard, one can observe that multilayered coatings incorporating many thin layers are an ideal way to ensure load-support properties. The alternation of layers with high/low shear modulus can also provide more benefits for multilayer coatings.

Apart from the thicknesses, the composition of the multifunctional coatings also affects the tribological characteristics. Multifunctional micro-diamond coatings have been prepared through chemical vapor deposition (CVD), and it was observed that only high-quality diamond coatings would enhance the tribological applications of stainless steel [21]. The enhancement could be achieved by surface scratching to obtain the structure of ultra-dense Q-carbon (a metastable allotrope of carbon) with high nucleation density, and the strains in the prepared diamond films were reduced. A different investigation revealed that transition-metal nitrides (TMN), such as Ti-based TMN coatings or Cr-based TMN coatings, can be used to take advantage of numerous thin layers to form a multilayer structure [10]. Generating nitrides and oxides on the layer was less complicated due to the Ti and Cr's ability to provide high adhesion between the many layers and metal substrates [22]. In addition, the crystallinity of Ti-based TMN phase is also increased, thereby decreasing the lattice strain with the increased layer thickness of Ti. Another investigation showed that the grain orientation of the TMN-based coatings could also influence the coefficient of friction (CoF) of the multifunctional coating. For instance, a (111) preferred grain orientation exhibited a lower CoF compared to that with (200) preferred grain orientation [23].

The performances of several TMN multifunctional coatings, including ZrN and MoN, have been investigated for many years. The rising need for industrial applications drives the ongoing development of TMN coating materials. Krysina et al. [24] reported that nanocrystalline protective (Zr,Nb)N coatings with different concentrations of niobium can be controlled by an electric arc evaporator using a zirconium cathode. It was observed that additional doping of the ZrN-based coating with niobium allows modification of the structural and phase composition of the coating. An increase in the concentration of niobium of up to 3 wt% results in a transition to a multiphase composition and finely dispersed structure, thereby increasing the hardness of the coating. Besides doping, the multilayered ZrN coatings can be improved by experimental parameter selection, such as applying substrate pretreatment upon vacuum-arc deposition with preliminary Ar^+ ion sputtering to allow dense and homogenous nanocrystalline coatings [25]. The adherence of the ZrN coatings has been improved by the preferential grains' orientation during the growth process and the formation of significant (111) texture.

Super-hard multifunctional coatings are considered technologically promising materials for tribology applications owing to their architectures, which act as crack inhibitors and fracture resistance. The super-hard multifunctional coatings of TiN/CrN, TiN/ZrN, and TiN/WN pairs were achieved by altering the lattice structure across the interface of the layers [26]. Meanwhile, the TiVN/TiSiN multilayered coating exhibited the same cubic NaCl-type structure as TiVN and TiSiN coatings, indicating that the alternate growth of TiVN and TiSiN in multilayered TiVN/TiSiN coatings resulted in periodic multilayered structure [27]. Most of the super-hard multifunctional coatings could expand the safe zone of cutting tools by minimizing the built-up edge (BUE) formation. Additionally, it is anticipated that the interface design, in conjunction with particular phases, will be able to control the wear rate and friction coefficient.

The mechanism of tribological protection by the multifunctional coatings containing hard nitride materials relies on three factors: (a) the structure and the solid distortion-induced strengthening effect caused by differently sized atoms [28,29]; (b) the formation of a multilayered structure with a coherent interface that improves performance owing to the combined effects of Hall–Petch strengthening, dislocation blocking, and epitaxial stability [30–32]; and (c) the incorporation of elemental carbon that improves the plastic toughness and tribological behavior of the metal nitride coatings [33,34]. The multi-element nitride coatings of ZrCrW(C)N (containing elemental carbon) have been proven to exhibit good plasticity and toughness, thereby improving tribological properties [35].

Transition-metal carbide (TMC)-based multifunctional coatings have also emerged as reliable coating materials for tribological applications. TMCs have interesting properties, including high hardness, high elastic modulus of up to 500 GPa, and modest fracture toughness [36]. In this case, elemental carbon (C) is added to the coating system, and the excess free carbon reduces the friction coefficient significantly without affecting the hardness or modulus [37]. In fact, the C element doping has a strengthening effect on the properties of the coating, especially when added into cemented carbide coating (CrN). Examples of TMC multifunctional coatings are (CrNbTaTiW)C [38] and (HfNbTaTiZr)C [39]. Such (HfNbTaTiZr)C coatings take the shape of nanocrystalline carbide grains produced by excessive carbon precipitation, which inhibits carbide grain growth and offers a large number of nucleation sites. In addition, similar (HfNbTaZr)C multimetal carbide coating was deposited at different substrate temperatures [40]. The prepared carbide coating exhibits nanocrystalline columnar morphology, with a slight increase in the grain size with increasing substrate temperatures. Furthermore, the addition of carbon to the metallic multicomponent coatings has led to improvement in indentation cracking resistance [41].

The TMC multifunctional coating of two-dimensional titanium carbide (Ti_3C_2) MXene has been prepared [42]. MXenes are new 2D materials that are produced by etching of the A element layers from bulk MAX phases, where M represents an early transition metal, A is a group IIIA or IVA element, and X can be either carbon (C), nitrogen (N), or both [43]. However, making MXene using the etching procedure and postprocessing is still challenging since these steps alter the chemical composition, defect, and flake dimension of $Ti_3C_2T_x$MXene. Until recently, several studies have been conducted on epoxy-based matrix composites with other 2D materials like graphene or boron nitride as the filler material for the multifunctional coatings [44]. Besides that, the transition-metal dichalcogenides (TMDCs) such as MoS_2 can also be used to fabricate multifunctional epoxy-based coatings [45]. The TMDCs material differs from the structure of graphene because the transition-metal atoms are sandwiched between two layers of chalcogen atoms.

Meanwhile, the tribological qualities have been improved by applying a tantalum carbide (TaC) composite of TMC coating [46]. It is worth noting that the C content varies as methane (CH_4) levels rise, as does the phase shift from hexagonal-Ta_2C to cubic-TaC, and to a typical nanocomposite structure composed of TaC nanocrystallites surrounded by an a-C matrix. The a-C matrix in the coating reveals that only an amorphous phase is present when the carbon content is at its greatest. In the meantime, a novel TaC@Ta core shell-like nanocomposite multifunctional coating was prepared via a self-assembly technique comprising TaC nanocrystalline enclosed with thin pseudocrystal Ta tissues by stimulating solid-state dewetting during layered deposition [47].

Transition-metal carbides (**TMC**) and nitrides (**TMN**), in the form of multilayer coatings, are successfully synthesized by the incorporation of carbon or nitrogen into the crystal lattices of the respective parent metals. They successfully demonstrate unique physical and chemical properties, which are found to be useful for tribologically demanding applications. Table 3.1 summarizes the advancement of multifunctional coatings for this specific use.

3.4 TRIBOLOGY EFFECT ON MULTIFUNCTIONAL COATINGS

Corrosion accelerated by tribological contact ensues when the passive film integrity is breached and the multifunctional coatings consistently lose their protective properties. A simultaneous effect of mechanical force (friction, abrasion, erosion, etc.) and corrosion when two surfaces come into contact while moving relative to one another (could be impingement, sliding, fretting, rolling, etc.) in a corrosive liquid is known as *tribocorrosion* (Figure 3.6a). Fretting wear occurs when adhesion,

TABLE 3.1
Multifunctional Coatings Used in Tribological Applications

Coating Material	Substrates	Deposition Method	Improvement	Ref.
Transition-Metal Nitride (TMN) Multifunctional Coatings				
(Zr,Nb)N	Titanium and steel	Vacuum-arc evaporation	Nb dopant altered the structure, low thickness (3–4 μm) did not have visible defects	[24]
ZrN	2024 Al-alloy	Vacuum-arc plasma with magnetic filter	High adhesion, homogenous, and dense nanocrystalline coating with thickness of ~1 μm	[25]
ZrCrW(C)N	AISI304 stainless steel	Vacuum-arc in N_2–C_2H_2 gas mixture	High nano-hardness and elastic modulus with a reduction in CoF of 14%	[35]
TiN/CrN, TiN/ZrN, and TiN/WN pairs	Si (100) and stainless steel (316LN)	DC magnetron sputtering	Strain-hardening by lattice mismatch across the TiN/WN interface	[26]
TiVN, TiSiN, and multilayered TiVN/TiSiN	Polished tungsten carbide	Cathodic-arc evaporation (CAE)	Higher resistance against plastic deformation compared to TiVN and TiSiN	[27]
TiAlN/TiAlCN multilayers	Steel AISI H11	DC magnetron sputtering / high-power impulse magnetron sputtering (HPiMS)	Coatings from DCMS exhibited higher adhesion to the substrates/coatings from HiPIMS, displayed higher resistance to abrasive wear	[48]
$Al_{0.7}Cr_{0.3}N$-$Al_{0.67}Ti_{0.33}$ N-based multilayer	High-speed steel (HSS) and cemented carbide (CC)	Industrial arc evaporation	Controlled residual stress state wear behavior at various temperatures with respect to their architecture	[49]
Transition-Metal Carbide (TMC) Multifunctional Coatings				
(CrNbTaTiW)C multicomponent	Stainless steel	Nonreactive DC-magnetron sputtering	Extremely hard multicomponent carbides	[38]
(HfNbTaTiZr)C	epi-polished c-plane sapphire	RF magnetron sputtering	High entropy carbide, broad range of carbon stoichiometries, highest hardness of 24 ± 3 GPa for $(Hf_{0.2}Nb_{0.2}Ta_{0.2}Ti_{0.2}Zr_{0.2})C$	[39]
TaC	Si and Ti6Al4V alloy	DC reactive sputtering	Controlled grain size with the width of matrix separation between TaC grain	[46]
TaC/Ta	Si and Ti6Al4V alloy	Alternately sputtering TaC and Ta targets	Ordered coherent TaC@Ta core–shell-like nanocomposite structure	[47]
(Ti_3C_2) MXene	SiO_2-coated silicon (Si)	Colloidal solution and spray-coated	Superlubricity against DLC with very low CoF	[42]

oxidation, and fatigue of materials happen consequently or simultaneously. The basic concept of tribocorrosion is the formation of a layer on the workpiece's surface as a result of the workpiece's contact with corrosive liquids or gases (Figure 3.6b). When sliding occurs between two metal surfaces or between a nonmetal and a metal surface in an aqueous environment, it results in a modification of the electrochemical processes occurring at the metal surface by the disturbance of layers on the surface, which may be an adsorbed monolayer or a thicker film of the reaction product. Long-term sliding of artificial joints, for instance, raises the temperature at the frictional surfaces, which is adverse to the adsorption of the synovial fluid on the sliding interfaces for lubrication [50].

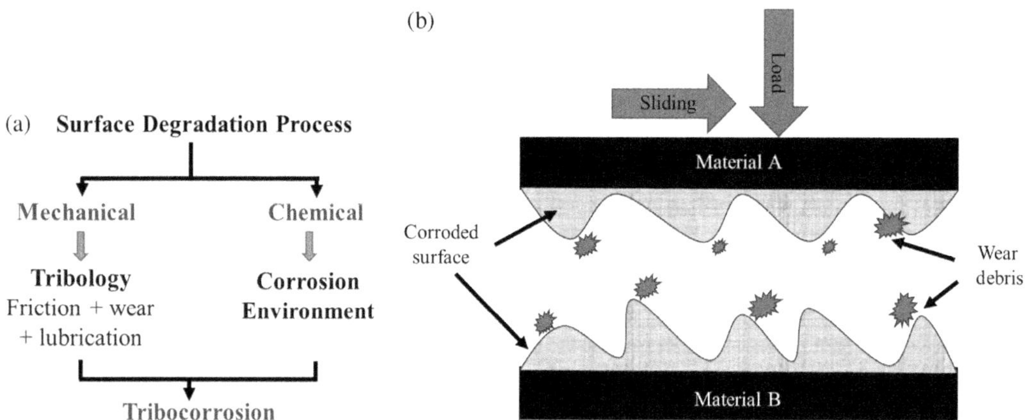

FIGURE 3.6 (a) Basic concept of tribocorrosion and (b) tribocorrosion between two sliding surfaces. (Courtesy of the author.)

The complexity of tribocorrosion lies in the fact that chemical and mechanical degradation mechanisms are not independent of one another; they influence each other, which leads to an increased rate of material loss [51]. Tribocorrosion is an effect of a pair of friction surfaces in synovial fluid, as exemplified by an implant of a hip joint endoprosthesis. In hip joint endoprostheses, friction occurs between the head and the acetabulum in which the surfaces in direct contact perform movements with less amplitude than the width of the contact surface at that point. Fretting in hip joint endoprostheses occurs in the case of stationary junctions. Friction is the most important factor in the artificial intervertebral disc, a spine implant used in total disc arthroplasty (TDA). The patient's motions result in friction between metallic surfaces. Tribocorrosive wear of the contact surfaces is a result of reciprocating friction, high contact pressure, and hostile surroundings. Inflammation of the tissues may develop after the implantation of orthopedic devices, and their chemical products may affect subsequent tribocorrosion behavior.

Most of the multifunctional coating systems work under conditions where there is constant mechanical interaction along with the corrosive environment. The material loss caused by wear is influenced by and is susceptible to material damage [52]. A corrosive environment can shorten the service life of the coatings by accelerating the rate of wear and, at the same time, enhancing or complementing corrosion damage. Tribological processes and frictional contacts contribute to the loss of about 23% of the total energy in the world. About 5% of this energy is used to overcome friction, and about 3% is used to build new parts or fix damaged ones. Therefore, good tribological practices allow for a 40% reduction in energy losses from friction and wear, which might add up to 1.4% of the gross domestic product (GDP) of any developed country [53]. Correspondingly, corrosion losses contribute to about 1% to 5% of the GDP of any nation. Additionally, this cost is increased by indirect corrosion-related losses such as production, repair, and breakdown expenses [54].

In many situations, such as in power generation, transportation, material processing, and turbine engines, high-temperature wear is a serious issue [55]. Faster surface oxidation kinetics, a decrease in the mechanical toughness and strength of the materials forming the contacting surfaces, and changes in adhesion between the coating and the materials underneath brought on by the combined effects of temperature and tribological factors all contribute to this wear. Tribocorrosion behavior depends on several parameters and can be classified into four major groups: (a) material properties, (b) electrochemical properties, (c) mechanical contact or tribological properties, and (d) working environment or solution [56] that affect wear or corrosion individually and eventually tribocorrosion as a whole.

The tribological coatings are generally designed to reduce friction between mating surfaces by means of decreasing interface wear [57]. Tribological coatings are also used as thermal barrier coatings (TBCs) that encounter heat generated in high-temperature environments to prevent engine surfaces from being oxidized and worn due to corrosion, thus reducing wear and eventually increasing the performance and component downtime [10]. At present, organic polymer materials used in metal surface coatings play a crucial role in drag reduction, enhancing wear resistance and corrosion resistance [10]. However, polymers and polymer composites are vulnerable to premature failure due to the formation of cracks and microcracks that arise due to changes in the environment or in mechanical properties that occur during coating processes. Therefore, the multifunctional coating materials listed in Table 3.1 seem more reliable, as discussed in the previous section.

The operation of tribological multifunctional coatings and their interfaces in moving machine components occurs more frequently under harsh contact conditions, especially those involving high temperatures primarily brought on by frictional heating. High-temperature tribology contacts operate in such applications as aerospace, power generation, automotive, and metalworking processes, sometimes at temperatures higher than 900°C. The term *high-temperature tribology* is ambiguous because there is no general limit as to what high-temperature means, and it is highly system dependent. A temperature considered high for a polymeric material will not be high for a metallic material such as steel or for a ceramic material. High-temperature tribology is considered to begin from a minimum temperature of 300°C to 350°C, where traditional lubricants like oils and greases begin to decompose and lose their lubricating effectiveness, up to a temperature of 1000°C. Furthermore, the operation of mechanical systems at high temperatures has detrimental effects on efficiency, performance, and reliability owing to the influence of temperature on friction and wear properties of contacting materials.

Noticeable effects induced due to the operation of tribological interfaces at elevated temperatures are the increased rate of tribochemical reactions (mainly oxidation) and degradation of the mechanical properties of the materials. The high-temperature friction and wear process is quite complex, involving changes in the physical, chemical, and mechanical properties of material surfaces. The need for lightweight materials in vehicles, for example, required complex shaped structural and safety components which can only be formed at elevated temperatures. Therefore, the interaction of the tool and workpiece at high temperatures gives rise to complex tribological phenomena. Optimization of friction in industry and transportation results in energy savings, improved quality and durability of products, and overall productivity.

3.5 CHALLENGES AND FUTURE PERSPECTIVES

Material deterioration associated with wear and corrosion has an incredible influence on the economics of engineering systems, which could result in material failure and high maintenance and replacement costs. The problem of tribocorrosion is encountered in numerous industries, for instance, nautical or marine [51], automotive [58], mining [59], biomedical (artificial joints, orthopedic plates and screws, dental implants) [58], and offshore industries [60]. In the fields of prosthetics and restorative dentistry, degradation by tribocorrosion leads to exposure to various microbial, biochemical, and electrochemical factors in the oral cavity, with consequent negative effects on human health [61].

Tribocorrosion systems are extremely complicated and depend on several factors, *viz.* mechanical, material, chemical, and electrochemical ones [52]. Tribocorrosion can be useful in certain applications, depending on the design and loading of the tribosystem. Unfortunately, the mechanisms of tribocorrosion are still not well understood. The factors that are interconnected make the tribocorrosion process more complex and difficult to monitor for each of the variables because a lot of factors vary with time, such as temperature and humidity.

In addition, the poor adherence to the substrate and the significant high residual stress induced by the fabrication process typically limit the performance of monolayer coatings. The performance of monolayer coatings is further hindered by the mechanical mismatch between substrates and coatings. Therefore, multilayer coatings emerged as a potential alternative to enhance coating performance through the multilayered structure that can control residual stress and enhance the toughness of the coated system. Commonly, multilayer coatings have multifunctional characteristics when "sandwich"-type structures are used. It is because each layer serves a variety of functions, including forming, antiwear, and anticorrosion.

Multifunctional coating systems can be developed through the rigorous growth of smart materials with desired tribological and physical and mechanical properties, such as a new generation of coatings with shape memory effects (SMEs). The SME consists of a multilayered surface composition with alternating layers of various functional applications and is expected to improve the operational properties of the working surfaces of products. The deformation behavior of samples with a composite surface layer of multicomponent SME materials, formed under high-energy impact during abrasive wear, significantly decreased the wear rate depending on the structural-phase state of the coatings and technology of their formation [62]. Likewise, using the SME approach, each layer in a multilayer coating would have significantly less stress than a thick multilayer if each layer slides over each other. Hence, alternating hard/soft layers or manufactured multilayer coatings with various ceramic materials are projected to improve adhesion and tribological protection with optimum layer thickness.

3.6 CONCLUSION

Research on multifunctional coatings necessitates a highly interdisciplinary approach encompassing many facilities and abilities. Many design strategies have been presented in this chapter to further enhance coating performance or to raise coatings' adaptability in various tribological situations. Among the numerous coating design concepts, multilayer coatings have attracted a great deal of attention in past years due to the added degree of freedom in tailoring the coating properties. They may be fabricated using various materials in several different forms and structures to meet specific requirements, and it is also suggested that molecular computer-based simulation be used to develop the multilayer formation and protection mechanisms associated with analytical tools for rapidly characterizing the performance in tribological applications. Overall, research on tribology is an emerging area for technologists in recent times, and the need for surface modifications with excellent performance is crucial. From this overview, we have also learned that multilayer coatings play an important role in the repair and remanufacturing of parts and components of the underlying metal, which efficiently reduces material loss and complies with industry standards for environmental protection.

LIST OF ABBREVIATION

Al Aluminum
BUE Built-up edge
C Carbon
CNT Carbon nanotubes
CoF Coefficient of friction
CVD Chemical vapor deposition
EHL Elastohydrodynamic lubrication
Hf Hafnium
HL Hydrodynamic lubrication
MEMS Micro-electromechanical systems
N Nitrogen
Nb Niobium
PUF Poly-urea-formaldehyde
PVD Physical vapor deposition
SME Shape memory effect
Ta Tantalum
TBC Thermal barrier coatings

TDA Total disc arthroplasty
Ti Titanium
TMC Transition-metal carbide
TMN Transition-metal nitride
W Tungsten
Zr Zirconia

REFERENCES

1. Anand, A., et al., *Role of Green Tribology in Sustainability of Mechanical Systems: A State of the Art Survey.* Materials Today: Proceedings, 2017. **4**(2): p. 3659–3665.
2. Blau, P.J., *Friction Science and Technology: from Concepts to Applications.* 2008: CRC press.
3. Blau, P.J., *Mechanisms for Transitional Friction and Wear Behavior of Sliding Metals.* Wear, 1981. **72**(1): p. 55–66.
4. Kennedy, F., J. Currier and B. Wong, *Tribotesting of Tibial Bearings of Knee Prostheses*, in *Tribology Series.* 2001, Elsevier. p. 371–379.
5. Brończyk, A., P. Kowalewski and M. Samoraj, *Tribocorrosion Behaviour of Ti6Al4V and AISI 316L in Simulated Normal and Inflammatory Conditions.* Wear, 2019. **434-435**: p. 202966.
6. Aibinder, W.R., et al., *Revisions for Aseptic Glenoid Component Loosening After Anatomic Shoulder Arthroplasty.* Journal of Shoulder and Elbow Surgery, 2017. **26**(3): p. 443–449.
7. Hirani, H., *Fundamentals of Engineering Tribology with Applications.* 2016: Cambridge University Press.
8. Vohra, K., et al., *Tribological Characterization of a Self Lubricating PTFE under Lubricated Conditions.* Materials Focus, 2016. **5**(3): p. 293–295.
9. Bhushan, B., *Introduction to Tribology.* 2013: John Wiley & Sons.
10. Sun, J., et al., *Tribological and Anticorrosion Behavior of Self-Healing Coating Containing Nanocapsules.* Tribology International, 2019. **136**: p. 332–341.
11. Liu, Y., et al., *Multilayer Coatings for Tribology: A Mini Review.* Nanomaterials, 2022. **12**(9): p. 1388.
12. Scharf, S., et al., *Multi-Functional, Self-Healing Coatings for Corrosion Protection: Materials, Design and Processing*, in *Handbook of Smart Coatings for Materials Protection.* 2014, Elsevier. p. 75–104.
13. Amiri, S., *Nano Coatings for Scratch Resistance*, in *Nanotechnology in the Automotive Industry.* 2022, Elsevier. p. 345–370.
14. Ali, S., et al., *Challenges and Opportunities in Functional Carbon Nanotubes for Membrane-Based Water Treatment and Desalination.* Science of The Total Environment, 2019. **646**: p. 1126–1139.
15. Blaiszik, B.J., N.R. Sottos and S.R. White, *Nanocapsules for Self-Healing Materials.* Composites Science and Technology, 2008. **68**(3): p. 978–986.
16. Madelatparvar, M., M.S. Hosseini and C. Zhang, *Polyurea Micro-/Nano-Capsule Applications in Construction Industry: A Review.* Nanotechnology Reviews, 2023. **12**(1): p. 20220516.
17. Mamat, M.F., et al., *A Characterization of Tung Oil-Filled Urea-Formaldehyde Microcapsules and Their Effect on Mechanical Properties of an Epoxy-Based Coating.* Malaysian Journal of Microscopy, 2023. **19**(1): p. 307–318.
18. Jakovljević, S. and D. Landek, *Tribological Coatings—Properties, Mechanisms, and Applications in Surface Engineering.* 2023, MDPI. p. 451.
19. Burakowski, T. and T. Wierzchon, *Surface Engineering of Metals: Principles, Equipment, Technologies.* 1998: CRC press.
20. He, Q., et al., *A Study of Mechanical and Tribological Properties as well as Wear Performance of a Multifunctional Bilayer AlTiN PVD Coating During the Ultra-High-Speed Turning of 304 Austenitic Stainless Steel.* Surface and Coatings Technology, 2021. **423**: p. 127577.
21. Joshi, P., et al., *Synthesis of Multifunctional Microdiamonds on Stainless Steel Substrates by Chemical Vapor Deposition.* Carbon, 2021. **171**: p. 739–749.
22. Cheng, Y., et al., *Internal Stresses in TiN/Ti Multilayer Coatings Deposited by Large Area Filtered Arc Deposition.* Journal of Applied Physics, 2008. **104**(9): p. 093502.
23. Azushima, A., et al., *Coefficients of Friction of TiN Coatings with Preferred Grain Orientations Under Dry Condition.* Wear, 2008. **265**(7): p. 1017–1022.
24. Krysina, O., et al., *Influence of Nb Addition on the Structure, Composition and Properties of Single-Layered ZrN-Based Coatings Obtained by Vacuum-arc Deposition Method.* Surface and Coatings Technology, 2020. **387**: p. 125555.

25. Vasylyev, M., et al., *Characterization of ZrN Coating Low-Temperature Deposited on the Preliminary Ar+ Ions Treated 2024 Al-alloy*. Surface and Coatings Technology, 2019. **361**: p. 413–424.
26. Kumar, D.D., et al., *Wear Resistant Super-Hard Multilayer Transition Metal-Nitride Coatings*. Surfaces and Interfaces, 2017. **7**: p. 74–82.
27. Chang, Y.-Y., et al., *Tribological and Mechanical Properties of Multilayered TiVN/TiSiN Coatings Synthesized by Cathodic Arc Evaporation*. Surface and Coatings Technology, 2018. **350**: p. 1071–1079.
28. Fu, Y., et al., *Structure and Tribocorrosion Behavior of CrMoSiCN Nanocomposite Coating with Low C Content in Artificial Seawater*. Friction, 2021. **9**: p. 1599–1615.
29. Wang, Q., et al., *Friction and Wear Performance of CrSiBCN Coatings Sliding Against Ceramic and Metal Counterparts in Sea Water*. Surface Engineering, 2021. **37**(6): p. 722–731.
30. Xu, Y.X., et al., *Effect of the Modulation Ratio on the Interface Structure of TiAlN/TiN and TiAlN/ZrN Multilayers: First-Principles and Experimental Investigations*. Acta Materialia, 2017. **130**: p. 281–288.
31. Fukumoto, N., H. Ezura and T. Suzuki, *Synthesis and Oxidation Resistance of TiAlSiN and Multilayer TiAlSiN/CrAlN Coating*. Surface and Coatings Technology, 2009. **204**(6–7): p. 902–906.
32. Liu, D., et al., *Structural, Interface Texture and Toughness of TiAlN/CNx Multilayer Films*. Materials Characterization, 2021. **178**: p. 111301.
33. Wang, Y., et al., *Improvement in the Tribocorrosion Performance of CrCN Coating by Multilayered Design for Marine Protective Application*. Applied Surface Science, 2020. **528**: p. 147061.
34. Ospina, R., et al., *Mechanical and Tribological Behavior of W/WCN Bilayers Grown by Pulsed Cacuum Arc Discharge*. Tribology International, 2013. **62**: p. 124–129.
35. Li, Y., et al., *Microstructure and Tribological p]Properties of Multilayered ZrCrW (C) N Coatings Fabricated by Cathodic Vacuum-Arc Deposition*. Ceramics International, 2022. **48**(24): p. 36655–36669.
36. Glechner, T., et al., *Structure and Mechanical Properties of Reactive and Non-Reactive Sputter Deposited WC Based Coatings*. Journal of Alloys and Compounds, 2021. **885**: p. 161129.
37. Meng, Q.N., et al., *Deposition and Characterization of Reactive Magnetron Sputtered Zirconium Carbide Films*. Surface and Coatings Technology, 2013. **232**: p. 876–883.
38. Malinovskis, P., et al., *Synthesis and Characterization of Multicomponent (CrNbTaTiW) C Films for Increased Hardness and Corrosion Resistance*. Materials & Design, 2018. **149**: p. 51–62.
39. Hossain, M., et al., *Carbon Stoichiometry and Mechanical Properties of High Entropy Carbides*. Acta Materialia, 2021. **215**: p. 117051.
40. Gopalan, H., et al., *On the Interplay Between Microstructure, Residual Stress and Fracture Toughness of (Hf-Nb-Ta-Zr) C Multi-Metal Carbide Hard Coatings*. Materials & Design, 2022. **224**: p. 111323.
41. Fritze, S., et al., *Hard and Crack Resistant Carbon Supersaturated Refractory Nanostructured Multicomponent Coatings*. Scientific Reports, 2018. **8**(1): p. 1–8.
42. Huang, S., et al., *Achieving Superlubricity with 2D Transition Metal Carbides (MXenes) and MXene/graphene Coatings*. Materials Today Advances, 2021. **9**: p. 100133.
43. Naguib, M., et al., *Two-Dimensional Transition Metal Carbides*. ACS Nano, 2012. **6**(2): p. 1322–1331.
44. Dong, M., et al., *Multifunctional Epoxy Nanocomposites Reinforced by Two-Dimensional Materials: A Review*. Carbon, 2021. **185**: p. 57–81.
45. Wang, X., et al., *MoS_2/Polymer Nanocomposites: Preparation, Properties, and Applications*. Polymer Reviews, 2017. **57**(3): p. 440–466.
46. Du, S., et al., *Optimizing the Tribological Behavior of Tantalum Carbide Coating for the Bearing in Total Hip Joint Replacement*. Vacuum, 2018. **150**: p. 222–231.
47. Ren, P., et al., *Self-Assembly of TaC@ Ta Core–Shell-Like Nanocomposite Film via Solid-State Dewetting: Toward Superior Wear and Corrosion Resistance*. Acta Materialia, 2018. **160**: p. 72–84.
48. Tillmann, W., et al., *Residual Stresses and Tribomechanical Behaviour of TiAlN and TiAlCN Monolayer and Multilayer Coatings by DCMS and HiPIMS*. Surface and Coatings Technology, 2021. **406**: p. 126664.
49. Spor, S., et al., *Evolution of Structure, Residual Stress, Thermal Stability and Wear Resistance of Nanocrystalline Multilayered $Al_{0.7}Cr_{0.3}N$-$Al_{0.67}Ti_{0.33}N$ Coatings*. Surface and Coatings Technology, 2021. **425**: p. 127712.
50. Xu, Y., et al., *Temperature-Sensitive Tribological Performance of Titanium Alloy Lubricated with PNIPAM Microgels*. Applied Surface Science, 2022. **572**: p. 151392.
51. Landolt, D., S. Mischler and M. Stemp, *Electrochemical Methods in Tribocorrosion: A Critical Appraisal*. Electrochimica Acta, 2001. **46**(24-25): p. 3913–3929.
52. Farooq, S.A., et al., *Tribo-Corrosion Behaviour of Composites and Coatings: An Overview of Influencing Factors, Evaluation Methods and Inhibitors*. Jurnal Tribologi, 2022. **35**: p. 92–116.
53. Holmberg, K., et al., *Global Energy Consumption due to Friction and Wear in the Mining Industry*. Tribology International, 2017. **115**: p. 116–139.

54. Javaherdashti, R., *How Corrosion Affects Industry and Life.* Anti-Corrosion Methods and Materials, 2000. **47**(1): p. 30–34.
55. Rose, S.R., *Studies of the High Temperature Tribological Behaviour of Some Superalloys.* 2000: University of Northumbria at Newcastle (United Kingdom).
56. Mathew, M., et al., *Significance of Tribocorrosion in Biomedical Applications: Overview and Current Status.* Advances in Tribology, 2009. **2009**.
57. Zia, A.W., Z. Zhou and L.K.-Y. Li, *Structural, Mechanical, and Tribological Characteristics of Diamond-Like Carbon Coatings*, in *Nanomaterials-Based Coatings.* 2019, Elsevier. p. 171–194.
58. Baba, Z.U., et al., *Towards Sustainable Automobiles-Advancements and Challenges.* Progress in Industrial Ecology, an International Journal, 2019. **13**(4): p. 315–331.
59. Katiyar, P.K., R. Maurya and P.K. Singh, *Failure Behavior of Cemented Tungsten Carbide Materials: A Case Study of Mining Drill Bits.* Journal of Materials Engineering and Performance, 2021. **30**(8): p. 6090–6106.
60. López-Ortega, A., R. Bayón and J. Arana, *Evaluation of Protective Coatings for Offshore Applications. Corrosion and Tribocorrosion Behavior in Synthetic Seawater.* Surface and Coatings Technology, 2018. **349**: p. 1083–1097.
61. Dini, C., et al., *Progression of Bio-Tribocorrosion in Implant Dentistry.* Frontiers in Mechanical Engineering, 2020. **6**: p. 1.
62. Blednova, Z.M., D. Dmitrenko and E. Balaev. *Tribological Properties of Multifunctional Coatings with Shape Memory Effect in Abrasive Wear.* in *IOP Conference Series: Materials Science and Engineering.* 2018. IOP Publishing.

4 Design and Fabrication of Smart Surface Coating and Thin Films for Future Industrial Applications

Muhammad Kaleem Shabbir, Javeed Akhtar, Zeeshan Ramzan, and Faiz Meeran

4.1 INTRODUCTION TO SMART THIN FILMS

Smart thin-film surfaces are the coating or deposited layers on the substrate's upper surface. By the addition of these surfaces, the properties of the substrate, such as thermal, electrical, environmental, and optical properties, can be changed. In smart thin-film surfaces, highly advanced materials are used. Smart thin films are also known as intelligent thin films because they respond automatically to changes in the environment. Smart coatings can sense and respond to their environment. Smart thin films that are placed on the surface of the substrate can be multifunctional. For the deposition of multifunctional layers, many types of charged and noncharged substances can be used. For smart thin-film surfaces, materials such as micelles, polyelectrolytes, nanoparticles, nanotubes, organic molecules with low molecular weight, clays, and self-healing agents can be used for multilayer functions [1].

4.1.1 Nature of Thin Films

In literature we mostly see that the thin films that are on the surface of the substrate are solid materials, but they can be solid, liquid, or gas. Similarly, the thickness of the thin films varies, starting from a few angstroms. The most effective thin films have thickness that ranges from 1000 Å to 5 to 10 μm. The thin films can be divided into three categories according to thickness: (1) Ultra-thin films that range in thickness from 50 Å to 100 Å, (2) thin films (very thin films) that range from 100 Å to 1000 Å thickness, and (3) thick thin films that can have comparatively thick surfaces. The thin films can also be divided on the basis of morphology, that is, they can be amorphous or crystalline. The thin films that are attached to the surface can also have different bonding attractions that can be ionic, covalent, or hydrogen bonding. Smart thin films can also be classified on the basis of conduction. On this classification the thin films are also divided into three types.

4.1.2 Conducting Thin Films

The conducting thin films have different conducting phenomena than the bulk material. The bulk materials can conduct electricity from all sides, but in the thin films, as size decreases the band gap increases, so conductivity lessens. The smart thin films are usually 1-D or 2-D materials. Due to this property they are extensively used in electronic devices and especially in modern chip technologies.

4.1.3 Semiconducting Thin Films

The semiconducting thin films usually have higher band gaps than the conducting thin films. In addition to this, these materials have some special characteristics like negative TCR, non-ohmic

 DOI: 10.1201/9781032635347-4

behavior with bois voltage, large thermoelectric power, and sensitivity to light, producing either photovoltaic or photoelectric current effects.

4.2 PROPERTIES OF THIN FILMS

4.2.1 Dielectric Properties

Smart thin films are basically materials that have properties to store and dissipate electrical energy when electromagnetic fields are applied. These properties of smart thin films show much resemblance to semiconductor material. The smart thin films provide insights into the electrical nature of the molecular and atomic species that are composed of the dielectric materials. Due to their energy-storing properties, the synthesis of the smart thin films are the crucial parts of the electronic circuits called capacitors. Capacitors consist of two electrodes with a dielectric material between them. Smart thin films act as dielectric materials and are placed between two electrodes. As the smart thin films have a high storage capacity, they are extensively used in memory devices, ultrasonic devices, and electronic circuits. Smart thin films are equally used in AC and DC devices.

4.2.1.1 Basic Concepts in Dielectrics

The dielectric phenomena occur when the electrical fields attract the electrons, protons, and nuclei of the dielectric materials. The opposite charge species are held together by the electrostatic force of attraction but are separated by the distance that makes two poles which have some moments as given by the product of the charge and the distance.

4.2.2 Optical Properties

Smart thin films have different optical properties from bulk materials. The optical properties of the smart thin films can be changed by controlling different parameters such as composition, thickness, conductivity, and porosity.

These films have several uses, including antireflective and high-reflective systems, multilayer filters, and polarizing systems. By controlling the twist angle of stacked layers, the optical properties of smart thin films can be modified. The smart thin films have countless optical properties. For example, they have the special ability to pass some wavelengths through them while reflecting others. This property makes them useful in LCD and display glass filters. Some properties like fluorescence are also optical properties of smart thin films [2]. Nowadays smart thin films are designed for use in fluorescent switching [3]. Controlled fabrication and aggregation can alter the optical properties of smart thin films.

The ability to adjust refractive indices is an intriguing use of smart thin films. For example, the refractive index of nonporous SiO_2 can be increased up to 15%. These properties are useful in lenses that are used in telecommunications [4]. Polymer thin films that have a high refractive index can be formed by charge-transfer complexation, which permits additional modulation of the refractive index. Generally, tuning of the refractive indices of smart thin films can be done by various methods, and every method has its own specifications, each with its own advantages and limitations. Among the limitations to tuning the refractive indices of smart thin films is the challenge of keeping the layer surface smooth and uniform [5]. Overall, there are many challenges to developing refractive indices that can be solved by new research technologies.

4.2.3 Self-Healing Properties

Smart thin films have the ability to self-heal, which makes them appealing for a variety of uses. For instance, excellent self-healing qualities have been developed in unconventionally self-healing flexible thin-film materials [6]. Self-healing polymer smart coating films can be used in a variety of applications, and researchers are exploring ways to develop smart and self-healing thin-film materials [7].

4.2.4 Surface Modification Properties

Surface modification is a process of modifying the surface properties of smart thin films to tailor their characteristics. This is accomplished by surface modification using radiation and chemical vapor deposition (CVD), which makes for simple tunability of the surface properties of the films [8, 9]. Surface modification methods like plasma immersion ion implantation (PIII) offer benefits over traditional beamline ion implantation, including a high dose rate, reduced sample charging and heating, and no vacuum. However, PIII has drawbacks, including no mass separation [10].

4.2.5 Biocompatibility Properties

Thin films have emerged as an excellent surface treatment for improving the biocompatibility of permanent implant materials. Polymer thin films can be employed as functional layers and conformal surface engineering materials. Indirect current magnetron sputtering was used to synthesize thin films of titanium dioxide (TiO_2), and its surface properties and biocompatibility were examined. After thermal activation, researchers produced Ti(1-x)Au(x) thin films that had increased hardness and biocompatibility. These films can be used as a biocompatible, super-hard covering to increase the lifespan of orthopedic implants [11]. Biocompatible thin films improve the biodegradability, biocompatibility, and wear resistance of permanent implant materials. Thin films can be utilized as surface treatments to improve the biocompatibility of implanted biomedical devices such as retinal prosthesis implantable chips. Biocompatible thin films can also be utilized to improve implant materials' physical qualities, such as wear resistance and coefficient of friction. With their potential to enhance the performance and lifespan of permanent implant materials, biocompatible thin films are well suited for biomedical applications [12].

4.3 MODERN TECHNOLOGIES FOR SYNTHESIS OF SMART THIN FILMS

The LbL approach is a new branch of materials science that serves as a foundation for the targeted creation of materials with specific architecture, qualities, and functions. LbL-deposited multilayers can have high self-healing properties. The properties of the LbL depend upon the structure of the material that is used for their synthesis.

In the process of depositing atomic inorganic smart thin films on the surfaces of substrates, a vacuum or a very low-pressure environment is usually required. Many other factors like substrate temperature, deposition rate, angle of incidence of the depositing particles, and the use of energetic particle bombardment can affect the properties of the thin films. The self-healing capacity of LbL-deposited multilayers is based on the mobility of the multilayer components, which is responsive to external stimuli. Due to changes in temperature, the movement of the particles that are deposited can change, which changes the shape of the films. The deposition rate also changes the composition of smart thin films. The film morphology, crystallinity, and stress can all be affected by energetic particle bombardment [13].

4.4 FABRICATION TECHNIQUES FOR SMART THIN SURFACES

Generally, thin films are fabricated by the installation of a tiny layer of a material (adsorbate) on the surface of any other material (adsorbent) to alter its chemical, environmental, optical, and thermal properties. Smart thin films are coatings that are intrinsically sensitive to environmental stimuli such as the pH level, temperature, light, and so on. They are sophisticated in terms of a self-repairing response to both exterior and internal stimuli, and the single/multiple functionality of their applications. This section includes an extensive summary of manufacturing processes for smart thin film deposition.

4.4.1 Chemical Vapor Deposition

Chemical vapor deposition (CVD) involves the introduction of a substance (adsorbate) in the vapor phase along with a carrier gas at an elevated temperature to initiate a chemical reaction to form a solid deposit. Unlike physical vapor deposition, this reaction does not require vacuum conditions [14]. During CVD, a heterogeneous reaction occurs at the surface of the substrate, rather than a homogeneous reaction in the gas phase. A variety of chemical reactions can occur, such as thermal decomposition (pyrolysis), hydrogen reduction, photolysis, oxidation, self-oxidation-reduction, carbonization, nitridation, and hydrolysis [15].

CVD encompasses a variety of process and reactor types, depending upon the intended applications. Deciding factors include the morphology of thin films, the thickness of the film, uniformity, cost, and requirements for the coating and substrate materials.

4.4.1.1 Thermal Activation

Thermal activation is the conventional method for triggering the chemical reaction at elevated temperature (900°C), although in the case of metal-organic substances, a temperature lower than 900°C is preferred. There are two types of CVD, atmospheric pressure CVD (APCVD) and low-pressure CVD (LPCVD). The overall chemical reaction involves the change of the reactant into the vapor phase, then allowing these vapors to enter the reactor where, depending upon the deposition condition, a homogeneous or heterogeneous chemical reaction takes place. Finally, crystallization of a thin film onto the substrate occurs. The kinetics of the reaction are either diffusion controlled or chemically controlled, depending upon the pressure employed. In the case of LPCVD <1 kPa, the speed of mass transportation across reactant, the rate of chemical reaction at the surface of the substrate and kinetic of the reaction is limited. In case of APCVD, the diffusion rate of the vapors is lower as compared to substrate, so diffusion cannot be done easily. LPCVD is advantageous over APCVD because it produces highly uniform thin films, and the introduction of unwanted particles can also be controlled. This technique is commonly employed to fabricate polysilicon and silicon nitride films [16].

4.4.1.2 Plasma Activation

Plasma activation take place at a low temperature, that is, around 300°C to 500°C. This approach is widely used for the deposition of organic, inorganic, and inert substances [17]. In this variant, electric power is used along with the thermal energy. The operating pressure is in the 10 to 100 Pa range. A reaction sample is enclosed between two electrodes. By using the electric field, a complex chemical reaction takes place in the precursor material under reduced pressure, which produces ions and electrons. These highly energetic bullets are directed to the surface of the substrate to initiate the deposition process. Formation of conformal films at lower temperatures makes this method superior to LPCVD and APCVD [14]. Plasma-enhanced CVD (PECVD) is an established methodology used at industrial scales to fabricate microelectronics, high-quality photovoltaics, transistors, and antireflective coatings [17].

4.4.1.3 Photon Activation

Photon activation involves the utilization of electromagnetic radiation, commonly ultraviolet radiation, to trigger the pioneer in or vapor phase or gas phase directly or to activate the intermediate reaction at low pressure, 0.1 to 3 torr [15]. The absorption of photons by the reactant molecules/atoms initiates the free radical mechanism. Mercury vapors are used as photosensitizers that are activated by the absorption of photons. These excited mercury atoms collide with the reactants and transfer the absorbed energy to generate free radicals. This variant has advantages over PECVD for the synthesis of SiO_2 and Si_3N_4 thin films because of the lower working temperature (150). Effective production equipment and need for the photo activation with mercury are the limiting factors to achieving an acceptable rate of deposition [18].

4.4.2 Atomic Layer Deposition

This technique involves the sequential deposition of a variety of materials in the vapor phase by using an activated substarter [19]. ALD is similar to CVD in that it involves consecutive self-limiting surface reactions of vapor phase precursor molecules and a substrate, but it is faster. This means that the reaction will come to a halt after all of the functional sites on the substrate have been occupied, resulting in a monolayer of material at the surface. This allows for precise conformational control of the deposited material's thickness [19].

A high temperature is applied to the substrate. One at a time, two distinct precursor molecules are delivered into the reaction chamber. The initial precursor molecule interacts with the substrate's surface, forming a reactive intermediate. To generate the required substance, the second precursor molecule interacts with the reactive intermediate. To eliminate any unreacted precursor molecules, the reaction chamber is purged with inert gas.

A thin coating of TiO_2 was created via ALD [20]. These are the defined steps: exposure of the precursor, purge, reactant exposure, and exposure of the reactant. TiO_2 layers are simple to manufacture and can have their thickness precisely regulated by repeating these steps. The chemical precursors are kept in a vapor phase by adjusting a certain temperature known as the "ALD temperature window" under a vacuum pressure of around 10 kPa or less in order to produce optimal reaction kinetics and prevent condensation or thermal decomposition of the precursor. The excellent conformity, atomic-level thickness control, and customizable film composition of ALD make it superior to conventional gas phase techniques for the deposition of thin films. ALD is a promising technology for forming thin films on complicated geometries of substrates. It is the most effective approach for producing highly conformal seed layers, which are required for applications like the integrated battery idea. Although metallic copper thin films may be made, their deposition is often limited by a shortage of precursors suitable for use with ALD techniques. Chemical precursor vaporization, managing their thermal breakdown, and a lack of effective self-limited reactivity with surfaces are some of the challenging components of ALD. Additionally, because the by-products of the suggested chemical reaction are probably going to contain dangerous or corrosive molecules, it is crucial to thoroughly study them while designing some chemical precursors.

4.4.3 Sol-Gel Coatings

The sol-gel technique is a wet-chemical approach for producing oxide materials. It is a flexible process for producing a wide range of materials such as ceramics, glasses, and films. The hydrolysis and condensation of a precursor molecule is the first step in the sol-gel technique. This technique produces a sol, which is a colloidal dispersion of extremely tiny particles (1–100 nm) into colloidal suspensions and solid solids, respectively. Molecular precursors undergo successive condensation reactions in a liquid medium to produce the oxide networks in the sol-gel process. Sol-gel coatings can be created using both organic and inorganic methods. The inorganic approach makes use of inorganic precursors, networks produced via a colloidal suspension (usually oxides), gelation of the sol, and network development in a continuous liquid phase. The organic approach makes use of functionally generated organic monomers containing double bonds. The sol-gel method is a wet-chemical method for producing oxide materials. To begin, dissolve metal or metalloid alkoxide precursors in alcohol or another low-molecular-weight organic solvent. The alkoxide groups in this solution are then hydrolyzed, meaning they are replaced by hydroxyl groups. The hydroxyl groups then combine to form a network of linked particles. This network gradually thickens to the point where it becomes a gel. Recent studies have demonstrated that coatings made of silica, ceria, vanadium, and molybdate can be modified using sol-gel technology to create a smart, self-healing coating with a functional gradient. This coating can function as a high-performance surface treatment, limiting the attack on the material's surface caused by water transport, and it can provide covalent bonding for reliable coating adhesion [21]. The sol-gel technique is used to make thin-film coatings with various features, such as photochromic, superhydrophobic, antireflective, bioactive, self-healing, micro- or nanocapsules, and nanocontainers. Solar-responsive smart windows, self-cleaning surfaces, optical gadgets, medical implants, and drug delivery systems are just a few of the uses for these coatings.

4.5 OTHER COMMON DEPOSITION TECHNIQUES

The three most prevalent procedures for depositing thin films on surfaces are dip coating, spin coating, and spray coating. Dip coating is a simple and adaptable process for coating planar and nonplanar materials. Spin coating is a more exact process for controlling the thickness of the film. Spray coating is a quick and effective approach for large-area substrates. In comparison to spray coating and spin coating deposition processes, the dip-coating approach takes longer due to the repeated adsorption, washing, and drying stages. Spin coating provides good control over layer thickness by varying the spinning speed and duration, but the substrate's size is limited by the spinning device's dimensions [22]. Spray coating is a contact-free technique for deforming low-process plastics at temperatures ranging from 100°C to 200°C. It may deposit conductive polymers including polyaniline, polypyrrole, and polythiophene on conductive and nonconductive surfaces. These methods have been used in a variety of applications, such as corrosion-resistant treatments, sensors and biosensors, semiconductors, electrical devices, water sterilization, and energy storage. [23].

Another cutting-edge method for the fabrication of multifunctional smart coatings for various purposes is layer-by-layer (LbL) self-assembly. Spraying, dipping, or spinning methods can be used to generate nano- or microscale multilayer thin films using polyelectrolytes, metal ions, and metal nanoparticles with alternating charges (Figure 4.1). The four sequential steps of the LbL self-assembly method are (1) coating the surface to be modified with the first solution via dip, spray,

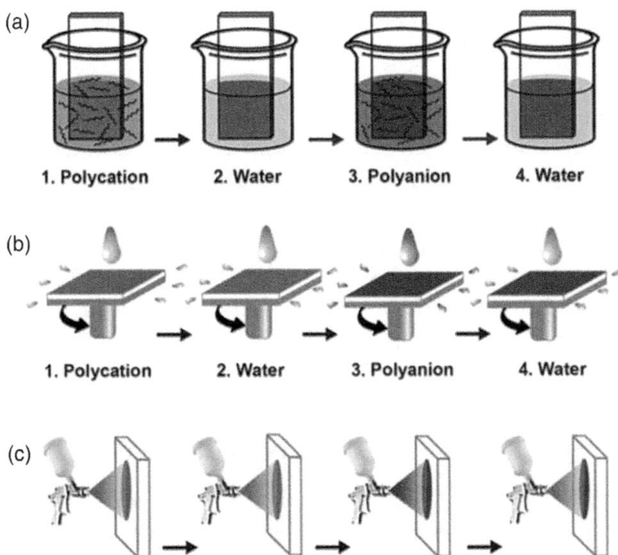

FIGURE 4.1 Steps involved in creating polyelectrolyte multilayer films using an LbL assembly. (a) Dipping LbL assembly. (b) LbL assembly with spin assistance. (c) Spray-aided LbL assembly. Steps a–c are iteratively repeated to create multilayer films. (Reproduced with permission from [24], © 2012 Royal Society of Chemistry.)

or spin for a predetermined amount of time, (2) rinsing the surface with clean solvent, (3) drying the surface, and (4) repeating steps 1–3 with the second solution comprising the oppositely charged moieties. The process is carried out repeatedly to reach the desired film thickness.

4.5.1 Deposition of Smart Coatings and Smart Thin Films

Smart thin films, surfaces, and interfaces can be made using smart coatings for a variety of applications. Corrosion-protective coatings with intelligent, active feedback and "on-demand" release of corrosion inhibitors are the most well-known examples of smart coatings. Superhydrophobic, superoleophobic, and superamphiphobic coatings, antibacterial coatings, self-cleaning coatings, antifogging coatings, antifreezing coatings, anti-icing coatings, bioinspired coatings, super-antiwetting coatings, and self-healing coatings made of shape-memory polymers are additional examples of smart coatings with specific functions [25].

4.5.1.1 Smart Anticorrosion Self-Healing Coatings

Recent advancements in the development of corrosion-protective coatings are based on active, self-healing polymers and intelligent, responsive coatings with autonomous protection of pores, crazes, scratches, and cracks through a variety of mechanisms, such as UV radiation, mechanical stresses, or thermal stress, or by external responsive mechanisms, such as pH, temperature, redox, or light. It is possible to develop self-healing, thin-film coatings by employing a variety of processes and deposition techniques. Sol-gel and chemical-conversion coatings are the two most popular coating types. One of the most promising research topics that uses the sol-gel process to generate intelligent coatings with active-feedback features is the development of stimuli-responsive, corrosion-protective coatings [26].

As illustrated in Figure 4.2, corrosion inhibitors can be embedded in a variety of nanostructured materials (such as nanocapsules or nanocontainers) and then integrated into the active-coating matrix with self-healing properties. For instance, corrosion inhibitors have been loaded into nanocontainers using straightforward physical adsorption, ion-exchange reactions, encapsulation by emulsion or condensation polymerization reactions, and LbL assembly by embedding the inhibitor between the alternating polyelectrolyte multilayers (PEMs) [28]. To further improve the stimuli-responsive functioning, polyelectrolyte multilayers can be placed on the external shell of nanocontainers. These layers become active when the pH, temperature, or redox potential of the surrounding environment change as shown in Figure 4.3.

Another method for creating self-healing coatings is to incorporate corrosion inhibitor–filled hollow capsules into the coating matrix. As shown in Figure 4.3, a three-stage procedure is used in this instance to generate polyelectrolyte capsules. A suitable capsule-core material is first chosen. Following that, multilayers of polyelectrolytes are deposited on the capsule shell wall. The metal oxide core material is finally dissolved by acid hydrolysis, leaving behind hollow, porous, and semipermeable polyelectrolyte capsules. Controlling the quantity of polyelectrolyte layers that are deposited allows for the best porosity of the manufactured capsules. Because it enables the capsule core to be filled with a variety of organic and inorganic corrosion inhibitors, this porosity is crucial [26].

4.5.2 Self-healing Nanocomposite Coatings In Microcapsules

One of the most current applications for the creation of self-healing nanocomposite coatings is microcapsules. Active compounds can be packaged inside another material using the microencapsulation technique to protect them from their environment, whether they are solid, liquid, or gaseous. The surrounding material is referred to as the shell, while the encapsulated active ingredient is referred to as the core materials. The concept of microencapsulation was first used in the 1950s to create microencapsulated dyes using coacervation for the production of carbonless copying materials. Numerous uses of microencapsulation technology have been developed, including self-healing

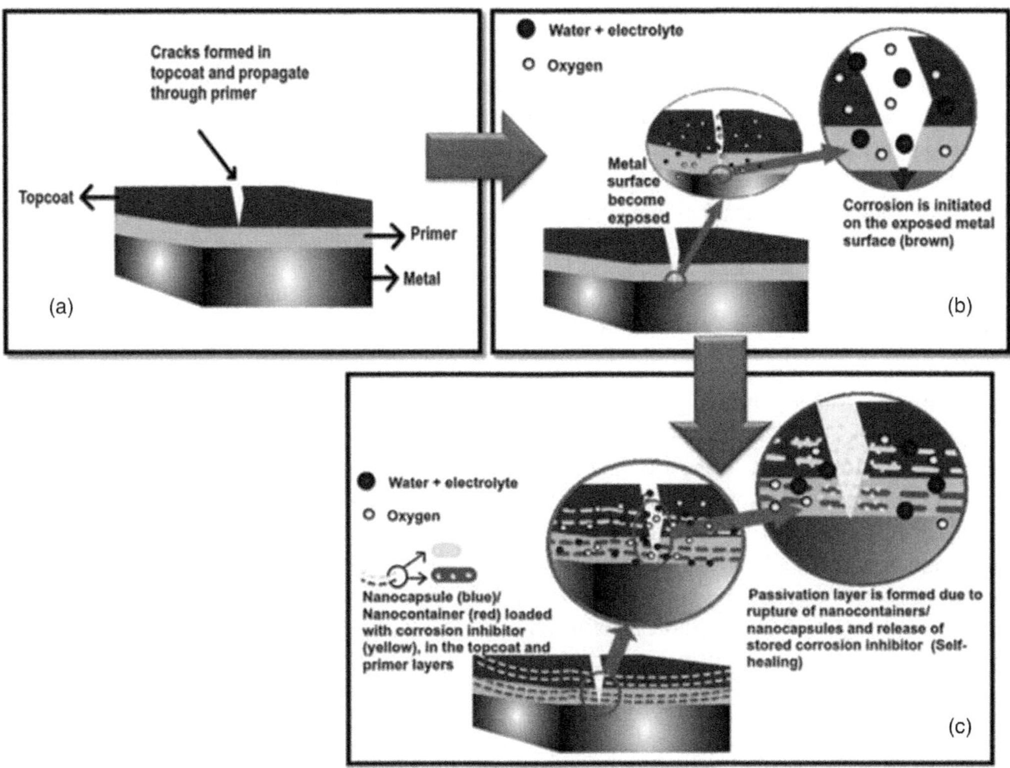

FIGURE 4.2 Schematic representation of the (a, b) corrosion process and (c) self-healing action for smart coatings embedded with inhibitor-loaded nanocapsules or nanocontainers. (Adapted with permission from [27].)

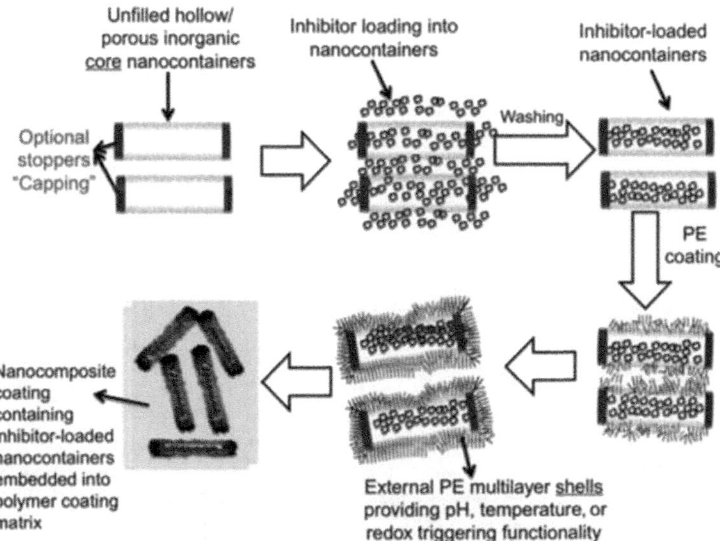

FIGURE 4.3 The development of functional inorganic, organic, core-shell nanotube-encrusted nanocomposite coatings is illustrated schematically. In order to achieve effective and regulated self-healing through the release of the inhibitor contained inside, inorganic nanotubes are first loaded with an appropriate corrosion inhibitor (internal core) and then coated with polyelectrolyte multilayers (external shell). (Reproduced with permission from [29].)

nanocomposites, cosmetics, catalysis, food applications, and medication delivery [30]. There are numerous ways to microencapsulate. Microencapsulation can be divided into two basic kinds. Monomers and prepolymers are employed as starting ingredients in the first category. Emulsion polymerization, suspension polymerization, and interfacial polycondensation are a few examples of this category. Similar to suspension cross-linking and solvent extraction/solvent evaporation, the second category of processes uses polymers as the initial raw material for the production of microcapsules. Other microencapsulation processes include coacervation/phase separation, spray drying, and polymer precipitation.

4.5.3 Smart Superhydrophobic and Superoleophobic Coatings

Another significant application for smart thin film is the emergence of surfaces with adjustable wettability (e.g., superhydrophobic, superoleophobic, or superhydrophilic) for various applications, including self-cleaning, antifreezing, anti-icing, and antifogging surfaces, and oil-water separation. Smart superhydrophobic coatings are another important application for smart, thin-film coatings. They can be fabricated using a variety of approaches, e.g., chemical etching, plasma etching, dip-coating deposition, spray coating, laser texturing, electrodeposition and LbL self-assembly. They are used in practical applications like extracting oil from oily wastes. Recently, interest in the development of controlled oil-water separation materials has increased. Smart responsive coatings have been developed to enable the synthesis of novel materials with adjustable surface wettability based on various triggering mechanisms, including pH and magnetic or electric fields [31].

For the production of materials with switchable, surface-oil wettability, such as nonwoven fabrics and melamine foam (MF), Ong et al. developed a two-step process. In the first stage, a very sticky polydopamine (PDA) layer was developed on the substrates using chemical oxidative polymerization of the dopamine monomer. Second, a pH-responsive block copolymer called poly(2-vinylpyridine-b-dimethylsiloxane) (P2VP-b-PDMS) was coated onto PDA-coated substrates, as shown in Figure 4.4 [32]. With pH values of 2 and 7, respectively, the functionalized melamine sponge showed switchable surface wettability.

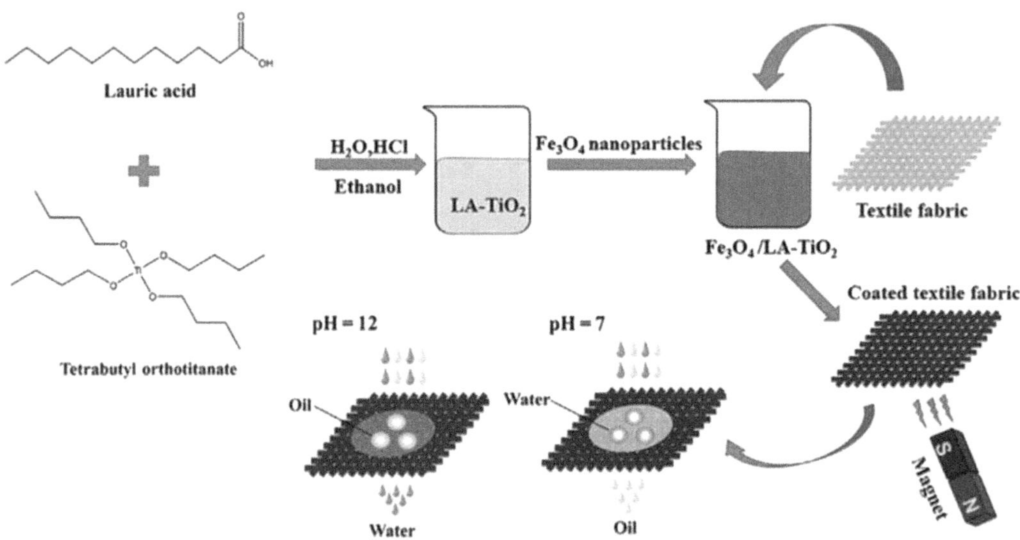

FIGURE 4.4 Fe_3O/LATiO_2 fabric preparation schematic illustration. (Reproduced with permission from [33], © 2018 Elsevier Ltd.)

4.5.4 Hybrid Nanocomposite Coatings

4.5.4.1 Nanocomposite Coatings Containing Nanoparticles and Nanofillers

In order to prevent corrosion, organic polymeric coatings frequently operate as a barrier between the metal and its surroundings. Any polymeric covering, however, is susceptible to corrosive species like water and ions like chloride. The characteristic of permeability shortens the lifespan of the polymeric coating and speeds up the early phases of corrosion. The longevity of a composite coating can be increased and the permeability can be decreased by adding inorganic fillers to the formulation of the polymeric coating. When compared to its counterpart in the conventional filler category, inorganic nanofillers, with their extremely large boundary volume along with the small gain size, offer improved barrier characteristics. Behzadnasab et al. developed epoxy-based nanocomposite coatings with varying quantities of zirconia nanoparticles and nano-clay that had undergone 3-aminopropyltrimethoxysilane (3-APTMOS) treatment. The optimal conditions for film formation and an improvement in corrosion protection performance were obtained by the addition of zirconia nanoparticles and nano-clay. In the event of the inclusion of spherical zirconia nanoparticles, 1 wt% better exfoliation of clay silicate layers in the resulting nanocomposite coating was associated with higher corrosion protection efficacy [34]. Elhalawany et al. created brand-new water-based coatings that comprise conductive polymer nanoparticles (CPNs) made of the corrosion-inhibiting polyaniline (PANI). A 1:1 mixture of nanohybrid particles (nHPs) of CPNs and nanosilica was used to create the hybrid coating [35]. The binder used was poly (vinyl acetate). The developed coating was applied to six different substrates to make painted films. The addition of silica nanoparticles enhanced the barrier properties of the nanocomposite painted films (PVAc/nHP), which were discovered to provide greater chemical, mechanical, and corrosion protection than the blank paint film (PVAc only). Similar outcomes were attained using water-based coatings using polyanisidine (PAns) and polytoluidine (PTol) with a styrene/butyl acrylate emulsion binder.

4.5.4.2 Nanocomposite Coatings

An amorphous film along with the nanocrystalline phase make up a nanocomposite covering. For the formation of nanocomposite coatings, various methods have been suggested. The most commonly used technique is reactive magnetron sputtering. Materials having hardness more than 80 GPa are considered ultra-hard materials, and nanocomposite coatings can be made to provide harder surfaces beyond what is permitted by the rule of mixing. For a range of industrial applications, increasing mechanical properties like the hardness of nanocomposite coatings is necessary. To achieve this, intricate theories have been created and numerous experiments have been conducted. However, while enhancing a material's hardness is given a lot of attention, its toughness receives insufficient attention [36]. Therefore, additional study is needed that takes corrosion behavior into account while also considering hardness and roughness. The design of nanocomposite coatings must take into account a number of variables and application requirements. The prevention of dislocation formation in the nanocrystalline phase, stopping the sliding of nanograins across grain boundaries, the impact of changing the lattice parameter, the significance of grain boundaries and crystal size, and finally the effects of the intermediate phases as well as the impurities should all be research topics in this field [37].

4.5.5 Conversion Coatings

A chemical or electrochemical reaction generates a layer of chemically bound substrate chromate, metal oxides, molybdate, vanadate, cerate, phosphates, or other chemicals at the metal surface, resulting in the emergence of the conversion coatings. Conversion coatings are frequently employed as inexpensive coating procedures that can prevent corrosion on metal substrates by serving as an insulating membrane across the metal surface to the environment [38].

4.5.5.1 Chromate Conversion Coating

The most popular conversion coating for enhancing the corrosion resistance of numerous metals and their alloys, including copper, aluminum, magnesium, and zinc, is chromate. The self-healing property of the coating, high electrical conductivity, and the high efficiency-to-cost ratio are the main factors influencing the widespread use of chromating. Due to its benefits, it is becoming the norm for corrosion protection. Additionally, chromating permits the application of additional finishing treatments and offers the highest level of under-film corrosion resistance. Hexavalent chromate, however, is listed as one of the most poisonous compounds by the U.S. Environmental Protection Agency (EPA) due to its ability to cause cancer as well as the fact that it is a waste product that is harmful to the environment. In coating processes, many attempts have been made to produce less hazardous or environmentally friendly alternatives as a result of current environmental legislation and growing requests for a complete ban on lethal hexavalent chromate. Although trivalent chromate was suggested as a potential substitute, it turned out to be less efficient than hexavalent chromate.

4.5.5.2 Chrome-Free Conversion Coatings

Recent years have seen the development of chrome-free conversion coatings based on salts, including cerate, stannate, vanadate, molybdate, silicate, and zirconate. These can function as a barrier coating to stop water from penetrating the material's surface and can create covalent bonds for strong coating adhesion.

4.5.5.3 Anodizing

The electrolytic process of anodizing is used to coat metals and alloys with a thick oxide layer. These films are used to increase paint adherence to the substrate and corrosion resistance. The processes involved in anodizing are as follows: (1) mechanical treatment; (2) degreasing, cleaning, and pickling; (3) electropolishing; (4) anodizing utilizing AC or DC current; (5) dyeing or post-treatment; and (6) sealing. The anodized films are made up of a relatively thick layer of a cellular structure with a thin barrier layer at the interface of metal and coating. Each cell has a pore whose size is influenced by the kind of electrolyte being used and the specifics of the experiment. The anodized film's quality is in turn determined by the density and size of the pores. One of the biggest obstacles to producing homogeneous anodic coatings is electrochemical inhomogeneity caused by phase separation in the material being coated [39]. Additionally, flaws, porosity, and inclusions from mechanical processing may cause uneven deposition, which in turn may make corrosion worse. During anodization, localized surface heating could have an impact on the materials' fatigue strength, which is another drawback, particularly in heavier films. Furthermore, the anodized metal that is produced is constructed of brittle ceramic, and it lacks the necessary mechanical qualities to meet the demands of some industrial applications.

4.5.6 Spray Coating

Spray coating involves impacting molten or softened particles onto a substrate to create a coating. In industrial settings, spray-coating techniques are frequently utilized to cover metals and glass that have unusual shapes as well as organic lacquers. The sections that follow include examples of some popular spray-coating processes [40].

4.5.7 Thermal Spraying

Melted coating ingredients are sprayed onto the substrate that is being coated during the thermal spraying process. A flame or an arc accelerates particles between 1 and 50 microns in size to high speeds while partially melting them. The particles deposit onto a surface to produce a coating [41]. Porosity, oxide concentration, and substrate adhesion all influence the coating's quality. Metal, ceramic, or polymer can be sprayed as the coating materials are typically heated chemically or

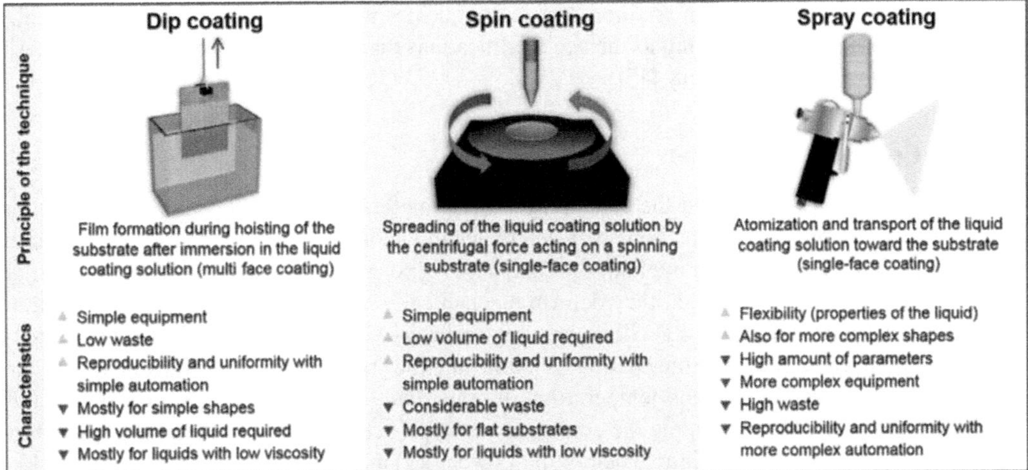

FIGURE 4.5 Principles and key features of the dip, spin, and spray coating systems are illustrated schematically. (Reproduced with permission from [43].)

electrically. When compared to other ordinary coating processes like electro- and electroless deposition techniques (i.e., CVD and PVD), the thermal spray technique has the ability to offer coatings for substrates with large surface areas that range in thickness from 15 μm to a few mm at a high deposition rate, as shown in Figure 4.5. Another benefit is the ability to spray different coating powders, such as ceramic, plastic, composite, or pure metal, onto the surface of the substrate [42].

4.5.8 High-Velocity Oxygen Fuel Spraying

In 1980, a new thermal spraying technique known as high-velocity oxy-fuel spraying (HVOF) was developed. In this method, oxygen gas along with a mixture of fuel are supplied to a chamber where combustion occurs. In this combustion chamber, they are ignited and react to each other. In HVOF, a variety of fuels are employed. Commonly utilized fuels include liquid fuels like kerosene and gaseous fuels like hydrogen, natural gas, methane, and propane. A jet of hot gas traveling at a high velocity of around 1000 m/s and a high pressure of about 1 MPa is created. Injecting a coating material powder into the hot gas stream causes the powder to accelerate to 700 to 800 m/s. The substrate that needs to be coated is in the path of the hot gas and powder stream. In the stream, the powder deposition occurs across the substrate after partially melting. The resulting coating, which is typically employed to increase corrosion and wear resistance, has a thickness of roughly 10 μm.

4.5.9 Plasma Spraying

In plasma spraying, coating material powders are injected into the plasma jet at a temperature of around 10,000 K, where they melt before being sprayed over the substrate to be coated. The type of the coating powders, the plasma gas's composition, and the gas flow all have an impact on the coating's final qualities because of the interaction between the substrate to be coated and the plasma coating substances. Factors affecting pace, energy input, proximity to the substrate, torch geometry, and final coating/substrate cooling are the differences between current and advanced coating processes [44].

4.5.10 Vacuum Plasma Spraying

The plasma spray process served as the foundation for the development of vacuum plasma spraying, which can work between 40°C and 120°C without overheating some coating materials like polymers, rubbers, or plastics. In addition, this technique has the potential to trigger nonthermally activated surface reactions, leading to surface modifications that are not possible under atmospheric pressure with molecular chemistry [45].

4.5.11 Cold Spraying

The HVOF spraying method and the cold spray coating method theoretically share the utilization of high velocity to promote interactions between the coating components and the coating substrate. However, with the cold spraying method, the carrier gas accelerates the particles to incredibly high speeds. When solid particles meet, they deform mechanically and plastically cling to the substrate to form a coating. In cold spraying, choosing a high velocity range is crucial because the velocity needs to be high enough to bind the coating substrate and the coating material. The material's characteristics, size of the powder, and temperature all affect the velocity. In 1990, a Russian research team made the first attempt to apply the cold-spraying approach in order to assess the particle erosion of an object subjected to high-velocity, fine-powder steam. The ability to use soft metals like Cu or Al as well as materials with high melting points like tungsten, tungsten carbide, cobalt, and titanium is a significant benefit of this process. Another benefit is the option to substitute an inert carrier gas, like helium or nitrogen, for oxygen. This method's drawbacks include its poor deposition efficiency and the need to use an extremely fine powder, which is unappealing from an industrial standpoint [46, 47].

4.5.12 Warm Spraying

The HVOF spraying method has lately undergone a revolutionary modification known as warm spraying. In this method, nitrogen is added to the combustion gas to lower its temperature, which is identical to the idea underlying the cold-spraying method. This technique has the advantage of having a coating that is more effectively coated than in cold spraying. In addition, compared to HVOF, the low temperatures that are used for warm spraying cause fewer chemical reactions in the feed powder. These advantages are therefore especially important for coating materials like polymers, titanium, and metallic glasses because they easily oxidize or deteriorate at high temperatures [48].

4.5.13 Roll-to-Roll Coating

A flat substrate is coated by being passed between two or more rollers during the roll-to-roll coating process. When using this method, following the proper setting of the gap between the top and the second roller, the coating material is applied to an application roll by one or more supplementary rolls. The application roller's coating is removed by the substrate as at the bottom it moves around the support roller. The coated substrate subsequently takes its final form after curing, but it has no impact on the properties of the coating. Direct roll and reverse roll coating are the two methods used in roll-to-roll coating. The substrate and the coating applicator roll rotate in the same direction when coating is applied directly to a substrate using a roll. The applicator roll rotates counterclockwise toward the substrate during the reverse roll coating process.

4.5.14 Meyer Rod Coating

The substrate is coated in the Meyer rod coating procedure while it travels over a roller that is only partially submerged in the coating substance. A wire-wound metering rod, also known as a Meyer rod, is used to control the amount (and occasionally the shape) of the coating to be applied to the substrate [49].

4.5.15 A Dip Coating

During the dip-coating procedure, the substrate is precisely controlled in terms of immersion time and withdrawal speed in a coating material bath with a predefined viscosity. After that, the substrate is taken out of the bath and allowed to drain. The substrate coated can then be forced to dry or roasted to complete drying. Dip coating is significantly influenced by the coating's viscosity. Throughout this coating process, the coating viscosity must remain consistent. This method is primarily used to apply primers before final coatings [50].

4.5.16 Advanced Polymers and Fillers

One important future development in coating technology is the use of improved polymers in coating structures.

4.5.16.1 Hyper-branched Polymers

Hyper-branched polymers that have small melt viscosities are produced through controlled radical polymerization; these are excellent for powder-coated materials and coatings having substantial solid contents. The addition polymerization of acrylic monomers in hypercritical fluids is one technique for producing solvent-free binders with constrained melting ranges for application as powder coatings with thin film thickness and reduced temperature curing capabilities [51].

4.5.16.2 Organic–Inorganic Hybrid Polymers

Developers of future coating technologies are likely to show a lot of interest in hybrid polymers. Because inorganic networks' stability and scratch resistance can be paired with organic polymers' flexibility, combinations of organic polymers and silicates can increase the coating's overall properties. Hybrid polymers have been successfully created using the sol-gel process. Most of the multiphase polymer systems that are currently in use do not produce translucent films, which has prevented a wider adoption of these systems as coatings. In this method, phase separation occurs in the nanofield after a block-and-comb polymer structure having intramolecular incompatibility. Brock reported a technique for creating polymers with enhanced transparent properties. This method employs use of a block-and-comb polymer structure with intramolecular incompatibility and nanoscale phase separation. In the future, the properties of the coating won't be negatively affected by better adhesion to the surface, thanks to advancements in the separation of phases and a greater comprehension of the energies of interface in polymer mixtures as shown in Figure 4.6 [52].

4.5.17 Conductive Polymer Coating

In recent years, conductive polymers based on polyanilines and polythiophene have been developed, and these materials may have increased corrosion resistance. These techniques may be used to produce electrostatic and electrochemically coated nonmetallic substrates, heat-conductive layers for electrically heatable surfaces, and other electro-related applications. Due to their capacity to absorb harmful gases, microporous coatings incorporating catalysts are currently the focus of substantial environmental science study. It will become more common to identify the owner of stolen goods using coatings that have a modest amount of active ingredients (such as photochromic coatings, which change transparency or opacity depending on the amount of light they receive) [54].

4.5.18 Water-Soluble Paint Coating

One of the major trends in the architectural paint industry for the foreseeable future is a rise in the usage of water-soluble latex paints. Compared to alkyd-based solvent-borne paints, water-soluble paints are less expensive, smell better, and don't create any harmful waste. Water-soluble paints

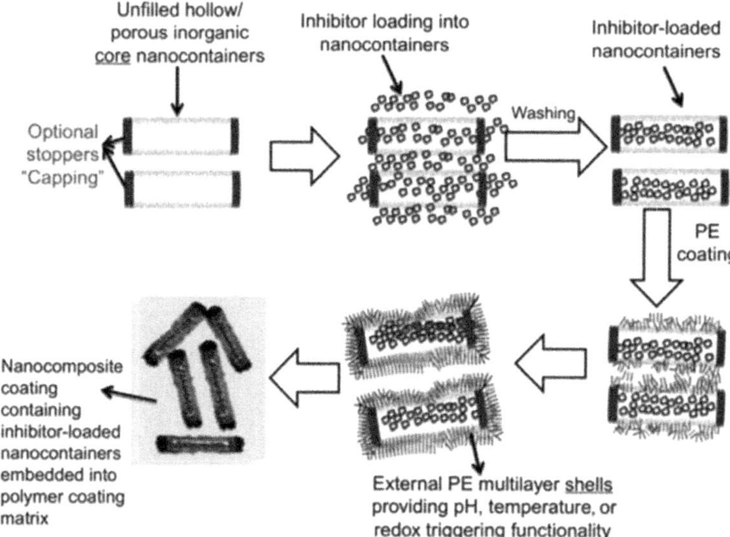

FIGURE 4.6 An example of a schematic pathway for the production of functional inorganic, organic, core-shell nanotube-encrusted nanocomposite coatings. In order to effectively and precisely self-heal, inorganic nanotubes are first loaded with an appropriate corrosion inhibitor (internal core), and then coated with polyelectrolyte multilayers (external shell) that carry certain responsive functions. (Reproduced with permission from [53].)

have faced fierce competition in the coatings and paints industry from better, more recently created polymers with distinctive properties [55].

4.5.19 Filler Coating

Although they can have a negative impact on coating quality, fillers are frequently employed in the coatings industry, mostly to cut costs. The fillers will have a specific function such as improving the coating quality for specific uses, ornamental effects, or enhancing the coatings' mechanical qualities. The design of ultra-thin films with nanolayers of specialty fillers or additives incorporated into the polymer matrix will be significantly affected by nanoscience and nanotechnology. Such fillers are anticipated to enhance mechanical strength, coating clarity, and corrosion resistance as a filler for custom coatings. Phyllosilicates with their distinctive leaflike structure will be the focus of more and more basic and applied study. In other research, phyllosilicates interlocked into the polymer matrix in nano form were employed to enhance the coating's barrier properties [56]. This strategy will gain popularity and be utilized more frequently for fire protection, electrical insulation, and particularly thermal insulation.

4.6 DEVELOPMENTS IN COATING PROCESSES

Nanocoatings can offer improved hardness and strength, and researchers will be interested in processing them. However, improvements in coating manufacturing and the availability of nanopowders limit the spread of nanocoatings. Nanocoating applications are still in their infancy because the procedure calls for extensive control over current. Furthermore, the technology calls for complex tools and multistep procedures. For thermal spray operations (such as plasma spraying or HVOF spraying), nanopowders are employed as feedstock materials. Thermal spraying offers the distinguishing advantage of a moderate to high rate of throughput and the ability to coat target materials with complex shapes using nanostructured feedstock powders generated using vapor, liquid, and solid processes. When compared to micro powder-made coatings, thermally sprayed nanocoatings with intermediate hardness demonstrated improved wear resistance. HVOF will attract the most research interest in the upcoming generation of nanocoatings due to its ability to create thick nanocrystalline ceramic coatings with wear characteristics superior to those created by plasma spraying [57]. HVOF also enables the production of nanocoatings with low porosity, high strength, and enhanced wear resistance. The electro-dip methods will undergo certain modifications to allow for lower curing temperatures, and new curing technologies utilizing radiation curing will expand their field of use to low-melting-point materials like plastics. The primary objective for coating industries will be to automate the coating process in order to quicken production lines, decrease the need for labor, and conserve energy. Future coating process improvements will result in fewer coating layers, complete automation of the coating process, module-based color management of the finished product, and automatic quality control [58]. Future developments in coating technology will often be driven by the following:

1. Modern lightweight materials include composites and magnesium alloys. In order to provide some desirable qualities, the study of rare-earth alloying elements like cerium and yttrium in magnesium alloys and composite materials will continue.
2. Specialized coating systems with unique chemistry and structures will become commonplace as a result of developments in polymers, new binders, and new fillers.
3. The future generation of coating technology will be significantly influenced by nanoscience and nanotechnology. More attention will be paid to self-healing coatings based on nanocontainers or nanocapsules that can be loaded with an inhibitor to stop corrosion on the substrate in the event that the coating layer is damaged or scratched. However, more research is still required to improve coating conditions because the existing process uses multiple phases (about 8–10 steps) and pricey raw ingredients, both of which are undesirable from an industrial standpoint.
4. Designing biocompatible nanocoating systems for use on medical implants and creating effective methods for creating high-performance hydroxyapatite coatings will be major foci of future work.
5. Industrial and basic research will be heavily focused on developments in coating processes. The next 10 years' worth of research will be devoted to developing new ways for surface modification as well as quicker and less expensive curing processes that use UV and electrostatic application.

4.7 SMART THIN-FILM SURFACE APPLICATIONS

The versatility and accurate control provided by thin-film technologies make them the eye-catching choice for promoting a wide diversity of challenges. As research is increasing day by day, in the future we can expect innovative and active biomedical applications of thin-film surfaces. For example, thin-film surfaces can be used for injection of exact dosages to the tissues and cells to reduce the side effects of medicines and improve efficiency. Some of the applications of smart thin-film surfaces are presented here.

4.7.1 Environmental Applications

4.7.1.1 Water Purification

Membranes that are formed from ceramic-based smart thin-film surfaces have the tendency to filter and distill water. These membranes are used in water purification because they have many applications, including excellent selectivity, high mechanical strength and durability, and excellent chemical and thermal stability [59]. These systems offer greater protection against entangling and scaling, which can increase the durability of systems receiving negligible maintenance. This is beneficial for those seeking reliable and efficient systems [60].

Smart thin-film surfaces provide protection against bacteria and dirt that make them suitable for water purification. For example, zinc aluminate films can inhibit the growth of some bacteria, including Gram-negative bacteria, *Escherichia coli*, and *Pseudomonas aeruginosa* [61].

TiO_2 has self-cleaning properties, so it can remove dust from water, and along with zinc aluminate it can be used to obtain bacteria- and dust-free water.

Smart thin films have great potential for use in water purification and antimicrobial activities. Some studies show that they can remove pollutants from water, but further advanced studies are needed to meet the new requirements.

Future study needs to reflect the types and concentrations of pollutants, the properties of smart thin films, and the cost-effectiveness of using this technology associated with other water purification methods. Smart thin films are biodegradable, but research is needed to make them effective for long terms.

4.7.1.2 Oil-Water Separation

Smart thin films are used both to separate oil from water and water from oil, so they can be used in direct practical applications in industry to solve common problems of oily wastewater and other oil/water pollutants [62]. Thin films that show a response to stimuli having wettability for water and oil makes them perfect for oil-water separation. For advanced technologies, polyaniline smart coatings are used. Ph-responsive devices are made from smart thin-film surfaces that are used for constant oil-water separation without ex situ treatments [62]. Superhydrophobic, superoleophobic, and superhydrophilic surfaces are made to adjust wettability. Special membranes are also made by using smart films for oil-water separation. The membranes are made on the concept of wettability, the growth of unique wet table systems and their coupling with membrane materials, and their process of manufacture and final usage in oil-water filtration [63].

There are two types of smart thin films, oil removal and water removal. The films that remove water have a hydrophobic and lyophilic nature. They can perform both functions, meaning they can attract oil and repel water simultaneously.

The oil removal films are only used for water removal, while the water removal films are used for oil removal from oil/water mixtures. The concentration of feed oil affects the removal time and purity.

The surface of the smart film is designed in a way that its structure is restricted to the nano scale that alters the energy and water absorption properties, enabling rapid and efficient separation of oil from water. The superhydrophobic and superhydrophilic properties allow water to roll off the surface while retaining oil droplets, and the reversible changing property makes them ideal for repetitive use and cleaning [62, 64].

4.7.1.3 Antifouling Coating

In biomedical applications such as biosensors, bioanalytical devices, and implants, fouling is a major issue that is caused by nonspecific proteins. Poly(dimethylsiloxane) (PDMS) is a material with many good properties, but it suffers from protein adsorption due to its hydrophobic nature. It has been a struggle to grow biocompatible materials for antifouling coatings of PDMS [65].

Smart thin films have numerous benefits for antifouling uses. They have the ability to change the surface of the material upon which they are placed to give it special properties like minimal surface energy and low surface tension, which can stop fouling and biofilm formation. They can respond quickly to various stimuli like changes in temperature, pH, and light to control the release of antifouling agents or to cause self-cleaning. Examples of smart thin-film technologies include stimuli-responsive polymers, surface-grafted brushes, and photocatalytic coatings. Smart thin films offer a promising approach to antifouling applications, providing protection against biofouling and other surface contaminants.

In marine engineering, smart thin films are used for antibacterial and antifouling purposes. The surfaces are made from materials that are environmentally friendly. Trisubstituted organotin compounds (TOCs) and heavy metals are widely used as antifouling agents, but they have bad impacts on the environment, so researchers are seeking alternatives like synthetic and natural biocides that have better antifouling activity and are environmentally friendly. Superhydrophobic coatings and nonstick coatings are being prepared that provide antifouling and self-responsive mediators in the coating medium. Capsaicin and chitosan have been used recently to prepare pH-responsive nanocapsules. Polyethylene glycol (PEG) is widely used to modify films in medical, sensing, and intelligent material areas and is highly biofriendly. Another material, sodium alginate (ALG), is nontoxic and is used in coatings that provide antibacterial and antifouling applications in marine engineering [66]. Moreover, the usage of nanomaterials in thin-film composite membranes can deliver better performance in terms of durability and selectivity [67].

4.7.2 Energy Applications

4.7.2.1 Solar Cell Applications

In the past two decades, there have been many changes in the environment all over the world, and countries have spent hundreds of billions of dollars in response. Solar photovoltaics (PV) is the most reasonable, widely available technology for producing electricity using sunlight shining through the solar cells packaged into a solar unit. Thin films (1 μm) are used in Si solar cells, thin-film solar cells, and solar modules as absorber, passivation, buffer, electron/hole transport, and antireflection coating (ARC) layers. Smart thin films have significant potential in these solar technologies. Many solar technologies make use of smart thin films made of cadmium telluride (CdTe), copper indium gallium diselenide (CIGS), and amorphous silicon (a-Si) [68].

The thin-film solar panels are much better than traditional solar panels because they are lighter, thinner, more flexible, and more durable. Due to these properties, they can easily be used on various surfaces such as walls, roofs, window glass, and other building materials.

As these thin-film solar cells are more durable, they do not need much maintenance, and they can also work in low temperatures. They are in high demand in the marketplace. Research continues in order to make them even more effective and less expensive using natural materials. Because they can work in low temperatures, they are favorable for use in shady areas where traditional solar panels are unable to function properly [68].

Solar cells convert solar energy into electricity using the photovoltaic effect [69]. Modern research is being done to make eco-friendly colored solar cells, for example, due to properties like high absorption coefficient, controllable bandgap energy (Eg), and sustainability in an outer space environment. Quaternary-chalcopyrite $Cu(In,Ga)Se_2$ (CIGS) is an appropriate material for solar cells. Moreover, white light soaking increases the effectiveness of CIGS layers [70].

Cadmium telluride and copper indium gallium selenide thin films are the most effective types of thin-film solar cells. CIGS cells have a greater absorption coefficient than non-silicon-based cells, and thus more solar energy is converted into electricity, and they have longer stability, making them the best materials for thin-film technology. CIGS cells have a conversion efficiency rate of 22.3%, which makes them superior to other materials [71].

4.7.2.2 Wind Turbines

Smart thin films are widely used in the energy sector, including wind energy. In wind turbines, smart thin films are used to check the damaged parts [72], and smart thin-film hybrid materials are used to protect the blades of turbines. Creating energy by air flow is attained by using many electronics. For example, a robust triboelectric generator is made up of thin-film membrane and electrode pairs on a single substrate. When membrane comes in contact with air flow, it generates a flow of electrons by successive courses of contact electrification and electrostatic induction. With smart film surfaces on electrodes, the output current can be boosted up to 500%. Thin-film-based generators can give much better rapid and nonstop light as compared to commercial light-emitting diodes. Due to their durability they can give constant electric output after millions of operational cycles. Due to thin films, we can take advantage of wind energy using the triboelectric effect [73]. Smart thin-film surfaces can increase the efficiency of turbines by detecting damages and maintaining structural health [74].

Smart thin films are used as sensors in turbine blades and energy storage systems. Embedded thin-film sensors are used in a new technology called SHM, which is used to detect damage in FRP-based wind blades. FRP-based wind blades are combined with carbon nanotube-based thin films for sensing. Using spray fabrication and thermal annealing, thin films were prepared, and when they were subjected to uniaxial tensile cyclic loading the films were piezoresistive [75]. Extra research and progress may be needed to determine the potential payback of using smart thin films in wind turbine technology.

4.7.2.3 Heat Exchangers

The patented thin-film heat exchanger improves heating and cooling energy from ventilation air with low-cost material. It can be retrofitted into existing buildings and can offer a potential national energy savings [76]. In heat exchangers, two fluids pass through different zones. This means there are two fluids, the primary fluid that is hot and the secondary fluid that is cold. In a heat exchanger, the heat is exchanged from hot fluid to cold fluid through heat transfer walls. In smart thin-film heat exchangers, secondary fluid is present between smart thin films along heat transfer walls that provide better results. Secondary fluid supply localized at the heat-transfer surface reduces the thickness of liquid film. An example of this type is the so-called falling-film exchanger. There is also research in which smart radiator devices use a smart, integrated thin-film structure based on V1-x-yMxNyOn to dynamically modify the thermo-optic features of the primary substrate in response to ambient temperature and a control voltage [77].

Thermodynamic thin-film technology is an advanced technique that is used to improve the efficiency of solar thermal collectors that convert solar energy into electrical energy. The heat that is obtained from these collectors can be used for heating water, producing electricity, and for many other purposes. Thermodynamic coatings are applied on the surface of the solar collectors that change color due to changes in light intensity. When light is low, the color of the collector will be dark; when light intensity increases, the collector changes to a lighter color that helps in the absorption of more light. Thermodynamic coatings improve the thermal emittance of collectors, which makes them able to radiate heat more efficiently, increasing their efficiency [78].

4.7.3 Aerospace Applications

4.7.3.1 Anti-icing Coating

Smart thin films can be used as anti-icing coatings for aerospace vehicles. Smart thin films can detect atmospheric conditions and respond automatically. In this way they can reduce ice adhesion, decrease the temperature of ice nucleation, and delay freezing time. Ice formation can occur when liquid water changes temperature and/or pressure, causing severe safety and operational issues in a wide range of settings, such as power transmission, telecommunications, pumps, and heat exchangers. Ice can also strongly adhere to exposed infrastructures, leading to serious safety and operational issues [79]. Anti-icing surfaces have the potential to reduce the interaction between the surface and water droplets, so water cannot attach to the surface, and the ice-forming mechanism becomes inactive [79].

4.7.3.2 Anticorrosion Coating

Smart thin-film sensors are widely used in other aerospace applications. Due to their ability to detect changes in environment they can provide anticorrosion protection to aerospace electronics. Corrosion is the primary change that can damage aircraft. Conditions like temperature, humidity, ultraviolet rays, and pressure can cause corrosion. Smart thin films can provide protection from corrosion in several ways. Smart thin films release corrosion inhibitors when the corrosive agent comes in contact with the surface of the aircraft [80]. Because smart thin films are on the surface of the material, they can act as a barrier between the surface and the environment, and the surface does not come in contact with the corrosive agent directly. Smart thin films have a very interesting self-healing property that can repair the surface of aircraft affected by corrosion [81].

4.7.3.3 Self-Cleaning Surfaces

The self-cleaning properties of smart thin films make them widely useful in aerospace electronics and aircraft. These self-cleaning properties help keep satellites and aircraft clean. TiO_2 has extensive self-cleaning properties and is used in aircraft interiors to perform self-cleaning functions for their surfaces [82]. Self-cleaning surfaces can also avoid the buildup of debris and bacteria, which can be important for maintaining a healthy environment in the aircraft [83].

4.8 CONCLUSION AND FUTURE OUTCOMES

Smart coatings have higher demand for the high industrial range. These cutting-edge coatings provide technological capabilities that traditional coatings do not. From this review, it becomes obvious that many of the methods which at the moment are only interesting research projects will soon be used in common coating applications. This chapter's goal is not to provide an exhaustive survey of the literature on the subject, but rather to highlight some of the characteristics of these coatings that might be of interest to industry. Many of the smart coatings comprise metal and metal oxide nanoparticles, which offer a variety of functional qualities and improved performance. New multifunctional coatings are expected to be created with their hybrid nanoparticles.

Smart or intelligent coatings have a variety of applications because of their potential for multifunctionality. According to recent breakthroughs, smart coatings for anticorrosive applications are one of the main self-healing coating techniques, owing to the use of nanocontainers that can be loaded with active agents and have shells with controlled permeability specific to various triggers. Nanoparticle-based smart coatings have a variety of formulations and uses. Nanocontainers made of various types of materials that are coupled together or loaded with different active centers are becoming more popular as inventions and applications. These sophisticated solutions include everything from corrosion prevention to the delivery of medication via bioactive surfaces.

The current efforts are concentrated on scaling up nanocontainer production and evaluation. Developing practical coating systems with industrial applications that are commercially viable based on foundational academic research on smart coatings remains a significant challenge. Furthermore, improving present smart coating technologies for internal and external applications in the various industries where they are being used is hampered by questions about durability. Another difficulty is that functional coatings can cost more when added. But efficient multifunctional smart coatings in real-time service will significantly increase market demand.

LIST OF ABBREVIATIONS

PIII	Plasma immersion ion implantation
CVD	Chemical vapor deposition
IcB	Ion duster beam
cntS	Carbon nanotubes
ALD	Atomic layer deposition
Lbl	Layer-by-layer
nHPs	Nanohybrid particles
APTMOS	Aminopropyltrimethoxysilane
PTol	Polytoluidine
PAns	Polyanisidine
HVOF	High-velocity oxy-fuel spraying

PDMS	Poly(dimethylsiloxane)
PEG	Polyethylene glycol
CPNs	Conductive polymer nanoparticles
PBS-TMOS	Polybenzoxazine-trimethoxysilane
P2VP-b-PDMS	Poly(2-vinylpyridine-b-dimethylsiloxane)
PDA	Polydopamine
PV	Photovoltaics
ARC	Antireflection coating
CIGS	Copper indium gallium diselenide
TOCs	Trisubstituted organotin compounds
MF	Melamine foam
CdTe	Cadmium telluride
a-Si	Amorphous silicon
APCVD	Atmospheric pressure CVD
PEMs	Polyelectrolyte multilayers
ALG	Sodium alginate

REFERENCES

1. K. Wasa, M. Kitabatake and H. Adachi. *Thin film materials technology: Sputtering of control compound materials*, Springer Science & Business Media, 2004, p. 518.
2. M.L. Alfieri, M. Iacomino, A. Napolitano and M. d'Ischia, Fluorescent film deposition from dopamine and a diamine-tethered, bis–resorcinol coupler, *International Journal of Molecular Sciences* 2019, *20*, 4532.
3. A.-C. Bas, X. Thompson, L. Salmon, C. Thibault, G. Molnár, O. Palamarciuc, L. Routaboul and A. Bousseksou, Bilayer thin films that combine luminescent and spin crossover properties for an efficient and reversible fluorescence switching, *Magnetochemistry* 2019, *5*, 28.
4. D. Huh, H.-J. Choi, K. Kim, J. Park and H. Lee, Refractive index tunable nanoporous SiO_2 thin film and its application to mechanically robust broadband anti-reflection, *Nanoscience and Nanotechnology Letters* 2018, *10*, 1101–1106.
5. N. Huo and W.E. Tenhaeff, High refractive index polymer thin films by charge-transfer complexation, *Macromolecules* 2023, *56*, 2113–2122.
6. S. Vu, G. Nagesh, N. Yousefi, J.F. Trant, D.S.-K. Ting, M.J. Ahamed and S. Rondeau-Gagné, Fabrication of an autonomously self-healing flexible thin-film capacitor by slot-die coating, *Materials Advances* 2021, *2*, 6676–6683.
7. X. Zhang and J. He, Hydrogen-bonding-supported self-healing antifogging thin films, *Scientific Reports* 2015, *5*, 9227.
8. N. Chen, D.H. Kim, P. Kovacik, H. Sojoudi, M. Wang and K.K. Gleason, Polymer thin films and surface modification by chemical vapor deposition: Recent progress, *Annual Review of Chemical and Biomolecular Engineering* 2016, *7*, 373–393.
9. T. Govindarajan and R. Shandas, A survey of surface modification techniques for next-generation shape memory polymer stent devices, *Polymers* 2014, *6*, 2309–2331.
10. B.P. Wood, Feedback: A key feature of medical training, *Radiology*, 2000, *215*(1).
11. A. Olejnik, K. Siuzdak, J. Karczewski and K. Grochowska, A flexible nafion coated enzyme-free glucose sensor based on au-dimpled Ti structures, *Electroanalysis* 2020, *32*, 323–332.
12. V.K. Vendra, L. Wu and S. Krishnan, *Nanomaterials for the life sciences, nanostructured thin films and surfaces,* Springer, William Andrew Inc publishing, 2010, 5.
13. K.J. Loh, J. Kim, J.P. Lynch, N.W.S. Kam and N.A. Kotov, Multifunctional layer-by-layer carbon nanotube–polyelectrolyte thin films for strain and corrosion sensing, *Smart Materials and Structures* 2007, *16*, 429.
14. K. Choy, Chemical vapour deposition of coatings, *Progress in Materials Science* 2003, *48*, 57–170.
15. H.O. Pierson, in *Handbook of chemical vapor deposition: Principles, technology and applications*, William Andrew, 1999.
16. N. Sharma, M. Hooda and S. Sharma, Synthesis and characterization of LPCVD polysilicon and silicon nitride thin films for MEMS applications, *Journal of Materials* 2014, *2014*, 1–8.
17. E. Acosta, Thin films/properties and applications, *Thin films/properties and applications*, IntechOpen, 2021.

18. K.K. Schuegraf, in *Handbook of thin-film deposition processes and techniques: Principles, methods, equipment, and applications*, William Andrew, 1988.
19. R. Johnson, A. Hultqvist and S. Doblado, A brief review of atomic layer deposition: from fundamentals to applications, *Materials Today* 2014, *17*, 236–246.
20. R. Vaidyanathan, S.M. Cox, U. Happek, D. Banga, M.K. Mathe and J.L. Stickney, Preliminary studies in the electrodeposition of PbSe/PbTe superlattice thin films via electrochemical atomic layer deposition (ALD), *Langmuir* 2006, *22*, 10590–10595.
21. A.S. Hamdy, D. Butt and A. Ismail, Electrochemical impedance studies of sol–gel based ceramic coatings systems in 3.5% NaCl solution, *Electrochimica Acta* 2007, *52*, 3310–3316.
22. Hosseini, Majid Haji and Abdel Salam Hamdy Makhlouf. Industrial applications for intelligent polymers and coatings. *Industrial applications for intelligent polymers and coatings*, Springer Cham, 2016.
23. K. Szymański, A. Hernas, G. Moskal and H. Myalska, Thermally sprayed coatings resistant to erosion and corrosion for power plant boilers-A review, *Surface and Coatings Technology* 2015, *268*, 153–164.
24. Y. Li, X. Wang and J. Sun, Layer-by-layer assembly for quick production of thick polymeric films. *Chemical Society Reviews* 2012, *41*(18), 59986009.
25. F. Xia and L. Jiang, Bio-inspired, smart, multiscale interfacial materials, *Advanced Materials* 2008, *20*, 2842–2858.
26. F. Zhang, P. Ju, M. Pan, D. Zhang, Y. Huang, G. Li and X. Li, Self-healing mechanisms in smart protective coatings: A review, *Corrosion Science* 2018, *144*, 74–88.
27. N. Abu-Thabit and A.S.H. Makhlouf, Recent approaches for designing nanomaterials-based coatings for corrosion protection, in *Handbook of Nanoelectrochemistry*, Springer, Cham, 2015, pp. 309–332.
28. E. Shchukina and D.G. Shchukin, Nanocontainer-based active systems: From self-healing coatings to thermal energy storage, *Langmuir* 2019, *35*, 8603–8611.
29. N.Y. Abu-Thabit and A.S.H. Makhlouf, Recent advances in nanocomposite coatings for corrosion protection applications, in *Handbook of Nanoceramic and Nanocomposite Coatings and Materials*, Elsevier 2015, pp. 515–549.
30. S. Ilyaei, R. Sourki and Y.H.A. Akbari, Capsule-based healing systems in composite materials: A review, *Critical Reviews in Solid State and Materials Sciences* 2021, *46*, 491–531.
31. Q. Shang and Y. Zhou, Fabrication of transparent superhydrophobic porous silica coating for self-cleaning and anti-fogging, *Ceramics International* 2016, *42*, 8706–8712.
32. Z. Sun, T. Liao, K. Liu, L. Jiang, J.H. Kim and S.X. Dou, Fly-eye inspired superhydrophobic anti-fogging inorganic nanostructures, *Small* 2014, *10*, 3001–3006.
33. T. Yan, X. Chen, T. Zhang, J. Yu, X. Jiang, W. Hu and F. Jiao, A magnetic pH-induced textile fabric with switchable wettability for intelligent oil/water separation, *Chemical Engineering Journal* 2018, *347*, 5263.
34. P. Nguyen-Tri, T.A. Nguyen, P. Carriere and C. Ngo Xuan, Nanocomposite coatings: Preparation, characterization, properties, and applications, *International Journal of Corrosion* 2018, *2018*.
35. S. Pourhashem, F. Saba, J. Duan, A. Rashidi, F. Guan, E.G. Nezhad and B. Hou, Polymer/Inorganic nanocomposite coatings with superior corrosion protection performance: A review, *Journal of Industrial and Engineering Chemistry* 2020, *88*, 29–57.
36. J. Musil, Hard and superhard nanocomposite coatings, *Surface and Coatings Technology* 2000, *125*, 322–330.
37. J. Patscheider, T. Zehnder and M. Diserens, Structure–performance relations in nanocomposite coatings, *Surface and Coatings Technology* 2001, *146*, 201–208.
38. T.S. Narayanan, Surface pretreatment by phosphate conversion coatings—A review, *Reviews in Advanced Materials Science* 2005, *9*, 130–77.
39. H. Umehara, M. Takaya and S. Terauchi, Chrome-free surface treatments for magnesium alloy, *Surface and Coatings Technology* 2003, *169*, 666–669.
40. F. Aziz and A.F. Ismail, Spray coating methods for polymer solar cells fabrication: A review, *Materials Science in Semiconductor Processing* 2015, *39*, 416–425.
41. L. Pawlowski, in *The science and engineering of thermal spray coatings*, John Wiley & Sons, 2008.
42. L.-M. Berger, Application of hardmetals as thermal spray coatings, *International Journal of Refractory Metals and Hard Materials* 2015, *49*, 350–364.
43. G. Barroso, Q. Li, R.K. Bordiab and G. Motz, Review of silicon-based polymeric and ceramic coatings, *Journal of Materials Chemistry A* 2019, *7*(5), 1936–1963.
44. P. Fauchais, Understanding plasma spraying, *Journal of Physics D: Applied Physics* 2004, *37*, R86.
45. S. Shankar, D. Koenig and L. Dardi, Numerical and experimental analysis of a solid shroud in multi-arc plasma spraying, *JOM* 1981, *33*, 13–20.

46. H. Assadi, H. Kreye, F. Gärtner and T. Klassen, Cold spraying–A materials perspective, *Acta Materialia* 2016, *116*, 382–407.
47. H. Assadi, T. Schmidt, H. Richter, J.-O. Kliemann, K. Binder, F. Gärtner, T. Klassen and H. Kreye, On parameter selection in cold spraying, *Journal of Thermal Spray Technology* 2011, *20*, 1161–1176.
48. J. Kawakita, H. Katanoda, M. Watanabe, K. Yokoyama and S. Kuroda, Warm Spraying: An improved spray process to deposit novel coatings, *Surface and Coatings Technology* 2008, *202*, 4369–4373.
49. K. Triyana and E. Suharyadi, High-performance silver nanowire film on flexible substrate prepared by meyer-rod coating, *IOP Conference Series: Materials Science and Engineering* 2017, p. 012055.
50. J. Puetz and M. Aegerter, *Sol-gel technologies for glass producers and users* 2004, 3748.
51. R.M. England and S. Rimmer, Hyper/highly-branched polymers by radical polymerisations, *Polymer Chemistry* 2010, *1*, 1533–1544.
52. T. Saegusa and Y. Chujo, An organic/inorganic hybrid polymer, *Journal of Macromolecular Science—Chemistry* 1990, *27*, 1603–1612.
53. N.Y. Abu-Thabit and A.S.H. Makhlouf, Recent advances in nanocomposite coatings for corrosion protection applications, in *Handbook of Nanoceramic and Nanocomposite Coatings and Materials*, Elsevier, 2015, pp. 515–549.
54. M. Angelopoulos, Conducting polymers in microelectronics, *IBM Journal of Research and Development* 2001, *45*, 57–75.
55. J. Yang, I. Bos, W. Pranger, A. Stuiver, A.H. Velders, A.C. Stuart and M. Kamperman, Blue AIEgens: approaches to control the intramolecular conjugation and the optimized performance of OLED devices, *Journal of Materials Chemistry A* 2016, *4*, 6868–6877.
56. M. Zhang, F. Xu, D. Lin, J. Peng, Y. Zhu and H. Wang, A smart anti-corrosion coating based on triple functional fillers, *Chemical Engineering Journal* 2022, *446*, 137078.
57. S.S. Behzadi, S. Toegel and H. Viernstein, Innovations in coating technology, *Recent Patents on Drug Delivery & Formulation* 2008, *2*, 209–230.
58. A. Makhlouf, in *Current and advanced coating technologies for industrial applications*, Elsevier, 2011, pp. 3–23, https://doi.org/10.1533/9780857094902.1.3
59. S. Bandehali, F. Parvizian, S.M. Hosseini, T. Matsuura, E. Drioli, J. Shen, A. Moghadassi and A.S. Adeleye, Planning of smart gating membranes for water treatment, *Chemosphere* 2021, *283*, 131207.
60. A.K. Nayak, Smart micro- and nanomaterials for drug delivery, *Micro and Nano Technologies,* 2022, 135–151.
61. H.-J. Choi, S.-Y. Seo, J.-S. Jung and S.-G. Yoon, Water-resistant and antibacterial zinc aluminate films: Application of antibacterial thin film capacitors, *ACS Applied Electronic Materials* 2021, *3*, 1429–1436.
62. J. Prakash, N. Singh, R. Mittal and R.K. Gupta, Stimuli-responsive smart surfaces for oil/water separation applications, *Stimuli-responsive Dewetting/Wetting smart surfaces and interfaces* 2018, 207237.
63. Y. Wei, H. Qi, X. Gong and S. Zhao, Specially wettable membranes for oil–water separation, *Advanced Materials Interfaces* 2018, *5*, 1800576.
64. M.H. José, J.P. Canejo and M.H. Godinho, Oil/water mixtures and emulsions separation methods—an overview, *Materials* 2023, *16*, 2503.
65. H. Zhang and M. Chiao, Anti-fouling coatings of poly (dimethylsiloxane) devices for biological and biomedical applications, *Journal of Medical and Biological Engineering* 2015, *35*, 143–155.
66. X. Hao, S. Chen, D. Qin, M. Zhang, W. Li, J. Fan, C. Wang, M. Dong, J. Zhang and F. Cheng, Antifouling and antibacterial behaviors of capsaicin-based pH responsive smart coatings in marine environments, *Materials Science and Engineering: C* 2020, *108*, 110361.
67. M. Borpatra Gohain, S. Karki, D. Yadav, A. Yadav, N.R. Thakare, S. Hazarika, H.K. Lee and P.G. Ingole, Development of antifouling thin-film composite/nanocomposite membranes for removal of phosphate and malachite green dye, *Membranes* 2022, *12*, 768.
68. N. Song and S. Deng, in *Thin film deposition technologies and application in photovoltaics*, IntechOpen, 2022, https://doi.org/10.5772/intechopen.108026
69. D. Mattox, Application of thin films to solar energy utilization, *Journal of Vacuum Science and Technology* 1976, *13*, 127–134.
70. W.-J. Lee, D.-H. Cho, J.M. Bae, M.E. Kim, J. Park and Y.-D. Chung, Ultrafast wavelength-dependent carrier dynamics related to metastable defects in Cu (In, Ga) Se_2 solar cells with chemically deposited Zn (O, S) buffer layer, *Nano Energy* 2020, *74*, 104855.
71. K. Chopra, P. Paulson and V. Dutta, Thin-film solar cells: An overview, *Progress in Photovoltaics: Research and Applications* 2004, *12*, 69–92.
72. A. Downey, S. Laflamme, F. Ubertini, H. Sauder and P. Sarkar, Experimental study of thin film sensor networks for wind turbine blade damage detection, *AIP Conference Proceedings* 2017, p. 070002.

73. X.S. Meng, G. Zhu and Z.L. Wang, Robust thin-film generator based on segmented contact-electrification for harvesting wind energy, *ACS Applied Materials & Interfaces* 2014, *6*, 8011–8016.
74. A. Kumar, S. Rudra, S. Thamizharasan, G. Pradhan, M. Rani, B. Sahu and A.K. Nayak, Crystal structure controlled synthesis of tin oxide nanoparticles for enhanced energy storage activity under neutral electrolyte, *Journal of Materials Science: Materials in Electronics* 2022, *33*, 13668–13683.
75. L.P. Mortensen, D.H. Ryu, Y.J. Zhao and K.J. Loh, in *Rapid assembly of multifunctional thin film sensors for wind turbine blade monitoring*, Trans Tech Publ, 2013, https://doi.org/10.4028/www.scientific.net/kem.569-570.515
76. J. De Lallee, G. Marie and R. Moracchioli, in Thin-film heat exchanger, Google Patents, 1979, Report no: FR 2341118.
77. R.V. Kruzelecky, E. Haddad, M. Soltani, M. Chaker and D. Nikanpour, Integrated thin-film smart coatings with dynamically-tunable thermo-optical characteristics, *SAE Transactions* 2002, 323330.
78. A. Krammer, O. Bouvard and A. Schüler, Study of Si doped VO_2 thin films for solar thermal applications, *Energy Procedia* 2017, *122*, 745–750.
79. M. Shamshiri, R. Jafari and G. Momen, Potential use of smart coatings for icephobic applications: A review, *Surface and Coatings Technology* 2021, *424*, 127656.
80. M. Saremi and M. Yeganeh, Application of mesoporous silica nanocontainers as smart host of corrosion inhibitor in polypyrrole coatings, *Corrosion Science* 2014, *86*, 159–170.
81. Y.J. Tan, J. Wu, H. Li and B.C. Tee, Self-healing electronic materials for a smart and sustainable future, *ACS Applied Materials & Interfaces* 2018, *10*, 15331–15345.
82. R. Chakraborty, K. Vilya, M. Pradhan and A.K. Nayak, Recent advancement of biomass-derived porous carbon based materials for energy and environmental remediation applications, *Journal of Materials Chemistry A* 2022, *10*, 6965–7005.
83. R. Blossey, Self-cleaning surfaces-virtual realities, *Nature Materials* 2003, *2*, 301–306.

5 Properties of Multifunctional Thin Films for High-Temperature Applications

Akash Mishra and Ajit Behera

5.1 INTRODUCTION

Surface functionalization has been instrumental in the invention of new industrial applications in a variety of fields, including electronics, biosensors, and the glass industry, during the past several decades. A thin coating of substance is often placed on a substrate to accomplish surface functionalization, which might alter how it subsequently interacts with its surroundings. The solid thin film, either organic, inorganic, or hybrid, sometimes known as "smart coating," is capable of carrying out an established array of activities in response to certain inherent properties or external stimuli [1, 2]. A thin film is a microscopically thin layer of material that is deposited or developed on a substrate, which is often a rigid surface made of materials such as glass, silicon, or another substance. Thin films, which may vary in thickness from a few atoms or molecules to many micrometers, are crucial elements in a variety of technologies and applications. Multifunctional thin films are specialized, microscopically thin material layers that exhibit a variety of unique and sometimes opposing characteristics and functions [3, 4]. These films are designed to fulfil several functions at once, thus rendering them extremely adaptable and beneficial across a variety of industries and technologies. The majority of polymer substrates have lower glass transition or decomposition temperatures and therefore are unable to tolerate any high-temperature treatment, thus restricting the practical uses of thin films on these substrates. Additionally, high transmittance and superhydrophobicity are primarily sacrificed in order to increase mechanical robustness. Thus, the creation of strong, antireflective, and superhydrophobic thin films on substrates that are susceptible to high-temperature treatment remains a difficulty.

Because of the polycrystalline structure and substantial leakage current involved in bismuth ferrite ($BiFeO_3$; BFO) bulks, it is challenging to attain the inherent ferroelectric characteristics. Consequently, the BFO-based thin films have received a lot of interest since, in broad terms, excellent thin-film samples may be successfully manufactured by selecting the right deposition processes and fine-tuning processing parameters [5]. Anatase and rutile structures, which diverge in optical characteristics and presumably photocatalytic efficacy, are the most frequent forms in which thin films of titanium oxide (TiO_2) were produced. TiO_2, which exhibits adjustable behavior based on the technique employed for manufacture, represents one of the materials in thin films that has been the subject of the most research. It is possible to produce various structural polymorphs and levels of crystallinity or amorphousness by adjusting the manufacturing conditions. If photocatalysis is examined, the crystal size becomes even more important while it is in its crystalline state. However, the many polymorphic phases of TiO_2 (such as rutile and anatase) play a significant role in the material's use [6]. La_2O_3 thin films that have a porous microstructure with tiny crystallite size are generated by burning a blend of lanthanum nitrate and urea solution [7]. Metastable NaCl-structure $Ti_{1-x}Al_xN$ layers grown by magnetron sputtering are the most studied group of TiN-based thin films, and these are suitable for use as protective coatings for cutting tools, which show good hardness (typically ~30 GPa), high wear and oxidation resistance (depending on Al concentration), and self-hardening effects at elevated temperatures up to about 900°C (resulting from spinodal

DOI: 10.1201/9781032635347-5

decomposition) [8]. Sputter-deposited (transition-metal) TM diboride thin films have far fewer commercial applications than TiAlN, principally because of their extreme brittleness and low oxidation resistance. According to B. Bakhit et al. [9], in TiN-based thin films, alloying TM diborides with Al enhances their oxidation properties. Superior hardness as well as ductility are necessary for TM diboride films to be able to prevent brittle breaking. Alloying ZrB_2 thin films with Ta may simultaneously boost the toughness and hardness of nanoindentation tests.

The addition of oxygen to the TiC matrix has been investigated, and it has been demonstrated that titanium oxycarbide (TiCxOy) thin films can be tailored to exhibit a wide range of optical and mechanical properties, ranging from those of metallic carbides to those of ionic oxides, by varying the oxygen/carbon ratio [10]. Furthermore, TiCxOy thin films have also been considered for applications in decorative tasks and tribology. However, the application of TiCxOy thin films as piezoresistive films for high-temperature pressure sensors has not yet been investigated [11]. There are issues with the vaporization of sulfur (S) atoms during the fabrication of copper gallium selenide (CGS) thin films at extreme temperatures, which lead to insufficient S content and defects in the crystal structure [12]. Post-sulfurization treatment may diminish the S deficit, but it is challenging to control S content in CGS due to its high vapor pressure. Therefore, further approaches are employed to stabilize the S content, and one of them is tellurium (Te) doping in CGS. The widespread interest in dimetal chalcogenide thin films has grown continuously and enormously because of their use in high-technology applications such as solar cells, photonics, phase change memories, sensors, and optoelectronic applications. The formation of a single-phase $AgSbTe_2$ was reported on AgSbTe thin films deposited by thermal evaporation, as described by Pinsker et al [13]. Nonetheless, Lakshminarayana et al. asserted the formation of two phases when they prepared the material in the form of a thin film (AgSbTe + Ag2Te). These facts confirm that the thin film is more thermodynamically stable and possesses a purer cubic phase than bulk materials.

Several types of diamond thin films have been synthesized and systematically studied, and they exhibit different microstructures, surface morphologies, and properties. There are many methods for obtaining thin films, including electroplating, anodic treatment, chemical vapor deposition (CVD), atomic layer deposition (ALD), and spin coating [14]. The discovery of superconductivity above 90 K in the rare-earth-alkaline-earth-copper-oxide system initiated an enormous amount of research activity, especially in technologies to prepare thin films of these high-temperature superconductors (HTS). Future innovations will need surface-integrated thin-film sensors, which can be flexibly applied to critical zones of various technical components to monitor and control the actual state of machines. The locations of interest are primarily tribological contacts, e.g., bearings, tools, and screws. For this purpose, hard and wear-resistant coatings must be developed [15].

5.2 DIFFERENT EXAMPLES OF THIN FILMS

Besides the examples shown in Table 5.1, let's discuss different thin films in detail. Ceramic coatings made of **titanium nitride** and applied to substrates made of ductile steel are frequently used as protective coatings. The rigid and useful ceramic substance titanium nitride (TiN) is reported to crystallize in the B1 NaCl structure. It is present as a solid solution having a nitrogen content of between 37.5% and 50%. TiN is frequently used on instruments for cutting to extend their life because of its excellent wear and corrosion-resistant characteristics. Titanium nitride is used for surgical implants such orthopedic and dental prostheses because of its biocompatible qualities and mixture of exceptional ductility and hardness [8]. **Zirconium diboride** (ZrB_2) is one of the most notable ultra-high temperature ceramics due to its exceptional mix of characteristics, which includes extremely high melting points (over 3000°C), great hardness, great oxidation resistance, and exceptional thermal and electrical conductivities. Because of its unique mix of qualities, ZrB_2 is a desirable material for application in harsh situations. The front edges of hypersonic spacecraft, crucibles, high-temperature electrodes, or parts for corrosive environments are a few examples [9].

TABLE 5.1
List of Various Categories where Thin Films Are Used

Category	Examples	Uses
Semiconductor	SiO_2	Insulating layer in integrated circuits (ICs).
	Si_3N_4	Provides electrical isolation and protection in ICs.
	Aluminum, copper metallic thin films	Interconnections in microelectronics.
Optical	Antireflection coatings	Reduce light reflection on lenses, camera lenses, eyeglasses, and solar panels.
	Dielectric mirrors	Used in optical systems, lasers, and telescopes.
	Beam splitters	Divide and redirect incident light; used in interferometers.
Solar cells	Amorphous silicon (a-Si)	Used in thin-film photovoltaic solar panels.
	Cadmium telluride (CdTe)	Used in solar panels.
	Copper indium gallium selenide (CIGS)	Used in flexible solar panels.
Magnetic	Permalloy	Magnetic recording media such as hard drives.
	Garnet films	Magneto-optical data storage.
Superconducting	Yttrium barium copper oxide (YBCO)	Superconducting devices, including quantum computing.
Barrier coatings	Titanium nitride (TiN)	A barrier and hard coating in cutting tools and microelectronics.
	Tantalum pentoxide (Ta_2O_5)	A moisture barrier in microelectromechanical systems (MEMS).
Transistors	Amorphous silicon thin-film transistors (a-Si TFTs)	Used in liquid crystal displays (LCDs).
	Oxide semiconductor thin-film transistors (oxide TFTs)	Used in some display technologies, e.g., OLED.

Regarding ultra-high magnetic recording media and soft magnetic thin films (such as a magnetic reading head) for high-frequency response magnetic devices, **cobalt-carbon** (Co-C) thin films containing cobalt nanocrystals enclosed by graphite-like carbon are being created using ion beam or DC sputtering processes [16]. There is a class of diamond thin films named **ultrananocrystalline diamond** (UNCD), which is grown using an Ar-rich/CH_4 chemistry in a microwave plasma-enhanced CVD process [Ar (99%)/CH_4 (1%)]. UNCD can be used as coatings for mechanical pump seals, field emission cold cathodes, RF MEMS (radio-frequency micro-electro-mechanical systems) and NEMS (nanoelectromechanical systems) resonators, switches for wireless communications and radar systems, biomedical devices, and biosensors [17]. With the goal to create glass materials containing photocatalytic and antibacterial capabilities with the least amount of transparency loss, transparent **titanium oxide**-based (TiO_2) thin films coated with Ag NPs were produced on glass. Numerous structural polymorphs and levels of crystallinity or amorphousness may be produced by altering the processing parameters [18]. **Vanadium dioxide** (VO_2) is a well-known thermochromic substance that exhibits rapid shifts in electrical and optical characteristics close to its phase shift temperature, T_C around 68°C. Under Critical temperature, TC that is around 68°C, VO_2 exhibits an insulation state (monolithic phase), but above Tc, it transitions to a metallic state (tetragonal phase). When VO_2 enters the metallic state ($T > T_C$), the electrical characteristics undergo a significant shift that causes the reflectance in the mid-infrared (IR) region to increase. Subsequently, its emissivity decreases, making VO_2 appropriate for smart window uses. In contrast, applications requiring spacecraft thermal control call for a reverse dynamic behavior (high emissivity at higher temperature and low emissivity at reduced temperature) [19]. Apparently it is possible for **zinc oxide**-based thin-film transistors (TFTs) to take the place of the common amorphous silicon channel layer in traditional TFTs. You may also use ZnO film, which has a high band gap of 3.3 eV, to create transparent TFTs. The ability of ZnO to identify ultraviolet (UV) rays is also well recognized [20].

5.3 MULTIFUNCTIONAL THIN FILMS (MTFs)

Environmental benefits can be obtained from using nanomaterials derived from nature, along with the remarkable mechanical properties and/or photonic crystal properties of biological composites like bone, nacre, wood, beetle scales, and butterfly wings, which also serve as significant sources of motivation for the creation of novel multifunctional materials. Piezoelectric, magnetostrictive, shape memory, and ferromagnetic shape memory thin films are examples of multifunctional thin films that are essential components for further miniaturizing technological gadgets. These may be referred to as MEMS (micro-electro-mechanical-systems) or perhaps NEMS (nano-electro-mechanical-systems) depending on the crucial dimension scale. Within a small area, miniature devices combine many functions, such as sensors and actuators. Multiple materials may be used to carry out these tasks; for instance, a piezoresistive microstructure made of B-doped Si can detect the deflection of a magnetostrictive actuator based on TbFe/FeCo multilayers. Yet, in order to achieve a greater degree of integration, both of these functions—or even more—should be combined into a single multifunctional material. This substance can conduct sensing and actuation at an innate level that reduces the difficulty of microscale devices' fabrication [17, 18]. Multifunctional substances are typically nanoscale multilayered thin films composed of a minimum of two distinct materials, preferably at least binary or ternary alloys. This promising and adaptable framework for creating multifunctional, mesoporous materials with photonic crystal characteristics on an extremely large surface area is provided by using a suspension of cellulose nanocrystals (CNC) to serve as a self-assembled blueprint during the creation of inorganic materials that embrace a regular internal structure obtained from CNCs. Table 5.2 presents different functions and applications of various types of multifunctional thin-film coatings.

5.3.1 PREPARATION OF MTFs

The creation of coatings that display many desirable qualities concurrently requires specialized methods and procedures for multifunctional thin films. The particular technique employed is

TABLE 5.2
Different Functions and Applications of Various Types of Multifunctional Thin-Film Coatings

Coatings	Functions	Applications
Antireflective and hydrophobic coatings	• Antireflective properties reduce glare and improve optical clarity. • Hydrophobic properties repel water and prevent water droplets from sticking to the surface.	Optical lenses, camera lenses, eyeglasses, and solar panels
Antireflective and self-cleaning coatings	• Antireflective properties reduce glare and improve light transmission. • Self-cleaning properties facilitate the removal of dust and dirt from the surface.	Architectural glass, automotive windshields, and solar panels
Transparent conductive films	• Transparency allows light to pass through. • Conductivity enables the transmission of electrical current.	Touchscreens, flat-panel displays, and solar cells
Heat-reflective and insulating coatings	• Reflects heat to maintain a cool surface. • Insulates against heat transfer.	Energy-efficient windows and building materials
Biocompatible and drug-eluting coatings	• Biocompatible surface interacts favorably with biological tissues. • Releases drugs or therapeutic agents in a controlled manner.	Medical implants, stents, and drug-delivery systems
Optical filters and anti-glare coatings	• Filters specific wavelengths of light for optical applications. • Reduces glare and reflections for improved visibility.	Camera filters, optical instruments, and displays

determined by the materials used along with the intended purposes. The **pulsed laser deposition** (PLD) technique creates thin films by ablating one or more targets that have been irradiated by a concentrated pulsed laser beam. The presence of droplets on the substrate surface, a high rate of deposition of around 0.1 nm per pulse, as well as the exchange of stoichiometry between target and generated film are the primary hallmarks of PLD. This deposition method has been extensively used to manufacture metallic systems, polymers, fullerenes, and various oxides, nitrides, or carbides [21]. The application of a very homogeneous film upon a flat substrate across a significant area (30 cm) and using a fully predictable and repeatable film thickness makes **spin coating** a special process. Inorganic, organic, and inorganic/organic solution combinations may all be treated using this approach. Microelectronics uses, coverings of polymeric photoresist, coverings in media for optical information mass storage, the use of a metal layer that reflects laser light, and the spin-coating of a thin film of acrylic plastic to prevent scratches are where this approach is frequently used. Sputter-deposited ZrBy thin films are frequently overstoichiometric, including B/Zr(y) ratios > 2 (excess B), which may affect their phase stability at high temperatures [22]. This issue may be successfully overcome by using **high-power impulse magnetron sputtering** (HiPIMS), which makes use of the variations in the ionization possibilities among TM and B atoms and gas rarefaction phenomena in order to direct ion fluxes onto substrates. HiPIMS has been shown to provide greater oversight over the characteristics of formed films, with enhanced sputtered species ionization fraction and significant amounts of active ions [23]. Over surfaces that are kept at room temperature, new thin layers of diamond-like carbon have been deposited using an **ion-beam deposition** process. The idea behind the method is that instead of warming the substrate, the energy needed for thin-film nucleation and development can be provided from the kinetic energy generated by the accelerated ion beam of the deposition material. The kinetic energy of impact on the substrate may be adjusted by altering the surface potential whenever ions that contain the substance to be coated are propelled in a narrow stream toward the substrate [3]. Simply varying the concentration of all dopant in the precursor mixture, **spray pyrolysis deposition** (SPD) offers low-cost deposition and straightforward in-vitro doping. SPD is a well-liked method that is easy to operate, inexpensive, and efficient for large-area deposition. The characteristics of the coating are primarily influenced by a number of factors, including the type of precursor solution, concentration of the solution, amount, operational pyrolytic decomposition, strongly uniform substrate/deposition temperature, local substrate cooling throughout spray, mono-dispersion of aerosols, spray rate, nozzle geometry, distance from substrate, and deposition time [24]. To attempt to offer a mechanism to separate the gas dissociation activity from the graphene substrate development process inside the system of **plasma-enhanced chemical vapor deposition** (PE-CVD), an innovative method of graphene deposition employing inductively coupled radio frequency plasma (ICRFP) is being researched. The ability to independently adjust the precursor gas dissociation and growth parameters renders this process more highly tunable than traditional CVD, which should improve management of the size and form of the final carbon nanostructures. Additionally, employing the precursor gas in a plasma state results in a higher density of chemically reactive carbon radicals by definition, which increases the rate of deposition, particularly in nonequilibrium low-temperature plasmas like the aforementioned ICRFP plasma, which also includes a merely ionized environment alongside an elevated electron energy [25].

5.3.2 Studies Related to MTFs

A useful method for analyzing the thermal characteristics of thin films with many functions is **thermogravimetric analysis** (TGA). Here, a little fragment of the multipurpose thin film is sliced and precisely weighed to determine the original mass. The TGA apparatus, which comprises a furnace and an accurate balance, is filled with the film specimen. The TGA equipment is set up for heating the sample gradually (often at a steady rate of temperature rise) while continually gauging its weight [10]. The TGA equipment constantly monitors the mass of the film specimen while the temperature rises. Every weight fluctuation is noted. These alterations could be a sign of a number of heat processes, such as oxidation, desorption, or breakdown. A useful method for determining the outermost area and porosity of materials, particularly multifunctional thin films, is the **Brunauer-Emmett-Teller** (BET) study. The film needs to have an established and precisely determined mass, be clean, and be devoid of impurities. A gas adsorption device is used with the specimen, usually employing nitrogen (N_2). An adsorption isotherm is produced by the gas that is absorbed onto the outer layer of the film under different relative pressures (P/P_0). During adsorption and desorption, the device monitors the total quantity of gas adsorbed at every corresponding pressure point [12]. The data that is gathered includes the relative pressures and the quantity of gas adsorbed, which is typically measured in micromoles or milligrames. The values from the adsorption isotherm are put through the BET equation. The BET equation connects the substance's specific surface area with the quantity of gas adsorbed at a particular relative pressure. Another electrochemical method called **cyclic voltammetry** (CV) is used to investigate the redox behavior, electrochemical action, and other electrochemical characteristics of materials. Although CV is frequently used to analyze electroactive substances, it may also be used to examine the electrochemical properties of multifunctional thin films. The versatile thin film is prepared for electrochemical examination. This may entail depositing or coating the film over an electrode's surface, like a working electrode within an electrochemical cell or a glassy carbon electrode. An electrochemical cell is filled with the produced film. A working electrode, a reference electrode (such as Ag/AgCl), and a counter electrode are normally included in a cell [17]. To speed up all electrochemical processes, an electrolyte solution is employed. In cyclic voltammetry, a voltage scan is done repeatedly, while the resulting measurements of current flow are recorded. Research allows us to look at the way the electrochemical behavior of the film changes over time. A **galvanostatic charge-discharge** study can be especially useful for analyzing the capacitance behavior, energy storage capability, and charge/discharge kinetics of thin films that might possess electrochemical or energy storage capabilities. The versatile thin film is ready for electrochemical examination. The film is frequently deposited or coated over an electrode substrate, which might be a functioning electrode or a component of an electrochemical cell. An electrochemical cell is filled with the manufactured thin sheet. A working electrode (the film), a reference electrode, and a counter electrode have all been present in the cell. Ionic conduction is aided by the application of an adequate electrolyte. The circuit receives a steady current to power the thin film [22]. Usually, during this procedure, energy builds up on or inside the film. The electrical current's direction is switched following a certain amount of charging. By pulling current out of the thin layer, the circuit is discharged, unleashing the accumulated charge. The electrical potential over the thin film is tracked as the system charges and discharges, and the electrical current that passes through the device is continually recorded. In order to learn more about electrical impedance and the electrochemical activities taking place at the film's surface, **electrochemical impedance spectroscopy (EIS)** is used. The conducting substrate, serving as the working electrode, is normally coated or deposited with the multifunctional thin film. It's possible to subject the film to an electrolyte solution. By introducing a modest potential change (often a sinusoidal voltage) into the circuit, the electrochemical cell gets initialized [24]. A possible scenario in a state of equilibrium has the perturbation overlaid on it. EIS includes varying the chosen sinusoidal voltage's frequency across a large spectrum, often between millihertz and megahertz. The system's temporal constants are used to determine the frequency range. The reaction of the structure to the disturbance is captured together with the voltage that exists across the working electrode, the current flowing through the cell, and their relative phases. This data is gathered according to frequency.

5.3.3 Tests Related to MTFs

A technique employed to gauge how effectively multifunctional films interact or resist water is the **water droplet impact test**. Develop the test sample of the multifunctional thin film. For evaluation, the film is often attached to a sturdy surface like a glass or silicon wafer. Apply a testing setup that enables regulated droplet release and impact event capturing. A high-speed camera is frequently used in this device to capture the droplet's impact. Make water droplets with a certain size and speed [18]. To guarantee accurate findings, the droplets must be discharged repeatedly. Place the droplet at a given altitude and angle over the sample. Allow the droplet to make contact with the surface of the film. For a complete record of the droplet impact event, use a high-speed camera. This covers the first point of interaction, droplet dispersal, including any further actions on the film's surface. The **sand impact abrasion test** is a technique designed to assess how resistant a substance is to abrasion and wear due to the impact of abrasive particles like sand, including multifunctional thin films. Prepare the multipurpose thin-film sample for testing. To achieve uniform conditions, the film is often attached to a substrate, like glass or a silicon wafer. Use a testing device intended for the sand impact abrasion test. A controlled abrasive material feeder, an impact chamber, and a target holder are typical components of this system [19]. The abrasive medium, which is often a certain grade of sand or abrasive particles, should be chosen. Essential variables for the evaluation include the abrasive media's size and composition. Place the film sample into the equipment's shock chamber. Verify that the surface of the specimen is pointing in the path of the discharge of the abrasive medium. A technique for determining how well coatings or thin films adhere to a substrate is the **tape adhesion test**. This method is frequently employed to gauge how well multifunctional thin films adhere to the substrates which support them. Create a specimen sample of the multifunctional thin film. Usually, a substrate material is applied to or deposited with the film. Verify that the film is thoroughly cured or given time to set according to the application. During the test, pick a suitable sticky tape. Tapes with pressure-sensitive adhesive (PSA) are often employed. The exact purpose and anticipated adhesion strength affect the tape choice. Tear the tape into uniformly sized and spaced pieces [25]. These pieces can be placed on the surface of the thin film in several places. Attach the tape strips to the film's surface. To establish proper contact, press the tape down using an appropriate instrument or device, like a roller or a predetermined weight. This must be done to make sure the tape sticks to the film tightly. The tape is then swiftly and gently detached from the film's surface after a certain amount of time. For uniformity, the pace and angle of removal are standardized.

The capacity of multifunctional thin films to prevent or destroy bacteria on contact is evaluated using **antibacterial activity tests**. The multifunctional thin film test samples should be ready; the films must be properly deposited, coated, cured, or set in accordance with the planned use. Choose the bacterial strains that will be tested. Gram-positive and Gram-negative bacteria, including *Escherichia coli* (*E. coli*) and *Staphylococcus aureus*, are common choices. For the bacterial strains, produce nutritional agar plates or another appropriate growing medium. The microorganisms for the test will be cultured on such plates [8]. Introduce a standardized microbial suspension to the plates of agar. This guarantees uniform initial circumstances for the antimicrobial test. Make sure that the multifunctional thin film samples are in close contact with the bacterial culture by placing them on the infected agar plates' surface. The plates should be incubated for a predetermined amount of time, often 24 hours, at a specific temperature (commonly 37°C). Determine the area around the film that is clear (the region of inhibition). This region denotes the region that the film has prevented or eliminated bacterial growth. Usually, a bigger zone denotes more potent antibacterial action. Colony-forming units (CFUs), which may be measured quantitatively along with visual inspection, may be counted in the region of restriction and in areas devoid of the film.

A technique to assess a film's flexibility and mechanical qualities, notably its ability to withstand bending or deformation under load, is the **flat-wise bending test** for multifunctional thin films. Create the test specimens of multifunctional thin films. Verify that all protective coatings and substrates have been eliminated and that the films are trimmed to uniform measurements. Use a testing device created especially for flat-wise bending tests. A force measuring system and a bending fixture or mandrel are commonly included in this device. Choose a mandrel or bending apparatus having a suitable radius of curvature. The desired use and required bending dictate the fixture choice [5]. For covering the curved surface, position the film specimen on the bending fixture. Check that the long axis of the film is parallel to the curve. Using a proper loading method, deliver a regulated force on the very top of the film. A certain pace must be used when applying the force. Twist the film slowly to fit the fixture's curve. To fit the fixture's shape, the film will bend. Calculate the amount of pressure used to bend the film. This test can reveal details about the flexibility and flexural strength of the film.

5.4 PROPERTIES OF MTFs

5.4.1 Wear Rate

The rate of wearing is a measurement of the amount of material that is lost from the film's surface within a certain time period, usually expressed as a volume loss per unit of time or distance (e.g., volume loss per meter). Using a confocal laser scanning microscope, the wear paths are obtained. The wear rates are calculated by the equation: Wear rate = $V/(F \times s)$, where V is the volume loss by wear (mm^3), F is the applied load (N), and s is the sliding distance (m). The monolithic pure carbon film offers the smallest coefficient of friction (0.1) but the highest wear rate, while pure Co film has the greatest coefficient of friction (0.6). Compared to pure Co and C films, Co-C nanocomposite films offer greater protection against wear. A greater quantity of carbon in the Co-C composite sheets led to increased wear resistance with a smaller friction coefficient. A high level of hardness (>30 GPa) with greater toughness work together to create an improved durability to wear. For applications at elevated temperatures, the wear rate characteristics of multifunctional thin films are crucial for determining the film's resistance to wear and material depletion. For repair and substitution planning, continuous wear rate tracking is essential, particularly for situations where longer service life is crucial. Corrosion and oxidation may influence wear rate in high-temperature situations, such as in the aircraft industry. To learn more about the wear processes affecting the film, tribological studies—which concentrate on the mathematics of friction, wear, and lubrication—are often conducted [26].

5.4.2 Elastic Modulus

Knowing whether multifunctional thin films react to mechanical stressors, such as temperature-induced distortion, requires knowledge of their elastic modulus characteristics. The elastic modulus, also known as Young's modulus, measures how rigid the material is and if it is prepared to withstand deformation in response to an exerted force. Elastic moduli for substances may differ depending on temperature. The elastic modulus can shift when the temperature rises in applications with elevated temperatures. It is possible to set up multifunctional thin films on surfaces with various thermal expansion coefficients. The elastic modulus of the film–substrate combination at elevated temperatures might be affected by the ensuing thermal expansion discrepancy. Materials may experience creep, which is a time-dependent distortion under constant stress, at high temperatures. Knowing the elastic modulus parameters is essential for comprehending the film's creep resistance. To comprehend how a substance will behave during mechanical loading, it is crucial to analyze stress–strain curves at different temperatures. These curves reveal information about the modulus, yield strength, and elongation characteristics of the film. Development and optimization may be aided by using FEA simulations to describe and forecast the behavior of multifunctional thin films at various temperature settings. When cross-linkers are added, the elastic modulus and fracture strain are often larger than in pure semiconductors as well as dielectrics [27].

5.4.3 Dielectric Properties

The capability of a substance to retain electrical energy within an electric field is measured by the dielectric constant, commonly abbreviated as ε. Higher insulating qualities are indicated by a high dielectric constant, whereas higher electrical conductivity could be indicated by a low value.

Temperature changes may affect the dielectric constant and dielectric strength. Developing and employing multifunctional thin films in high-temperature settings requires a comprehension of such temperature relationships. Substances' coefficients of thermal expansion (CTE), particularly in layers bonded to substrates with differing CTEs, can affect a material's dielectric characteristics. Along with the high photoluminescence characteristics, doping La^{3+} ions could be a useful method for enhancing ferroelectric and dielectric capabilities. Each thin film has decreasing dielectric constant values due to more frequent occurrences. Dielectric rise was ascribed to the progression of microstructural deformation. Thin films' dielectric characteristics can change over multiple frequencies. For situations using high-frequency electrical impulses, like those in electronic gadgets and communication systems, this becomes especially important. Large values of dielectric constant along with dielectric strength are required for multifunctional thin films that serve as protection in electrical components like high-temperature capacitors and sensors. Certain films are appropriate for either sensors or actuators because they have ferroelectric or piezoelectric capabilities. The dielectric features of multifunctional thin films may be characterized at different temperatures and frequencies using testing techniques like dielectric spectroscopy or impedance analysis [28].

5.4.4 Thermochromic Properties

Thermochromic thin films are multipurpose materials that may alter their color or spectral properties in accordance with temperature changes. A number of processes, including liquid crystal phase transitions, bidirectional chemical reactions, or structural modifications, can cause thermochromic films to alter color. Both reversible and irreversible color variations are possible with thermochromic films. If the temperature shifts, reversible films revert to their initial condition, but irreversible films never will. The film's reaction time indicates how rapidly it alters color according to shifts in temperature. Certain thermochromic films are made for smart windows. Regulating the transfer of both light and heat depends entirely on their transmission characteristics. Methods like spectrophotometry and calorimetry can be employed to investigate the optical properties, including color changes of thermochromic coatings at elevated temperatures. Due to its outstanding temperature-responsive behavior at a critical transition temperature (τc) of 341 K (68°C), which is close to room temperature, solar energy–regulating vanadium dioxide (VO_2) has emerged as an established thermochromic substance as well as an excellent option for intelligent architectural glazing [29].

5.4.5 Film Thickness and Surface Morphology

Multifunctional thin films' width and surface structure constitute crucial factors in high-temperature uses because they affect its functionality, performance, and endurance. Thin films with several uses must possess accurate and regulated thicknesses. To produce certain optical, electrical, or mechanical characteristics, such regulation is necessary. The precise regulation of film thickness on either the nanoscale or microscale is possible using a variety of deposition processes, including physical vapor deposition (PVD), chemical vapor deposition (CVD), and atomic layer deposition (ALD). It is crucial to address this problem because an imbalance in the thermal expansion coefficient in the film along with the substrate might cause stress and delamination at elevated temperatures. The coating's surface texture may have an influence on both its appearance and substrate adherence. Surface roughness can be affected by high temperatures owing to alterations in material along with thermal expansion. Nanostructures, along with other surface topographic characteristics, may affect how the film interacts with environmental conditions such as temperature, humidity, and chemical environments. Surface anatomy may affect the film's capacity to stay moist. If the coating is subjected to hot fluids or gases, this feature is crucial. Methods including scanning electron microscopy (SEM), atomic force microscopy (AFM), and profilometry are frequently used to describe surface morphology [30].

5.4.6 Electrical and Piezoresistive Properties

Multifunctional thin films are intended to conduct electricity in certain situations. The resistivity of substances might be affected by extreme temperature settings. To retain consistent electrical features in use at elevated temperatures, films that have a lower thermal coefficient of resistance (TCR) are preferred. It is critical to comprehend the maximum voltage and current ratings of the film to be able to operate high-temperature electrical systems safely. The shift in electrical resistance of an object in reaction to mechanical strain is measured by the piezoresistive coefficient. To guarantee whether the film produces precise strain or pressure readings in high-temperature applications, piezoresistive detectors must be calibrated precisely. One of the first teams to report the electromechanical activation of CNTs was Baughman et al. They discovered that the application of an electrochemical voltage to a single-walled carbon nanotube (SWCNT) sheet results in sheet distortion, or a piezoelectric reaction. The researchers used an atomic force microscope to selectively deform a metallic SWCNT and discovered broad, bidirectional modifications in electrical conductivity. The reversible alteration of the resistance to electricity of poly (3,4-ethylenedioxythiophene) doped with poly(4-vinylbenzenesulfonate) (polystyrene sulfonate) (PEDOT:PSS)/multi-walled carbon nanotubes (MWCNTs) indicates that such thin films exhibit piezoresistive behavior when a pulling force is applied to them. The large aggregation in CNT packages determines the films' piezoresistive behavior [31].

5.4.7 CO_2 Gas-Sensing Properties

The long-term durability of CO_2-sensing components might be affected by elevated temperatures. Once subjected to CO_2, certain CO_2-sensing films depend on modifications to electrical conductivity. Variations in the dielectric characteristics of optical CO_2 sensors could be a sign of the CO_2 concentrations. Certain optical properties may be employed in optical sensors for identifying CO_2 by changing their wavelength. For CO_2 detection, materials including metal-organic frameworks (MOFs), polymers, and ceramics are frequently employed. In systems demanding repeated cycling, the amount of time required for the film to return to its initial condition following contact with CO_2 is crucial. In the Gartner et al. (2022) study, the use of a gas needle valve permitted CO_2 gas to flow into the steel chamber. For sensing, an industrial CO_2 gas tank was utilized. High CO_2 gas sensing performance with reasonable response and recovery times was demonstrated using a ZnO electrode. The working temperature had an impact on the degree of sensitivity of the CO_2 gas sensor, so tests were conducted between 250°C and 400°C to determine the optimal temperature. With the intense heat transfer caused by the high temperature, a certain amount of the gas molecules exited from the surface before the reaction, which caused the reaction to diminish. As a result, it was found that for optimal responsiveness, the ideal operating temperature is necessary [32].

5.4.8 Supercapacitive Properties

Multifunctional thin films' supercapacitive qualities across high-temperature uses have significance for power transfer and energy storage in a variety of sectors, notably aerospace, automotive, and energy storage systems. When making high-temperature supercapacitors, the electrolyte must be carefully considered. Analyze the periodic stability of the supercapacitor under high temperatures. Evaluate the extent to which the capacitance holds up to multiple charge-discharge operations. The supercapacitor's resistance inside is gauged using equivalent series resistance (ESR). High temperatures may cause ESR to grow, which may decrease a device's effectiveness and ability to supply power. Galvanostatic charge-discharge along with cyclic voltammetry are two techniques used to measure the supercapacitor's endurance at high temperatures. Electrodes that feature a permeable microstructure, including a honeycomb-like porosity of La_2O_3 sheet, make outstanding supercapacitors. The supercapacitive properties of the electrode are enhanced by the use of polyvinylidene fluoride (PVDF) to act as a binder [33].

5.4.9 Reflectance and Integral Luminous Transmittance

Multifunctional thin films may be made that carry or reflect light in a particular spectrum of wavelengths. It is crucial to understand how such films behave during high-temperature conditions over the desired wavelength region. Spectrophotometry, a technique that gauges the amount that light bounces back at different wavelengths, is often used to characterize the reflectance qualities of materials. Various temperatures can be used to do this study. Improved transmission (in the visible area) and reflection (in the near-infrared region) could be a useful property for clear conductive oxide, heat mirror, and other energy-efficient applications. Increasing the transmittance in the visible light spectrum and having a high reflectivity in the infrared region of the multifunctional thin films is therefore a beneficial pattern to conserve energy in manufacturing and structures. Consequently, by using opaque heat reflector coatings to limit infrared radiation, buildings may reduce their energy use. Sb_2O_3 films with Ni and Mn doping exhibit improved NIR transmittance along with antireflectance. Although transition-metal-doped Sb_2O_3 thin films have more UV absorbance along with a wider wavelength spectrum, they have much lower UV reflectance than undoped films. The quantity of visible light that penetrates a substance is often referred to as luminescent transmittance. Multifunctional thin coatings are often made to limit the transmission of visible light. Heat-reflective films that permit the transmission of light with visible wavelengths while inhibiting infrared radiation are necessary for several high-temperature uses. The qualities of thin films for light transmittance may be affected by surface shape. Si-Al gel film enhances solar variation (Tsol) and luminous transmission. Despite certain inherent issues like low luminous transmittance (Tlum) and weak oxidation resistance, vanadium dioxide (VO_2) has significant promise for use as solar energy–shifting glazing [34].

5.4.10 Wetting Properties

Surface imperfections may boost the flat surface's wetting characteristics. Severe wetting properties of coverings, such as superhydrophilicity and superhydrophobicity, may allow for novel ways of influencing and regulating how fluids behave on surfaces. The development of surfaces that entirely oppose wetting with water (superhydrophobic state; droplet of water contact angle >150° with little contact angle hysteresis) or are entirely and instantly wet by water (superhydrophilic state; water droplet contact angle 5° within 0.5 s or less) might allow for uses such as self-cleaning, antifogging, and bacteria-resistant surfaces. By adding irregularity at the proper length scale, it may be feasible to dramatically improve the wetting of a surface using water. Based on this fundamental idea, it has been shown that microporous surfaces or surfaces with lithographic patterns may both be made superhydrophilic. The emergence of superhydrophilic wetting features (water droplet contact angle 5° within 0.5 s or less) is directly responsible for the antifogging qualities of these surfaces. Light-scattering drops of water are prevented from developing on the outermost layer through the superhydrophilic multilayer's practically immediate, sheet-like wetting. It wasn't until a specific quantity of bilayers had been placed over a surface that persistent superhydrophilic wetting qualities emerged [35].

5.5 APPLICATIONS OF MTFs

Multifunctional thin films find applications in a wide range of high-temperature environments due to their versatility and unique properties; see Table 5.3.

In this section we cover additional noteworthy uses. For a range of high-frequency equipment uses in various fields connected to the rapidly expanding information technology sector, **high-temperature superconducting** (HTS) thin films provide special features that may be used. One crucial characteristic is a remarkably low degree of microwave absorption at temperatures made possible by low-powered cryocoolers. In the future, advances in integrated high-temperature superconductor circuit technology could create a significant opportunity to create digital devices that can operate at clock frequencies up to 100 GHz. Around 1988, yttrium-barium-copper-oxide

TABLE 5.3
List of Various Applications of Multifunctional Thin Films in Diverse Industries

Industry	Applications
Aerospace and aviation	• Thermal protection coatings for spacecraft and aircraft • Antireflective coatings for optical sensors in high-altitude environments • Heat-resistant coatings for engine components
Energy generation and storage	• Heat-resistant films for concentrated solar power systems • High-temperature supercapacitors for energy storage • Thin films for thermoelectric generators
Power generation	• Coatings for gas turbine components, improving efficiency and durability • High-temperature materials for molten salt energy storage systems
Electronics and sensors	• Thin films for high-temperature sensors, such as pressure and gas sensors • Thin-film resistors for high-temperature electronics
Optics and photonics	• Smart windows with adjustable luminous transmittance for energy-efficient buildings • Optical coatings for high-temperature lenses, prisms, and mirrors
Oil and gas industry	• Coatings for drilling tools and components to resist high-temperature and corrosive environments • High-temperature sensors for well monitoring and control
Nuclear industry	• Radiation-resistant and high-temperature coatings for nuclear reactors and fuel assemblies
Semiconductor industry	• Thin films for high-temperature annealing and processing in semiconductor manufacturing

($YBa_2Cu_3O_{7x}$), became one of the most notable examples of the oxide superconductors; having a transition temperature Tc of 92 K, it provided the basis for the first effective creation of epitaxial thin films. The invention of flat HTS passive microwave devices, which are currently the HTS thin film gadgets with the greatest economic potential, together with the growth of wireless communication, was spurred on by this scientific work as well as by subsequent findings of extremely low microwave damages in YBCO thin films in 1989. Cryogenic RF pickup coils constructed from HTS thin films are already standard equipment in industrial high-sensitivity nuclear magnetic resonance spectrometers, and comparable advancements are being designed for low-field magnetic resonance imaging devices for use in health care.

Pressure sensors made of piezoelectric materials show promise across a range of commercial uses, including the automobile and petroleum industries. Lead zirconate titanate ($Pb[Zr_xTi_{1-x}]O_3$, PZT) transducers are widely used in conventional piezoelectric sensors. The technique is applicable to particular high-temperature and harsh environment operations. Piezoelectric devices require no external power sources, making them suitable for the production of small, power-efficient devices. Piezoresistive and capacitive SiC pressure sensors have both been described. SiC still has modest yields and is sensitive to a small spectrum of target pressures.

A **heat mirror** is a system that was originally invented for reflecting solar heat in tropical regions and stopping interior warmth from escaping in colder regions. A heat mirror displays high transmittance at shorter wavelengths together with elevated reflectance at longer wavelengths. A thin layer of a noble metal, such as Au, Ag, Cu, or Al, or a metal-like nitride, such as TiN and ZrN, or a doped oxide semiconductor, such as In_2O_3:Sn and ZnO:Al, is often layered within two antireflection coatings, TiO_2 being one of them. One of the most prominent uses of multifunctional thin films is as the antireflection layer of a TiN-based heat mirror. A spectrophotometer along with an FT-IR system were used to measure the transmittance/reflectance of the heat mirror over a variety of wavelengths.

It is crucial to reduce the presence of certain colors in wastewater before it is transported into aquatic ecosystems since water containing hazardous dyes has substantial detrimental impacts on human health and aquatic life. The **photocatalytic activity** of pristine and Zn-substituted SnS nanostructured thin films may be evaluated and analyzed beneath UV and solar illumination for photodegradation of methylene blue. In particular, Sn1-xZnxS (x = 1%) has demonstrated the greatest photocatalytic efficiency across every film specimen because of its better exhibited properties. It is possible to efficiently construct a light-concentrated solar generator having the benefits of self-powered characteristics, high sensing performance, cheap cost, and simple manufacture. In the realm of solar energy, the application of polymer-based luminescent solar concentrators (LSCs) provides a viable method for concentrating light with no need for costly tracking apparatuses and to enhance the diffuse light sensitivity of traditional photovoltaic (PV) devices. Thin, multipurpose films may be used to create all of these materials [30–33].

5.6 SUMMARY

The crucial importance of multifunctional thin films in high-temperature uses is addressed in this chapter. Such adaptable films, renowned for having outstanding thermal stability, have uses in an array of sectors, including electronics, energy, and aerospace. The chapter analyzes the characteristics of multifunctional thin films for high-temperature uses in the aerospace, energy, electronics, and optical sectors. It describes the obstacles to producing strong, antireflective, and very hydrophobic thin coatings on delicate substrates. The chapter includes the adjustable behavior of titanium oxide (TiO_2) as well as the intrinsic ferroelectric characteristics of bulk bismuth ferrite ($BiFeO_3$). The two structures that produce TiO_2 most often are rutile and anatase, which have different optical characteristics and might behave differently as photocatalysts. Among the most desirable class of TiN-based thin films, suitable as protective coatings for cutting tools, is TiN-based Ti1-xAlxN layers generated by magnetron sputtering. Yet, as a result of their low oxidation resistance and elevated brittleness, sputter-deposited (transition-metal) TM diboride thin films have relatively few practical uses. Knowledge about these films' complex characteristics becomes more important in a future where high-temperature conditions become more common. This short yet thorough introduction emphasizes the revolutionary potential of multifunctional thin films in developing science and supplying cutting-edge solutions that succeed even in extremely demanding high-temperature environments.

REFERENCES

1. N. Kaiser, "Review of the fundamentals of thin-film growth," *Appl. Opt.*, vol. 41, no. 16, p. 3053, Jun. 2002, doi: 10.1364/AO.41.003053.
2. D. A. Hardwick, "The mechanical properties of thin films: A review," *Thin Solid Films*, vol. 154, no. 1–2, pp. 109–124, Nov. 1987, doi: 10.1016/0040-6090(87)90357-9.
3. X. Xiang, Z. He, J. Rao, Z. Fan, X. Wang, and Y. Chen, "Applications of ion beam irradiation in multifunctional oxide thin films: A review," *ACS Appl. Electron. Mater.*, vol. 3, no. 3, pp. 1031–1042, Mar. 2021, doi: 10.1021/acsaelm.0c01071.
4. G. Subramanyam *et al.*, "Challenges and opportunities for multi-functional oxide thin films for voltage tunable radio frequency/microwave components," *J. Appl. Phys.*, vol. 114, no. 19, Nov. 2013, doi: 10.1063/1.4827019.
5. J. Wu, Z. Fan, D. Xiao, J. Zhu, and J. Wang, "Multiferroic bismuth ferrite-based materials for multifunctional applications: Ceramic bulks, thin films and nanostructures," *Prog. Mater. Sci.*, vol. 84, pp. 335–402, Dec. 2016, doi: 10.1016/j.pmatsci.2016.09.001.
6. S. Mathew Simon *et al.*, "Recent advancements in multifunctional applications of sol-gel derived polymer incorporated TiO2-ZrO2 composite coatings: A comprehensive review," *Appl. Surf. Sci. Adv.*, vol. 6, p. 100173, Dec. 2021, doi: 10.1016/j.apsadv.2021.100173.
7. E. A. Cochran, K. N. Woods, D. W. Johnson, C. J. Page, and S. W. Boettcher, "Unique chemistries of metal-nitrate precursors to form metal-oxide thin films from solution: Materials for electronic and energy applications," *J. Mater. Chem. A*, vol. 7, no. 42, pp. 24124–24149, 2019, doi: 10.1039/C9TA07727H.
8. H. de Sousa e Silva *et al.*, "Morphological analysis of the TiN thin film deposited by CCPN technique," *J. Mater. Res. Technol.*, vol. 9, no. 6, pp. 13945–13955, Nov. 2020, doi: 10.1016/j.jmrt.2020.09.080.

9. B. Bakhit *et al.*, "Multifunctional ZrB2-rich Zr1-xCrxBy thin films with enhanced mechanical, oxidation, and corrosion properties," *Vacuum*, vol. 185, p. 109990, Mar. 2021, doi: 10.1016/j.vacuum.2020.109990.
10. A. C. Fernandes *et al.*, "Property change in multifunctional TiCxOy thin films: Effect of the O/Ti ratio," *Thin Solid Films*, vol. 515, no. 3, pp. 866–871, Nov. 2006, doi: 10.1016/j.tsf.2006.07.047.
11. L. Marques, H. M. Pinto, A. C. Fernandes, O. Banakh, F. Vaz, and M. M. D. Ramos, "Optical properties of titanium oxycarbide thin films," *Appl. Surf. Sci.*, vol. 255, no. 10, pp. 5615–5619, Mar. 2009, doi: 10.1016/j.apsusc.2008.08.022.
12. S. A. Vanalakar *et al.*, "Effect of post-annealing atmosphere on the grain-size and surface morphological properties of pulsed laser deposited CZTS thin films," *Ceram. Int.*, vol. 40, no. 9, pp. 15097–15103, Nov. 2014, doi: 10.1016/j.ceramint.2014.06.121.
13. R. M. Imamov and Z. G. Pinsker, "Determination of the crystal structure of the hexagonal phase in the silver-tellurium system," *Sov. Phys. Crystallogr.*, vol. 11, no. 2, pp. 182–188, 1966, [Online]. Available: https://rruff-2.geo.arizona.edu/uploads/SPC11_182.pdf
14. R. E. Tressler, and V. S. Stubican, "Preparation of thin films of sulforspinels," *Mater. Res. Bull.*, vol. 2, no. 12, pp. 1119–1124, 1967, doi: 10.1016/0025-5408(67)90141-9.
15. O. Shekhah, J. Liu, R. A. Fischer, and C. Wöll, "MOF thin films: Existing and future applications," *Chem. Soc. Rev*, vol. 40, no. 2, p. 1081, 2011, doi: 10.1039/c0cs00147c.
16. T. J. Konno, and R. Sinclair, "Crystallization of amorphous carbon in carbon—Cobalt layered thin films," *Acta Metall. Mater.*, vol. 43, no. 2, pp. 471–484, Feb. 1995, doi: 10.1016/0956-7151(94)00289-T.
17. S. Bhattacharyya *et al.*, "Synthesis and characterization of highly-conducting nitrogen-doped ultrananocrystalline diamond films," *Appl. Phys. Lett.*, vol. 79, no. 10, pp. 1441–1443, Sep. 2001, doi: 10.1063/1.1400761.
18. M. Kitano, M. Matsuoka, M. Ueshima, and M. Anpo, "Recent developments in titanium oxide-based photocatalysts," *Appl. Catal. A Gen.*, vol. 325, no. 1, pp. 1–14, May 2007, doi: 10.1016/j.apcata.2007.03.013.
19. C. Wan *et al.*, "On the optical properties of thin-film vanadium dioxide from the visible to the far infrared," *Ann. Phys.*, vol. 531, no. 10, Oct. 2019, doi: 10.1002/andp.201900188.
20. K. Kandpal, and N. Gupta, "Perspective of zinc oxide based thin film transistors: A comprehensive review," *Microelectron. Int.*, vol. 35, no. 1, pp. 52–63, Jan. 2018, doi: 10.1108/MI-10-2016-0066.
21. H.-U. Krebs *et al.*, "Pulsed Laser Deposition (PLD) – A Versatile Thin Film Technique," in: Kramer, B., Ed., *Advances in Solid State Physics*, Vol. 43, Springer, 2003, pp. 505–518. doi: 10.1007/978-3-540-44838-9_36.
22. Y. Zhang *et al.*, "Rapid and selective deposition of patterned thin films on heterogeneous substrates via spin coating," *ACS Appl. Mater. Interfaces*, vol. 11, no. 23, pp. 21177–21183, Jun. 2019, doi: 10.1021/acsami.9b05190.
23. T.-T.-N. Nguyen, Y.-H. Chen, M.-Y. Chen, K.-B. Cheng, and J.-L. He, "Multifunctional Ti-O coatings on polyethylene terephthalate fabric produced by using roll-to-roll high power impulse magnetron sputtering system," *Surf. Coatings Technol.*, vol. 324, pp. 249–256, Sep. 2017, doi: 10.1016/j.surfcoat.2017.05.082.
24. N. L. Tarwal *et al.*, "Growth of multifunctional ZnO thin films by spray pyrolysis technique," *Sensors Actuators A Phys.*, vol. 199, pp. 67–73, Sep. 2013, doi: 10.1016/j.sna.2013.05.003.
25. F. Böke, I. Giner, A. Keller, G. Grundmeier, and H. Fischer, "Plasma-enhanced chemical vapor deposition (PE-CVD) yields better hydrolytical stability of biocompatible SiOx thin films on implant alumina ceramics compared to rapid thermal evaporation physical vapor deposition (PVD)," *ACS Appl. Mater. Interfaces*, vol. 8, no. 28, pp. 17805–17816, Jul. 2016, doi: 10.1021/acsami.6b04421.
26. O. H. Auciello *et al.*, "Science and technology of ultrananocrystalline diamond (UNCD) thin films for multifunctional devices," D. K. Sood, R. A. Lawes, and V. V. Varadan, Eds., Mar. 2001, pp. 10–20. doi: 10.1117/12.420857.
27. P. Liu, A. Lam, Z. Fan, T. Q. Tran, and H. M. Duong, "Advanced multifunctional properties of aligned carbon nanotube-epoxy thin film composites," *Mater. Des.*, vol. 87, pp. 600–605, Dec. 2015, doi: 10.1016/j.matdes.2015.08.068.
28. K. Naveen Kumar, J. L. Rao, and Y. C. Ratnakaram, "Optical, magnetic and electrical properties of multifunctional Cr3+: Polyethylene oxide (PEO) + polyvinylpyrrolidone (PVP) polymer composites," *J. Mol. Struct.*, vol. 1100, pp. 546–554, Nov. 2015, doi: 10.1016/j.molstruc.2015.07.066.
29. L. Hu, S. Lyu, F. Fu, J. Huang, and S. Wang, "Preparation and properties of multifunctional thermochromic energy-storage wood materials," *J. Mater. Sci.*, vol. 51, no. 5, pp. 2716–2726, Mar. 2016, doi: 10.1007/s10853-015-9585-9.
30. H. S. Kim, B. H. Sohn, W. Lee, J.-K. Lee, S. J. Choi, and S. J. Kwon, "Multifunctional layer-by-layer self-assembly of conducting polymers and magnetic nanoparticles," *Thin Solid Films*, vol. 419, no. 1–2, pp. 173–177, Nov. 2002, doi: 10.1016/S0040-6090(02)00779-4.

31. S. Biehl, H. Lüthje, R. Bandorf, and J.-H. Sick, "Multifunctional thin film sensors based on amorphous diamond-like carbon for use in tribological applications," *Thin Solid Films*, vol. 515, no. 3, pp. 1171–1175, Nov. 2006, doi: 10.1016/j.tsf.2006.07.143.
32. M. Gartner *et al.*, "Multifunctional Zn-doped ITO Sol–Gel films deposited on different substrates: Application as CO2-sensing material," *Nanomaterials*, vol. 12, no. 18, p. 3244, Sep. 2022, doi: 10.3390/nano12183244.
33. G. Durai, P. Kuppusami, S. Arulmani, S. Anandan, S. Khadeer Pasha, and S. Kheawhom, "Microstructural and electrochemical supercapacitive properties of Cr-doped CuO thin films: Effect of substrate temperature," *Int. J. Energy Res.*, vol. 45, no. 14, pp. 20001–20015, Nov. 2021, doi: 10.1002/er.7075.
34. G. Wei, D. Yang, T. Zhang, X. Yue, and F. Qiu, "Fabrication of multifunctional coating with high luminous transmittance, self-cleaning and radiative cooling performances for energy-efficient windows," *Sol. Energy Mater. Sol. Cells*, vol. 202, p. 110125, Nov. 2019, doi: 10.1016/j.solmat.2019.110125.
35. F. Ç. Cebeci, Z. Wu, L. Zhai, R. E. Cohen, and M. F. Rubner, "Nanoporosity-driven superhydrophilicity: A means to create multifunctional antifogging coatings," *Langmuir*, vol. 22, no. 6, pp. 2856–2862, Mar. 2006, doi: 10.1021/la053182p.

6 Hydrophobic and Hydrophilic Behavior of Multifunctional Thin Film

Elmira Khanmamadova, Rashad Abaszade, and Rasoul Moradi

6.1 INTRODUCTION

Thin-film technologies offer the incredible opportunity to use materials in diverse ways that are incredibly well-suited for a broad spectrum of uses. The skill of altering the traits of these materials to make them appropriate for various applications and crafting thin films with multiple functions is of utmost importance. The ability to make these materials either repel water (hydrophobic) or embrace it (hydrophilic) plays a vital role in customizing them for a wide range of applications, including protection [1–5]. On the other hand, hydrophilic properties indicate the tendency of materials to attract water and other polar liquids. Hydrophilic thin films are ideal for applications that involve interaction with water or biological fluids, especially in biomedical applications. These films are typically made of water-attracting polymers, such as polyethylene glycol (PEG), polyvinyl alcohol (PVA), or hydrogel materials [6–10].

This will delve into the in-depth examination of the impact of hydrophobic and hydrophilic behavior on the design and application of multifunctional thin films (Figure 6.1). The versatility provided by these properties of thin-film materials significantly enhances their adaptability to a wide range of applications [11–14]. This diversity underlies the significant influence of thin films

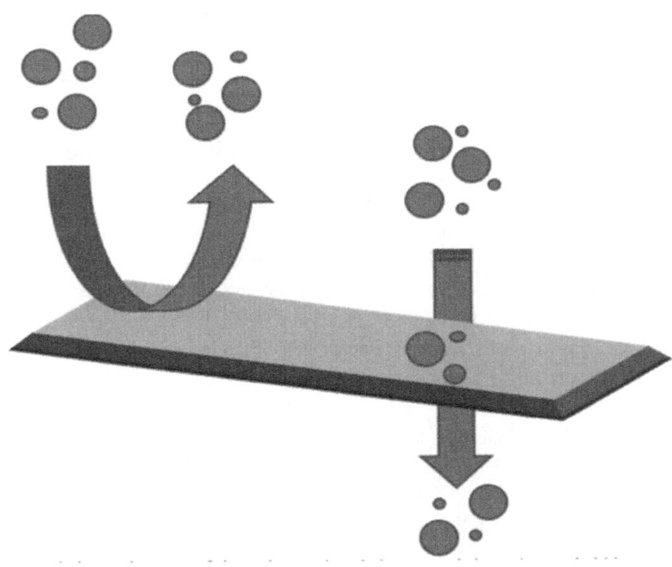

FIGURE 6.1 Combination of hydrophobic and hydrophilic properties of a multifunctional thin film.

DOI: 10.1201/9781032635347-6

in various sectors, ranging from electronics to biomedicine, energy production to environmental protection.

6.2 MATERIAL SELECTION

Science has come a long way in a short time, especially when we talk about the cool stuff they're doing in materials science. One of the standout inventions is this thing called thin-film technology [15–20].

Imagine super, super thin layers—like really thin, almost as fine as a strand of hair—and they stick these onto different surfaces [21]. People are finding all sorts of uses for them, from gadgets and health equipment to new energy sources and green tech solutions [20–26].

Some materials attract water and allow it to spread on their surface and called hydrophilic materials (Table 6.1). This is very important for things like medical uses, especially when you think about growing cells or delivering medicine. Common materials that have this water-loving property are polyvinyl alcohol (often called PVA), polyethylene glycol (or PEG for short), and materials like hydrogels that soak up water [23, 24, 27–35].

Hydrophobic material properties and their use in specific applications has been given in Table 6.2.

TABLE 6.1
Describing Hydrophobic Material Properties and Their Use in Specific Applications

Features	Description	Applications
Water repellent	Hydrophobic materials attract water and other polar liquids. This property is measured by the material's ability to remove water.	Waterproof coatings, outdoor equipment, clothing
Oil repellent	Hydrophobic materials also attract oils and other nonpolar liquids. This is measured by the material's ability to remove oils.	Oil-proof coatings, kitchenware, industrial equipment
Anti-adhesive	Hydrophobic materials minimize their interaction with most materials, making them anti-adhesive.	Nonstick coatings, pots and pans, biomedical devices
Anticorrosion	Hydrophobic materials reduce interaction with corrosive substances such as water and air.	Marine equipment, auto parts, buildings
Easy to clean	The surface of hydrophobic materials does not attract water and other liquids, making them easy to clean.	Home appliances, bathroom utensils, kitchen utensils

TABLE 6.2
Hydrophilic Material Properties and Their Use in Specific Applications

Features	Description	Applications
Water hammer	Hydrophilic materials attract water and other polar liquids. This property is measured by the material's ability to absorb water.	Absorbent pads, hydrogels, moisturizing products
Steam permeability	Hydrophilic materials allow the passage of water vapor, which creates perspiration-breathable products.	Sportswear, band-aid, breathable materials
Biocompatibility	Hydrophilic materials generally interact well with biological systems and are suitable for biological applications.	Biomedical devices, drug delivery systems, tissue engineering
Surface energy	Hydrophilic materials have high surface energy and provide the ability to adhere to the material and spread.	Adhesives, coatings, printing
Resolution	Materials that have a liking for water, known as Hydrophilic materials typically have the ability to dissolve in water and other substances that are polar in nature. This quality of being soluble in water and polar solvents is a way these materials shield themselves.	Pharmaceutical formulations, detergents

These two basic surface properties of thin films determine how the films will perform in a given application [23, 26, 34]. Material selection is a critical factor when optimizing a film for a particular application. A good material selection ensures the expected performance of the film in a particular application, thus enabling thin films to be used effectively in a wide range of applications [21].

6.3 EVALUATING WAYS TO APPLY COATINGS

There are several techniques for applying coatings, from using chemicals to spraying or even dipping. Each method changes the smoothness, thickness, and evenness of the final coat in its own way. Thin films are like versatile "stickers" made from materials that either repel or attract water. The way we apply these films can change how they behave on surfaces. Just think of them as either being "water-loving" (hydrophilic) or "water-avoiding" (hydrophobic) [36–39]. This trait is pretty crucial because it affects how these films do their job in different situations. Let's look at some popular ways we put these films on surfaces.

Chemical Vapor Deposition (CVD): Chemical vapor deposition, often known as CVD, is like a super-efficient kitchen technique for applying thin coats to surfaces. Imagine turning an ingredient into vapor and letting it settle evenly on a plate. That's how CVD works, but with high-tech precision in hot, vacuum-like conditions. This ensures that the final coating is smooth, high-quality, and just the right thickness. A cool thing about CVD? It can make surfaces behave differently around water. Some CVD-coated surfaces shoo away water like an umbrella. Water beads up and rolls off. On the other hand, some CVD surfaces love water and spread it evenly, like spreading jam on toast.

Why does this matter? Well, imagine a surface that pushes water away, making it self-cleaning or waterproof. That's the magic of hydrophobic coatings. On the flip side, a surface that embraces water can keep things moist or speed up certain chemical reactions, thanks to hydrophilic coatings.

Physical Vapor Deposition, or PVD for short, is like giving an object a brand-new, thin outer skin. Imagine turning a solid material into vapor (like how water turns into steam) and then letting it settle onto something else, sticking to it. This happens in a special vacuum chamber and doesn't need super high temperatures [30]. Now, what's cool is that PVD lets us tweak the new "skin" to either love water (hydrophilic) or repel it (hydrophobic). Think of it like setting up a room: Depending on the furniture, lighting, and colors, you can make it either cozy and welcoming or sleek and minimalist. For instance, if we want our object's new skin to shoo water away, we'd choose materials that naturally don't like water, like certain metals or organic stuff. We'd also keep the temperature on the cooler side and pump up the gas pressure in our special chamber. It's like dressing up for a rainy day—you'd choose a raincoat and waterproof boots [27, 31]. On the other hand, if we want our object to attract water, we'd pick materials that are friendly with water, like certain metal oxides. We'd also raise the temperature a bit and keep the gas pressure low. Think of it as setting up a beach-themed room with sand-colored cushions, seashells, and sunny yellow walls. PVD is super versatile. Depending on how we mix and match everything, we can use it for loads of different purposes to make surfaces either befriend or fend off water.

Spray Coating: Spray coating is a quick and economical thin-film coating method. In this method, a material is atomized into fine droplets using a nozzle or atomizer and then sprayed onto a substrate [28]. This technique offers the advantage of rapidly coating large surfaces. Spray coating allows for adjusting the composition of the coating solution and the application process to determine the thin films' surface properties. The composition affects the properties of the material in the coating solution. For example, a hydrophobic material solution can be used to create a hydrophobic thin film. The application process also affects film properties. The spray speed, nozzle-to-substrate distance, spray angle, and gas pressure used during the coating process can be adjusted to control the film's surface properties. For example, the spray speed can be reduced or a gas flow can be used during the coating process to achieve a smoother surface [16, 31].

Spray coating can be used, depending on the choice of suitable materials and coating parameters, to achieve either hydrophobic or hydrophilic thin films [37]. This technique has found applications in various industries, including surface protection, coating, paint, electronics, solar energy, and medicine.

Immersion Coating: Immersion coating is a cool technique that involves coating surfaces with special thin films that can either love or hate water—kind of like how some of us enjoy rainy days and others don't. How to make these surfaces relate to water in a certain way? It all comes down to the mixture of chemicals we use and the speed at which we extract the surface from the mixture. For example, if we need a surface that is not too fond of water (hydrophobic), we choose a mortar with the right water-repellent ingredients. And here's a fun trick: Slowly pulling the surface out of the solution gives the chemicals a better chance to spread well and make sure they're doing their water-repellent job well.

On the other hand, if we want our surface to be water-loving (hydrophilic), we again choose our solution carefully, this time with water-attracting ingredients. And a quick exit from the solution helps it to remain a thin layer, enhancing its moisture-loving nature.

Simply put, dip coating is like giving a surface a new character—whether it's waterproof or not (based on a study by Zhang, X. and Shen, Z. from 2000).

6.3.1 Effect of Hydrophobic and Hydrophilic Methods on Thin Film

Hydrophobic and hydrophilic thin-film coating methods can have different impacts on the quality, thickness, and homogeneity of the film. These factors directly influence the performance and functionality of the film in application.

The above table summarizes the impacts of hydrophobic and hydrophilic thin film coating methods on film quality, thickness, and homogeneity. (Table 6.3) Each coating method provides certain advantages and limitations and affects the film characteristics in various ways.

Chemical vapor deposition (CVD) has the potential to create high-quality and homogeneous films. It provides controllable film thickness at an atomic level and offers high homogeneity [17]. Physical vapor deposition (PVD) likewise has the ability to produce high-quality and homogeneous films and can be used to control film thickness.

The spray coating method may produce films of lower quality. Although the film thickness can be controlled, there are some limitations in terms of homogeneity. The dip coating method can be effective for creating homogeneous films, and the film thickness can be adjusted [19].

Table 6.3 provides an overview of the effects of film coating methods. However, the most suitable film for a particular application can be obtained using more specific parameters and optimized conditions for each coating method. The coating methods and parameters are continuously researched and developed to improve film quality, thickness, and homogeneity.

TABLE 6.3
Effects of Hydrophobic and Hydrophilic Thin-Film Coating Methods

Coating Method	Film Quality	Film Thickness	Homogeneity
Chemical vapor deposition (CVD)	High quality, homogeneous film	Atomically controllable	High homogeneity
Physical vapor deposition (PVD)	High quality, homogeneous film	Controlled thickness	High homogeneity
Spray coating	Lower quality film	Controlled thickness	Lower homogeneity
Dip coating	Homogeneous film	Adjustable thickness	High homogeneity

6.3.2 Choice of Film Parameters in Fabrication Technique

Optimizing the chosen method is a critical step to achieving the desired film properties in hydrophobic and hydrophilic thin-film coatings. This process involves adjusting and optimizing the coating parameters. Coating parameters include solution composition, coating speed, temperature, pressure, withdrawal speed, gas environment, etc. Coating parameters are of critical importance in the formation of film layers with hydrophobic or hydrophilic properties. Here are some significant coating parameters:

1. **Solution Composition:** When you mix up a coating solution, what you put in it really matters. It can change how the final layer feels and acts. If you want it to repel water or attract it, you've got to use the right ingredients.
2. **Coating Speed:** When we're applying a solution onto a surface, think of it like painting a wall. The speed at which we move our brush (or coating speed) can determine how thick the paint layer will be and whether it will have an even look. So we need to choose the right speed to get the perfect, consistent finish we want.
3. **Temperature:** When you're making a film layer, think of it like baking. The temperature you use can change how quickly things dry and how the final layer turns out, similar to how cookies can be crunchy or chewy depending on how they're baked. Just like some recipes work best at a specific oven temperature, some materials give the best film results at certain temperatures. So it's very important to keep an eye on the temperature.
4. **Pressure:** When you're applying a coat, the amount of gas pressure you use can change how the liquid breaks up and spreads out. Think of it like adjusting the nozzle on a garden hose; it determines how the water comes out. This pressure can make the coating smooth and even or bumpy and uneven. It's essential to find the just-right pressure for the best results.
5. **Withdrawal Speed:** In dip coating, imagine dunking a cookie into milk. The speed at which you pull out the cookie (similar to the withdrawal speed in dip coating) determines how much milk sticks and the texture it leaves on the cookie. To get just the right amount of milk and the perfect texture, you need to find the best pull-out speed. In our world, getting this speed right helps achieve the perfect thickness and feel for the coated layer.
6. **Gas Environment:** In certain coating techniques, the type of gas used in the process plays a crucial role. The specific makeup of this gas environment can influence the texture, structure, and chemical makeup of the film layer that's created.

Experimental studies and parameter optimization play a significant role in determining the most suitable parameters for the coating method [12, 18]. This involves systematically changing various parameters and evaluating their effects on film characteristics. Subsequently, the most suitable parameter combinations that will provide optimal film properties can be determined.

In this process, characterization and analysis methods are also important. Below is a table containing some analysis techniques commonly used to evaluate properties such as film thickness, surface morphology, chemical composition, and hydrophobicity/hydrophilicity of hydrophobic and hydrophilic thin films (Table 6.4).

TABLE 6.4
Appropriate Analysis Techniques for Evaluation of Film Properties

Film Property	Analysis Techniques
Film thickness	Profilometry, ellipsometry, effective medium measurements, TEM, SEM
Surface morphology	SEM, AFM, profilometry
Chemical composition	XPS, FTIR, Raman spectroscopy, EDX
Hydrophobicity/Hydrophilicity	Water contact angle measurements, contact angle measurements

Table 6.4 shows us how to test different characteristics of films, like how thick they are, their surface details, their chemical makeup, and how they interact with water. Think of films as being like the protective screens on your phone or the coatings on sunglasses.

To figure out how thick these films are, we can use tools like profilometry and others. Imagine taking a very detailed close-up picture of the edge of the film to see its depth. That's sort of what these tools do. When we want to get a good look at the surface of the film—like its texture and tiny details—we use tools like SEM and AFM. To get the nitty-gritty on what the film is made of chemically, we turn to things like XPS and FTIR. They're like detectives that can tell us about every tiny piece of a puzzle in the film. And when we're curious about how the film interacts with water, we do simple tests to see if water beads up or spreads out [38]. It helps us know if a film is more like a raincoat (repelling water) or a sponge (absorbing water).

In short, all these fancy techniques are like our toolkit to understand everything about these films. By using them, we can make sure films do exactly what we want them to, whether it's protecting our gadgets or making sure our sunglasses are just right.

6.4 QUANTITATIVE MEASUREMENT

Hydrophobicity/hydrophilicity levels are quantitatively measured. Major parameter to be used is given below:

Water Contact Angle (θ): Imagine a tiny droplet of water resting on a surface. The angle where the droplet meets the surface tells us if it is hydrophilic or hydrophobic. A big angle means the surface isn't a fan of water, but a small angle means it loves it. This angle helps us understand if the water droplet will glide or stick around on that surface.

To determine the angle at which water touches a surface, you'd need a few simple tools as below: Something to create a water droplet, like a sprayer.

1. A device, often called a profilometer, to check how smooth the surface is down to tiny details.
2. A way to closely watch and snap a picture of the exact moment the water droplet touches the surface, perhaps using a camera or even a microscope.
3. Some handy software and tools to crunch the numbers and get the measurements right.

When you want to figure out how water interacts with a certain surface, you'd usually go through this simple process:

- Start by getting your surface ready. This means cleaning it up and sometimes tweaking it a bit to make it either repel or attract water.
- Gently place or spray a tiny drop of water onto it.
- Watch closely! See if the water drop spreads out or rolls away. Capture this moment with a picture.
- Now, using some fancy software or tools, determine the angle where the water touches the surface. This angle helps us understand the surface's relationship with water.
- Think of it this way: Imagine dropping water on a raincoat versus on a paper towel. On the raincoat, the water might just roll off, forming a sort of "steep hill." On the paper towel, the drop might spread out, forming a "gentle slope." These hills and slopes are called water contact angles.

Setting up the water angle measuring tool and Measurement Device: First off, make sure your water angle measurement device is set up just right. Think of it like tuning a guitar—you've got to get things like the size of the water drop, how fast it falls, and the temperature just right.

a. **Dropping the Water:** Picture this: You're holding a delicate pipette and you let a single water droplet fall gracefully onto a surface. It should either sit there peacefully or spread out in an instant.

b. **Capturing the Moment:** Snap a picture or use your measurement device to get an image of this droplet.
c. **Understanding the Droplet's Shape:** Now, dive deep into that image. Look for things like how wide it is, how round, and the angle where it meets the surface. It's all in the details!
d. **Doing the Math:** Finally, with the data in hand, you get to play detective. Use the right formula (often, it's the Young equation) to figure out the water's contact angle.

Imagine you have a glass of water. Now, there are three places where tension occurs due to the interaction between different substances (equation 6.1):

$$cos(\theta) = \frac{(\gamma sv - \gamma sl)}{\gamma lv} \tag{6.1}$$

γsv: This is the tension at the top layer of the water, where it meets the air. Think of it as the "skin" on the surface of the water when you touch it gently.
γsl: This is the tension between the glass and the water. It's why the water might curve up slightly at the edges where it touches the glass.
γlv: This is a unique connection between water and the air right above it. It's similar to the resistance you feel when you swiftly move your hand through water.

Imagine a water droplet landing on a table. The way it spreads out or beads up tells us a lot about how hydrophobic or hydrophilic that table's surface is. The water contact angle is like measuring the slant or tilt of that droplet. We use it to figure out if the surface likes water or pushes it away. Basically, we look at the point where the droplet touches the table and measure the angle it makes. This angle (θ) is worked out using a special formula, but in essence, it's like seeing how comfy the water droplet feels on that surface.

$$\theta = cos^{-1}\left(\frac{P - S}{P}\right)$$

In simpler terms, think of P as the force that makes the surface of a liquid act like a stretched elastic sheet, while S is like the force between that surface and the liquid below it.

Think of the contact angle as the way a drop of water might stick or slide out on a surface. By looking at how steep or shallow this hug is, we can tell a lot about the surface. If the drop stays rounded and doesn't spread out much, like on a raincoat, it's because the surface is hydrophobic and the angle is usually above 90 degrees. On the other hand, if the drop spreads out flat, like on a paper towel, it's because the surface is hydrophilic and the angle is typically below 90 degrees (Figure 6.2).

Some basic equipment is needed for measuring the contact angle. Imagine you're working with a tool that sprays a tiny drop of liquid. Next, there's an arrangement that lets you watch closely, even

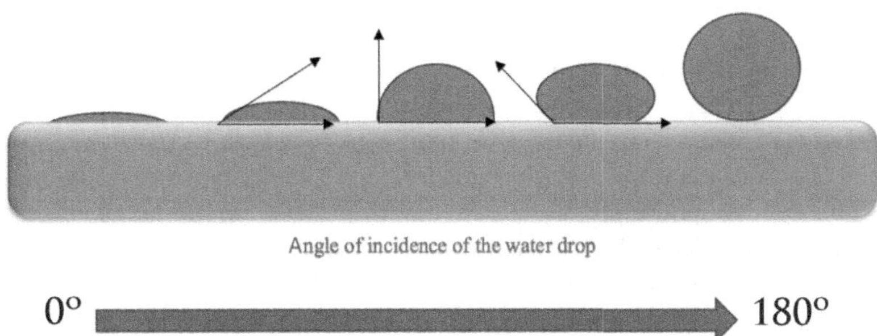

FIGURE 6.2 The water contact angle represents the angle of a water droplet falling on the surface.

take photos, of how this liquid drop spreads or maybe gathers up on a surface. Kind of like using a super magnifying glass or camera. And to make sense of what you see, you've got some handy software and tools for crunching numbers.

When people want to find out how much a liquid likes or dislikes a surface (that is, measure the contact angle), they do these things:

1. They get the surface all set up first. Think of it as preparing a canvas for painting. They clean it up, and sometimes they treat it so that it either attracts water like a sponge (we say it's hydrophilic) or repels it like a raincoat (that's hydrophobic).
2. After that, they spray or drop a tiny bit of liquid onto this prepared surface. It's like watching a drop of rain fall onto a leaf.
3. They then watch closely, capturing how the liquid behaves on that surface. It's like taking a snapshot of that raindrop either spreading out or beading up.
4. Finally, from those snapshots, they figure out the contact angle. It's the angle where the liquid drop touches the surface. It gives them a clue about how much the liquid and the surface get along!

Imagine you've spilled a drop of water on a table, and you notice the shape it forms. This shape can actually tell us a lot about how the table's surface interacts with water. Some surfaces make water spread out, while others make it bead up. This angle of water on the surface helps us understand if something is water-loving (hydrophilic) or water-repelling (hydrophobic). By measuring this angle, scientists can choose the best materials for specific tasks. And yes, there's a formula they use to get that angle, but it's all about how that drop interacts with its landing, as in equation 6.2:

$$\text{spot } \alpha = 180° - \theta \tag{6.2}$$

where θ represents the water contact angle.

Hydrophobicity Index (HI): The hydrophobicity index (HI) is a value that indicates how much a material repels water. This measurement helps us understand whether a material is more hydrophilic or more hydrophobic based on its surface properties.

The process of calculating this index involves a few key steps:

1. Collecting data on water contact angles: This step involves measuring how a water droplet interacts with the material's surface. We do this by placing a droplet of water on the material and observing the angle formed where the droplet touches the surface. These measurements are made under various conditions, such as with different materials, surface treatments, and environmental factors.
2. Analyzing the data: The collected water contact angle data is then analyzed to calculate the hydrophobicity index. This is typically done using a mathematical formula or procedure that generates a number indicating the material's hydrophobicity or hydrophilicity based on the water contact angle data.
3. Interpreting the results: The calculated HI value helps us understand the material's interaction with water. A higher HI value means the material is more hydrophobic, while a lower value suggests it's more hydrophilic.

The hydrophobicity index is a valuable tool that helps scientists and engineers make informed decisions when choosing materials, modifying surfaces, or applying coatings to enhance a material's hydrophobic or hydrophilic properties. This index is based on measurements like water contact angle and surface tension, and it's often used to compare the water-repelling or water-attracting properties of different materials.

When we talk about the HI value, it's a way to measure how a drop of water behaves on a surface. Think about rain falling on a leaf or on a car's windshield. Sometimes the water forms beads, and other times it spreads out. This behavior is influenced by the water's angle of contact with the surface and the force that the surface exerts on the water, which we call surface tension.

The formula we use to get this HI value is shown as equation 6.3:

$$HI = \frac{(\theta - 90)}{\gamma} \tag{6.3}$$

Here's what the symbols mean:

- θ is the angle of the water drop with the surface, like the tilt of a raindrop on a window.
- γ is the surface tension. You can think of it as the force that the surface uses to either pull the water close or push it away.

The interesting thing is, the HI value can be negative or positive. If it's negative, it means the surface is hydrophilic. If it's positive, the surface is hydrophobic. So, the bigger the HI value, the more the surface acts like a raincoat, pushing the water away.

The essential ingredients of water contact angel parameters can be surface tension of the liquid.

Surface Tension (γ): Think about how a liquid, like water, behaves on that surface. Whether it spreads smoothly or gathers into drops depends on what is called its energy state. This energy state is similar to the mood of a liquid, which is influenced by how much it likes the surface and how "sticky" it is to it.

$$\gamma = \gamma lv + \gamma sv \cdot cos\, \theta \tag{6.4}$$

In this context: γ represents the surface tension. γlv denotes the liquid-surface tension resulting from the liquid-surface interaction. γsv denotes the liquid-environment tension resulting from the liquid-environment interaction. θ represents the contact angle.

The contact angle depends on the surface tension and how the liquid is spread over the surface. This whole relationship between surface tension and how liquids behave on surfaces is described by the Young equation or the Young-Laplace equation.

6.5 SURFACE PRETREATMENT: ENSURE CLEAN, SMOOTH, AND PROPERLY TREATED SURFACES FOR EFFECTIVE FILM ADHESION

Surface pretreatment is a step that involves properly cleaning the surface, controlling its roughness, and optimizing its chemical composition before the application of a film layer [13]. These pretreatment processes aim to ensure the film adheres properly to the surface, exhibits desired behavior, and has long-term durability.

Surface cleaning involves removing any dirt, oil, dust, or other contaminants from the surface. These contaminants can hinder the film layer's adherence to the surface and negatively affect the performance of the film. Suitable cleaners, solvents, or surfactants are generally used for surface cleaning.

Roughness control involves determining the surface's smoothness and roughness. A rough surface can make it difficult for the film layer to spread or adhere evenly. Devices such as surface profilometers or microscopes may be used for roughness control [14]. If necessary, to reduce or optimize roughness, surface processes such as sanding or polishing may be applied.

Chemical composition determines the structural and chemical properties of the surface. The film layer's surface chemistry can influence its adhesive strength, hydrophobic or hydrophilic behavior, and chemical resistance. Chemical agents, primers, or coating solutions can be used to optimize surface chemistry. These steps ensure that the film layer is compatible with the surface and exhibits the desired performance.

TABLE 6.5
Methods and Materials for Surface Treatment

Component	Example Methods	Chemical Formula
Surface cleaning	Washing the surface with detergent or soapy water	*NaOH* (Sodium hydroxide)
	Cleaning the surface with isopropyl alcohol or ethyl alcohol	$CH3CH(OH)CH3$ (Isopropyl alcohol)
	Cleaning the surface with acidic or alkaline cleaning solution	*HCl* (Hydrochloric acid)
		NH4OH (Ammonium hydroxide)
Roughness control	Sanding or brushing the surface	*SiC* (Silicon carbide)
	Surface profilometer or microscope for roughness measurement	
Chemical composition	Treating the surface with acid	*H2SO4* (Sulfuric acid)
	Coating the surface with primer or precoating material	
	Subjecting the surface to passivation process	*CrO3* (Chromium trioxide)

Surface pretreatment aims to ensure that the film adheres correctly to the surface, exhibits desired behavior, and has long-term durability. Different surface properties may be required for each application, so appropriate cleaning, roughness control, and chemical composition optimization methods should be selected. Methods and materials are presented in Table 6.5.

In this table, examples of cleaning methods such as washing the surface with detergent or soapy water and cleaning with isopropyl alcohol or ethyl alcohol are provided for surface cleaning. Additionally, cleaning the surface with acidic or alkaline cleaning solutions is also an option [15]. Mechanical processes such as sanding or brushing can be used for roughness control, and roughness measurement can be performed using a surface profilometer or microscope. For chemical composition optimization, methods such as treating the surface with acid, coating with primer or precoating material, or subjecting the surface to a passivation process.

The chemical formulas represent examples of some chemical substances that can be used in the surface preparation process. Different chemical formulas and compositions can be used for different applications and materials.

These examples are just some of the methods that can be used for surface preparation. Suitable methods should be determined for each application and material, and appropriate chemical formulas should be selected and optimized. Proper surface cleaning, roughness control, and chemical composition ensure the adhesion of the film to the surface, exhibit the desired behavior, and provide long-term durability.

Below is an example table providing a more extensive description of some techniques and methods used for film characterization (Table 6.6).

These methods and techniques play a crucial role in the comprehensive characterization of films and the evaluation of their hydrophobic and hydrophilic behavior. They provide valuable information about the surface properties, spreading properties, morphology, and chemical composition of the film, helping us to understand its performance and properties.

6.5.1 Preparation and Characterization of High-Performance Hydrophobic Films

In this section, the preparation and characterization processes of hydrophobic films are examined, providing important insights into hydrophobic behavior and interactions with water. Contact angle measurements determine the angle formed by a water droplet placed on the film surface. A high contact angle indicates that the water droplet has limited contact area with the film surface and is repelled. This reflects the hydrophobic properties of the film. Water spreading tests evaluate the spreading speed and pattern of a water droplet on the film. In hydrophobic films, the water droplet

TABLE 6.6
Film Characterization Techniques

Characterization Technique	Methods	Description
Contact angle measurements	Tilt method: This method involves gauging the angle at which a droplet makes contact with a surface, serving as a way to safeguard the material.	This approach enables us to gauge the level of water resistance or water attraction exhibited by a surface. When the desired contact angle is substantial (exceeding 90°), it suggests that the surface repels water effectively. Conversely, if the contact angle is minor (below 90°), it signifies that the surface has an affinity for water.
	Captive bubble method: Determining the contact angle involves the act of positioning an air bubble onto the surface and then observing and measuring the angle at which the bubble makes contact with the material.	This technique is employed to examine how surfaces interact with water, determining whether they repel or attract it. If an air bubble remains compact (with a steep contact angle), it signifies that the surface is not fond of water (hydrophobic). On the other hand, if the bubble spreads out (with a shallow contact angle), it signifies that the surface is water-friendly (hydrophilic)
Water spreading tests	Drop shape analysis: Analyzes the spreading pattern and speed of a liquid droplet.	Used to evaluate the hydrophobic or hydrophilic behavior of the surface. A hydrophobic film repels the water droplet, while a hydrophilic film spreads and absorbs the water droplet.
	Washburn method: Evaluating the way water spreads by gauging the absorption of liquids in a capillary material	This method is employed to assess how water-repellent or water-attracting a surface is. A surface that repels water will soak up and disperse minimal amounts, whereas a surface that attracts water will soak up and spread out larger amounts.
Surface morphology analysis	Atomic force microscopy (AFM): Creating detailed images of the texture found on the surface of the film using high-resolution imaging techniques.	This technique, known as Atomic Force Microscopy (AFM), is employed to examine the physical structure of a thin layer. AFM captures detailed pictures of the surface, revealing insights about its bumpiness, surface characteristics, and how its structures are arranged.
	Scanning electron microscopy (SEM): In-depth visualization of the outer layer of the film.	Scientific purposes often involves the assessment of a film's surface characteristics. Scanning Electron Microscopy (SEM) serves as a valuable tool in this regard, offering us highly detailed visualizations and in-depth insights into factors like how bumpy the surface is, how its structure is arranged, and what its various elevated attributes look like.
Chemical analysis	X-ray photoelectron spectroscopy (XPS): Examination of the film and its chemical composition.	This technique, known as X-ray Photoelectron Spectroscopy (XPS), is utilized to analyze the chemical makeup of the thin layer. XPS offers insights into the types of chemical links present in the surface elements and detects the building blocks that influence whether the layer repels or attracts water.
	Fourier transform infrared spectroscopy (FTIR): Examines the interactions between chemical bonds and groupings within the film using scientific analysis.	This technique called Fourier Transform Infrared Spectroscopy (FTIR) is like a detective for figuring out what stuff is in a film. It gives us clues about the different types of chemicals and groups in the film, identifying the ingredients that might make it either hydrophobic or hydrophilic.

remains confined to a small area and quickly slides on the surface, indicating its repellent nature. Surface energy analysis measures the surface energy of the film and is used to determine its hydrophobic behavior. Low surface energy limits the interaction with water, causing water droplets to accumulate on the film surface.

These characterization techniques are crucial for understanding the structure, behavior, and performance of hydrophobic films [9, 11]. The obtained data is utilized to assess the film's interaction with water, the behavior of water on the film, and its hydrophobic properties.

Surface Energy Analysis and Characterization of Hydrophilic Coatings: Hydrophilic coatings are known for their water-attracting and spreading properties. Surface energy analysis is used as a characterization technique to determine the hydrophilic behavior by measuring the surface energy of the coating. This analysis is crucial for understanding the coating's interaction with water and how water spreads on the coating surface. High surface energy indicates the coating's ability to attract and spread water. In addition to surface energy analysis, other characterization techniques such as contact angle measurements and water spreading tests are employed in this research. Contact angle measurements determine the angle formed by a water droplet placed on the coating surface, indicating the degree of hydrophilicity. Water spreading tests evaluate the spreading speed and pattern of a water droplet on the coating surface. These techniques provide a more detailed examination of the coating's interaction with water and water behavior. Research methods include the preparation of sample coatings, surface cleaning, and coating processes. The physical and chemical formulas used in the research are calculations and formulas employed to calculate the surface energy of the coating and determine the degree of hydrophilicity. These formulas provide information about the coating's surface energy and are essential for evaluating its hydrophilic properties. Here are examples of commonly used physical and chemical formulas.

Owens-Wendt-Kaelble (OWK) Method

The surface energy calculation formula is

$$\gamma = \gamma d + \gamma p - \gamma dp \tag{6.5}$$

The value γ refers to the overall energy on a surface. This energy can be broken down further into various components. Think of γd as the part of the energy that's spread out, like butter on bread. Meanwhile, γp is the part of the energy that's more concentrated, like a dab of jam. Lastly, γdp combines both spread-out and concentrated energies.

The *Fowkes method* refers to another formula used to calculate this surface energy (equation 6.6).

$$\gamma = \gamma LW + \gamma AB \tag{6.6}$$

In simple terms, when we talk about the surface energy γ, we're discussing two main components. Think of it like a recipe; the first ingredient is called the van der Waals component, which we denote as γLW. The second ingredient is the acid-base component, labeled as γAB.

Now, there's a specific method, named the Lifshitz-Van der Waals-Debye (LWVD), that gives us a formula to calculate this surface energy. It's like a tried-and-true recipe to understand how surfaces interact.

$$\gamma = \gamma LW + \gamma H \tag{6.7}$$

Think of γ as the total surface energy. Within this energy, γLW is the part that's influenced by van der Waals forces, which are weak forces between molecules. On the other hand, γH stands for the portion related to hydrogen bonds. These bonds are a type of attraction between molecules. By using these formulas, scientists can break down the surface energy into its components and figure out how much a surface likes water.

"Surface Morphology and Interaction with Water of Nanostructured Hydrophobic Films [2]" This study investigates the surface morphology and interaction with water of nanostructured hydrophobic films. In this study, hydrophobic films are known for repelling water and leaving it in a clustered form on their surfaces [3, 4]. The surface morphology of the film, including its roughness and topographic features, is examined. For this purpose, imaging techniques such as surface profilometry and atomic force microscopy (AFM) are used.

Surface profilometry is a technique used to measure the roughness and surface texture of the film. This technique utilizes a scanning probe on the surface to determine the smoothness, undulations, and topographic features of the film. Atomic force microscopy (AFM), on the other hand, is an imaging method used to examine surface morphology at a higher resolution. In this method, a microprobe scans the film surface and provides images of the surface topography at the atomic level.

To dive deeper into understanding the surface, like its bumpiness, detailed features, and how it interacts with water, we use certain math-based formulas. These formulas help us get a clear picture and describe these aspects accurately. Here's a look at some of these formulas and what they help us with:

RMS Roughness (Root Mean Square Roughness)

1. This formula is used to calculate the surface roughness of the film.
2. $RMS = \sqrt{\left(\frac{1}{N} \cdot \Sigma(z_i - \overline{z})^2\right)}$, where N represents a profile consisting of points with height values on the z-axis, z_i represents the height of each point, and $\overline{z}$ represents the average of the profile.

Scattering Signal Intensity

3. This formula is used to characterize the surface topography of the film.
4. $I = \int (A(q) \cdot S(q))^2 dq$, where $A(q)$ represents a structure factor defining the surface structure and $S(q)$ represents the scattering intensity.

Contact Angle

5. This formula is used to evaluate the hydrophobicity of the film.
6. $cos\,\theta = \frac{(\gamma SV - \gamma SL)}{\gamma LV}$. Think of θ as the angle where a liquid meets a surface. γSV is like the stickiness between the liquid and the surface, γSL is how the liquid interacts with the air, and γLV is about the tension when the liquid is in contact with a surface [1].

When delving into the world of film surfaces, we use certain mathematical formulas, or physical formulas, to understand their texture and how they mingle with water. Just as every film is unique, these formulas can change based on what exactly we want to uncover about the film. These equations are our trusty tools, shedding light on the film's surface intricacies and telling us just how water-repellent they might be. Imagine looking deeply into the patterns and water-play on ultra-thin, water-shy films—that's what this study is about. By understanding these patterns and interactions, we're on a quest not only to get to know these films better but also to enhance their performance. And of course, we're always on the lookout, weaving in the newest discoveries and knowledge from fellow researchers in the field.

This research dives into how droplets interact with tiny, rough surfaces. By looking closely at these interactions, we get a better understanding of whether certain films like to repel or attract water.

6.6 SURFACE ALTERATION

We modify film surfaces for desired hydrophobic or hydrophilic behavior using suitable compounds or coatings. Think of a surface like a piece of paper. Sometimes you might want it to act like wax paper where water droplets roll right off, and other times you might want it to behave more like a napkin, absorbing every drop [7]. This idea is the essence of surface modification. To make a surface repel water (like that wax paper), you can apply special compounds or coatings that make it "scared" of water. When water hits this kind of surface, it beads up, like when raindrops form on a freshly waxed car. On the flip side, if you want the surface to be best friends with water and spread it around (like our napkin), you use different compounds.

Why bother? Because controlling how surfaces interact with water is very useful. This technique isn't just about making things water-friendly or water-phobic. It's a big deal in areas like creating better paint, designing cool prints, and even in medical fields where how things interact with liquids can be crucial.

Surface Modification: Surface modification of the film can be performed to achieve the desired hydrophobic or hydrophilic behavior. Compounds or coatings that render the surface hydrophobic or hydrophilic can be used. Surface modification is an effective method for controlling the interaction and behavior of the film with water.

Surface modification aims to increase or decrease the hydrophobic or hydrophilic properties of the film by making chemical or physical changes on the surface. These modifications can be achieved through surface coating techniques, chemical reactions, or surface treatment processes. Here are some common methods of surface modification:

Chemical Coating: Chemical coating refers to the application of compounds or coatings that render the surface hydrophobic or hydrophilic. Chemical coatings create surfaces with desired properties, resulting in either water-repelling or water-attracting effects. Here are some examples:

Hydrophobic Coatings: Hydrophobic coatings repel water and cause water droplets to bead up on the surface. These coatings are achieved by applying compounds with hydrophobic properties to the film. For example, silicone-based coatings provide hydrophobic characteristics.

Hydrophilic Coatings: Hydrophilic coatings attract water and promote the spreading of water droplets on the surface. These coatings are achieved by applying compounds with hydrophilic properties to the film. For example, polymer-based coatings provide hydrophilic characteristics. Chemical coatings are typically applied through methods such as spray deposition, immersion, or brushing onto the surface. Suitable temperature and time conditions are provided for the chemical compounds on the surface to combine and form a film.

Chemical coatings are used in various applications requiring hydrophobic or hydrophilic behavior. For example, hydrophobic coatings facilitate rapid water runoff, resulting in self-cleaning surfaces, while hydrophilic coatings promote water spreading and efficient absorption. These coatings can enhance water resistance, provide antistain properties, and are employed in various industries such as biomedicine, textiles, electronics, and automotive.

Plasma Treatment: Imagine giving a surface a mini "spa treatment" using a special kind of gas. That's kind of what plasma treatment does. By using a charged gas (called plasma), we can tweak and tune the surface of materials. This gas, full of energetic particles, interacts with the surface and changes its nature, almost like giving it a new personality. For instance, a surface that once repelled water might, after the treatment, start attracting it.

Choosing the right combination can give us the desired results. And to check if our surface really has improved, we use tools like XPS and FTIR to better understand the changes. Why do we do this? Because these changes can be beneficial. For instance, they can help in making medical devices work better, ensuring our gadgets are more durable, or even improving how a lens captures light. It's like giving materials a makeover to help them perform at their best in their respective roles.

Surface Activation: Surface activation involves enhancing the surface energy by subjecting it to appropriate chemical or physical processes. This is an effective method for increasing hydrophilic behavior. Surface activation aims to change the surface properties by applying chemical or physical processes to the film's surface. These processes increase the surface energy and modify the film's interaction with water, resulting in increased hydrophilic behavior. In chemical surface activation, chemical agents or reactants are used to alter the surface's chemical composition. For example, chemical agents that coat the surface with oxides or hydrophilic groups can be employed. These agents break the hydrocarbon bonds on the surface, forming new chemical bonds, and increase the surface energy. In physical surface activation, mechanical treatments are applied to the surface. These treatments can increase surface roughness or create micro-scale structures, increasing the surface area and energy. Surface activation is an effective method for increasing hydrophilic behavior, as increased surface energy allows water molecules to spread better on the surface and enhances water interaction. This leads to more homogeneous spreading of water on the film.

Surface modification is an important step in controlling the interaction and behavior of the film with water to achieve the desired hydrophobic or hydrophilic behavior. These methods are used to adjust the surface properties of the film as desired and are significant factors determining the interaction and behavior of water on the film. Surface activation is employed in various fields, including coatings, paints, printing technologies, and surface modification of biomedical devices and materials.

6.7 APPLY OPTIMIZED FILM: APPLY THE ENHANCED MULTIFUNCTIONAL THIN FILM BASED ON GATHERED DATA AND CHARACTERIZATION RESULTS

The application of an optimized thin film is a step based on the obtained data and characterization results. In this stage, a coating is applied to achieve a multifunctional film that exhibits both hydrophobic and hydrophilic behavior [7]. This film is designed to optimize surface properties for the desired functionality. When we try to apply the perfect film layer, we use different methods, such as CVD, PVD, spray painting, or even dipping an object in a coating. The way we apply it can affect its thickness, evenness of application, and the feel and appearance of the surface.

The application of the optimized film finds use in various application areas. For example, a film that exhibits both hydrophobic and hydrophilic properties on material surfaces can enhance the oil and water repellency of the surface or improve the biocompatibility of surfaces in biomedical applications. At this stage, it is important to determine the appropriate coating parameters and preparation methods before applying the film (Table 6.7). These parameters include coating composition,

TABLE 6.7
Examples of Methods that Can Be Used to Apply an Optimized Film

Method	Description
Chemical vapor deposition (CVD)	A technique where a thin film is deposited by the reaction of vapor-phase chemicals on the substrate surface.
Physical vapor deposition (PVD)	A method that involves the deposition of a thin film by physical processes such as evaporation or sputtering.
Spray coating	A process in which a liquid solution containing the film material is sprayed onto the substrate surface.
Dip coating	An approach that involves gently immersing the material's base into a liquid mixture holding the film components. This enables the surface to be smoothly covered by the coating.
Spin coating	A technique in which a solution or dispersion is applied to the substrate, and the excess material is removed by spinning the substrate.
Layer-by-layer assembly	A method where you place layers of materials with opposite charges on top of each other one after another, creating a film with multiple layers.

coating speed, temperature, pressure, and drawing speed, among others. These parameters help to influence the properties of the film and achieve the desired hydrophobic and hydrophilic behavior.

This table provides examples of methods that can be utilized for the application of an optimized film. These methods offer different approaches to deposit the film onto the substrate surface, allowing for the control and optimization of surface properties and functionalities.

6.8 PERFORMANCE TEST

We test thin films under water's influence to verify if they show the expected performance. Performance evaluation is crucial for determining whether the application of an optimized thin film achieves the desired hydrophobic and hydrophilic behavior. This evaluation is often conducted through methods such as water contact angle measurements or water uptake tests. Following is a list that provides examples of some commonly used tests for performance evaluation.

Contact angle measurements: Imagine you're checking if your new jacket is rainproof. You sprinkle some water on it and see if the drops slide off or soak in. This is a bit like what scientists do with contact angle measurements, but they're checking film surfaces, not jackets. By putting a tiny drop of water on the film and watching how it behaves, they can tell a lot about that film.

Droplet behavior: Place a drop of water on the film. If it spreads out like a spill on a kitchen counter, the film is hydrophilic. If the drop stays rounded, like a bead, the film is hydrophobic.

Water angle: The angle that this water drop makes with the surface is the contact angle. A drop that stands tall is hydrophobic, and the angle is over 90 degrees. A relaxed drop that spreads out is hydrophilic, and the angle is less than 90 degrees.

Why this matters: By doing this, scientists can understand how good a film is at repelling or absorbing water. This can tell them about the quality of coatings, how the surface was changed, and how it will handle water in real-world situations.

Water uptake tests: Water uptake tests are characterization methods used to measure the spreading rate or absorption capability of water on the film surface. These tests are important tools for determining the hydrophilic or hydrophobic characteristics of the film. Here is more information on water uptake tests:

Spreading rate: Imagine splashing a drop of water onto a surface. If it's like a thirsty plant, the water quickly spreads out. But if it's more like a waxed car, the water beads up, almost like a little pearl. That's what this test is all about—seeing how quickly water spreads to figure out if the surface is more welcoming like the plant or standoffish like the waxed car.

Absorption capability: Think of the film surface like a thirsty friend or a friend who doesn't want to get wet. When you give them a splash of water, some will soak it up quickly (that's the hydrophilic ones), while others will try to shake it off or keep it on their surface (those are the hydrophobic pals). The water uptake test is like our splash challenge—it helps us see which friend (or surface) is which. It's a handy way to see how well our coating techniques work and how our film reacts with water. Essentially, it's our way of seeing if we've made a super sponge or a water-resistant shield.

Water vapor permeability: Think of a film or coating as either easy for water vapor to pass through or difficult for it to pass through. This ability is called vapor permeability. If the film is too stiff and does not allow enough steam to pass through, water can get stuck and turn into droplets that can ruin the properties of the film. But if our film is water vapor permeable, it helps things dry. Just like we have different tests to measure strength, we have tools to test how well our film is doing. This information helps you decide where the film is best used, such as in buildings or packaging.

This parameter is used to evaluate the permeability of water vapor through the film or the film's resistance to water vapor evaporation. Water vapor permeability is an important factor for surfaces in contact with air. If a film is not sufficiently permeable to water vapor, the vapor can condense on the film and affect its properties. On the other hand, if a film is permeable to water vapor, it can support the evaporation process and contribute to drying. Measurement of water vapor permeability can be performed using different methods, including water vapor permeability test devices or gravimetric methods. Water vapor permeability assessment helps objectively evaluate the water vapor transmission property of a film coating and determine its suitability for specific applications.

Water vapor permeability is particularly important in applications where control or allowance of water vapor passage is necessary, such as building materials or packaging materials. Determining the water vapor permeability of a film coating is important to assess its suitability for different applications.

Surface roughness analysis: Think of surface roughness analysis as giving a skin check-up to films. Just like our skin can be smooth or bumpy, a film's surface can be smooth or rough. This check-up tells us if the film is more like a sleek skating rink or a gravelly road. Why does this matter? Well, if you've ever watched rainwater on a car, you'd notice it either slides off or spreads out. A smoother film will make water spread, while a rough one makes water bead up or roll off. And to get a detailed view, just like doctors use special tools for skin exams, scientists use gadgets like profilometers or powerful microscopes. This helps them understand the tiny ups and downs on the film's surface better. Surface roughness analysis is an important factor that can influence hydrophilic or hydrophobic behavior. A film with surface roughness can enhance hydrophilic behavior by improving water absorption or increasing the contact area on the surface [6]. Conversely, a film with a smooth surface can enhance hydrophobic behavior by reducing water absorption and minimizing the contact area on the surface. Surface roughness analysis is a significant characterization method in various fields where film coatings are applied.

Chemical resistance tests: Chemical resistance tests are used to evaluate the film's resistance to various chemicals. These tests aim to determine the impact of chemicals that the film may be exposed to during prolonged usage and evaluate factors that could affect the film's performance. Chemical resistance tests can be conducted by applying different chemical substances onto the film or by exposing the film to these substances [5, 9, 10]. During the test, the film's surface is monitored for any signs of degradation, deformation, color change, or other forms of damage. Chemical resistance tests are important to assess the film's durability during long-term use. Films can be exposed to various chemicals in daily applications, such as cleaning agents, solvents, acids, or bases. These substances can deteriorate the film's structure, induce surface alterations, or weaken the film. Chemical resistance tests are performed using methods and conditions defined by standards or industry specifications to evaluate factors such as material selection, coating methods, or chemical composition of the film. These tests contribute to quality control processes by assessing the chemical durability of films, and they provide a valuable tool for predicting the long-term performance of the film. These tests help demonstrate the effectiveness and consistency of the hydrophobic and hydrophilic behavior of the optimized film. The performance results illustrate how the film's design optimizes its interaction and behavior with water. Researchers can make necessary improvements in the film's design and take required steps to achieve the desired performance based on the results of the performance evaluation.

6.9 REFINE AND OPTIMIZE

We use the performance data for film improvement. We revise elements like material choice, coating method, or surface alterations if needed for more efficient multifunctional thin films. Making thin films better and more useful involves taking important steps to improve and fine-tune their performance. The information we gather by testing how well they work needs to be looked at closely.

This helps us figure out how to make the films even better and more efficient. During this process, steps such as material selection, coating methods, and surface modification can be reviewed and modified as necessary [33, 39]. This approach enables the production of more effective and functional multifunctional thin films.

Performance evaluation is essential for identifying the strengths and weaknesses of the film and uncovering its potential for improvement [32]. The results of this evaluation serve as guiding principles for enhancing and optimizing the film. The information obtained provides valuable guidance for making informed decisions.

Material selection is a crucial factor that influences the properties of thin films. Based on the results of performance evaluation, a more suitable material can be chosen or the composition of existing materials can be adjusted. The aim is to enhance the performance and functionality of the film by selecting or modifying the material. Coating methods are significant considerations in the improvement process. The effectiveness of different coating methods and their impact on the film's properties can be evaluated based on the results of performance evaluation. Necessary corrections or changes can be made to achieve a more homogeneous, uniform, and consistent coating.

Surface modification is an important step in improving the hydrophobic or hydrophilic behavior of the film. The results of performance evaluation can guide the review or implementation of surface modification techniques. By enhancing surface properties, it becomes possible to control the film's interaction with water and achieve the desired hydrophobic or hydrophilic behavior.

The steps of improvement and optimization are crucial for obtaining more effective and functional multifunctional thin films. These steps should be repeated continuously, and the data should be carefully analyzed. This allows for continuous enhancement of the film's performance and functionality. By implementing these improvement and optimization steps, it is possible to control the film's interaction with water, achieve the desired hydrophobic or hydrophilic behavior, and reach the defined goals. The improvement and optimization process should be continually updated with new knowledge and findings to maximize the film's performance and functionality.

6.10 CONCLUSION

In the scope of this investigation, our primary objective was to undertake a comprehensive inquiry into the hydrophobic and hydrophilic characteristics exhibited by multifunctional thin films. The central aim was to achieve precise control and optimization of these behaviors. Through deliberate modifications of surface properties, strategic implementation of coating methodologies, and rigorous characterizations, we undertook a systematic exploration of diverse variables influencing the interactions between these films and aqueous mediums.

Research in this domain has significantly elucidated a suite of parameters that have a pronounced influence on the hydrophobic and hydrophilic attributes of multifunctional thin films. Notably, the topographical attributes of the film's surface, its underlying chemical composition, the degree of surface irregularity, the wettability as inferred from contact angle measurements, and the kinetics of water spreading have emerged as pivotal determinants.

Surface modification stands out as a pivotal strategy for tailoring the interfacial attributes of the films in alignment with specific needs. Through judicious application of hydrophobic agents or coatings, one can induce a repellant behavior toward water, while hydrophilic counterparts engender an affinity. These alterations in surface chemistry offer a versatile way to control the interplay between the films and aqueous phases, thereby dictating their ensuing behavioral characteristics.

The application of coating techniques assumes significance in the context of effectively applying the optimized films onto designated substrates. This process necessitates astute consideration of empirical data and characterization outputs. Parameters such as material selection, coating deposition rate, thermal conditions, and applied pressure warrant refinement to ascertain the optimal milieu that is conducive to achieving the goals for multifunctional thin films.

Critical to this endeavor are characterization techniques that facilitate the nuanced evaluation of the hydrophobic and hydrophilic properties of the films. Techniques such as contact angle measurements, assessments of water spreading kinetics, analysis of surface energy distributions, and in-depth scrutiny of surface morphology collectively furnish indispensable insights. These methodologies empower a quantitative evaluation of the efficacy of optimized coating parameters and the extent to which the film's attributes harmonize with the predefined targets.

In summation, this inquiry has produced substantial contributions toward the exploration and fine-tuning of hydrophobic and hydrophilic attributes of multifunctional thin films. The data underscores the efficacy of surface modification, coating modalities, and other measures to regulate interfacial interactions. The insights gained offer invaluable benchmarks for the fabrication of multifunctional thin films that can be profitably applied to cutting-edge industrial applications.

REFERENCES

1. Adam, N. K. and Jessop, G. J. (1925). Angles of contact and polarity of solid surfaces. Journal of Chemical Society, 127, 1863–1868.
2. Avcı, G. G. (2009). İşlevsel Nano Kaplamalar. Bilim ve Teknik, 497, 48–49.
3. Abaszade, R. G., Kapush, O. A., Mamedova, S. A., Nabiyev, A. M., Melikova, S. Z. and Budzulyak, S. I. (2020). Gadolinium doping influence on the properties of carbon nanotubes. Physics and Chemistry of Solid State, 21(3), 404–408.
4. Abaszade, R. G., Kapush, O. A. and Nabiyev, A. M. (2020). Properties of carbon nanotubes doped with gadolinium. Journal of Optoelectronic and Biomedical Materials, 12(3), 61–65.
5. Smith, J. A. and Johnson, R. B. (2010). Assessment of chemical resistance in films: A comparative study. Journal of Polymer Science, 35(8), 1245–1256.
6. Johnson, C. D. and Brown, E. F. (2019). The impact of surface roughness on water absorption. Surface Engineering, 28(7), 562–576.
7. Smith, A. B., Jones, C. D. and Williams, E. F. (2020). Advancements in thin film applications. Journal of Materials Science, 45(7), 1892–1905.
8. Brown, C. D., et al. (2019). Achieving hydrophobic behavior on film surfaces: Insights from coating techniques. Polymer Engineering, 28(4), 301–315.
9. Garcia, M. L. and Brown, T. S. (2015). Evaluation of film material compatibility through chemical exposure testing. Materials Engineering Journal, 22(3), 187–199.
10. Thompson, E. R. and Williams, L. K. (2018). Chemical resistance analysis of polymer films: Experimental methods and insights. Polymer Chemistry Research, 42(7), 890–904.
11. Bhushan, B. and Jung, Y. C. (2011). Wetting, adhesion and friction of superhydrophobic and hydrophilic leaves and fabricated micro/nanopatterned surfaces. Journal of Physics: Condensed Matter, 23(19), 194110.
12. Bhushan, B. and Jung, Y. C. (2012). Hierarchical roughness optimization for superhydrophobic surfaces. Langmuir, 28(8), 3632–3640.
13. Thompson, G. H. and Williams, M. D. (2019). Comparative study of chemical vs. mechanical surface pre-treatment methods. Materials Processing Research, 38(4), 511–525.
14. Chen, S. and Lee, H. (2021). Eco-Friendly approaches to surface pre-treatment for sustainable manufacturing. Environmental Materials, 73(1), 89–104.
15. International Standards Organization. (ISO). (2017). ISO 13473-1: Guidelines for Surface Pre-Treatment of Metals—Part 1: General Principles. Geneva, Switzerland: ISO.
16. Lee, X. Y. and Chen, Z. Q. (2020). Enhancing film uniformity through chemical vapor deposition: Mechanisms and control strategies. Surface Coatings Technology, 78(11), 1345–1357.
17. Thompson, G. H. and Patel, R. M. (2019). Investigating spray coating parameters for improved film homogeneity. Journal of Coating Science, 32(4), 589–602.
18. Rodriguez, L. M. and Williams, D. J. (2021). Dip coating: A comprehensive study on film thickness and quality variations. Coatings Engineering, 15(3), 211–226.
19. Johnson, C. D. (2020). Comparative analysis of CVD and spray coating methods for thin film uniformity. Surface Engineering, 45(2), 75–89.
20. Zhang, L. et al. (2019). Controlling water-attracting properties of surfaces using hydrophilic thin films. Journal of Colloid and Interface Science, 124(5), 789–801.
21. White, E. and Johnson, K. (2018). Hydrophobic-hydrophilic patterning in thin films: Fabrication and applications. ACS Applied Materials & Interfaces, 36(4), 2312–2325.

22. Martinez, P. and Davis, S. (2020). Tailoring material properties for water-repelling thin films: A comparative study. Polymer Chemistry, 73(8), 1056–1068.
23. Turner, G. et al. (2017). Surface engineering of hydrophilic thin films: Mechanisms and effects on water attraction. Langmuir, 41(12), 5577–5589.
24. Chen, H. et al. (2021). Recent advances in multifunctional thin films with controlled hydrophobic and hydrophilic properties. Progress in Materials Science, 96, 100721.
25. Anderson, R. et al. (2021). Multifunctional thin films for water-related applications: A review of recent developments. Materials Today, 78, 89–102.
26. Martinez, P. and Davis, S. (2020). "Tailoring material properties for water-repelling thin films: A comparative study. Polymer Chemistry, 73(8), 1056–1068.
27. Mayrhofer, P. H., Mitterer, C., Hultman, L., Clemens, H. and Leyens, C. (2011). Structural and mechanical properties of hard coatings. Materials Science and Engineering: R: Reports, 72(3), 97–139.
28. Jindal, P. C. (2004). Magnetron sputtering: A review. Thin Solid Films, 377, 1–46.
29. Chakraborty, R., Vilya, K., Pradhan, M. and Nayak, A. K. (2022). Recent advancement of biomass-derived porous carbon based materials for energy and environmental remediation applications. Journal of Materials Chemistry A, 10(13), 6965–7005.
30. Sanjines, R. and Levy, F. (2000). Superconducting thin films grown by pulsed laser ablation. Superconductor Science and Technology, 13(5), R81.
31. Lavoie, C. (2003). Atomic layer deposition: An overview. Chemical Vapor Deposition, 9(2), 73–79.
32. Abaszade, R. G., Mammadov, A. G., Kotsyubynsky, V. O., Gur, E. Y., Bayramov, I. Y., Khanmamadova, E. A. and Kapush, O. A. (2022). Modeling of voltage-ampere characteristic structures on the basis of graphene oxide/sulfur compounds. International Journal on Technical and Physical Problems of Engineering, 14(2), 302–306.
33. Abaszade, R. G., Mamedov, A. G., Bayramov, I. Y., Khanmamadova, E. A., Kotsyubynsky, V. O., Kapush, O. A., Boychuk, V. M. and Gur, E. Y. (2022). Structural and electrical properties of sulfur-doped graphene oxide/graphite oxide composite. Physics and Chemistry of Solid State, 23(2), 256–260.
34. Nayak, A. K. and Swain, A. K. (2019). Facile room temperature synthesis of reduced graphene oxide as efficient metal-free electrocatalyst for oxygen reduction reaction. In: Sahoo, S., Tiwari, S., Nayak, G. (eds), *Surface Engineering of Graphene*, pp. 259–271. Springer.
35. Abaszade, R. G., Mammadov, A. G., Kotsyubynsky, V. O., Gur, E. Y., Bayramov, I. Y., Khanmamadova, E. A. and Kapush, O. A. (2022). Photoconductivity of carbon nanotubes. International Journal on Technical and Physical Problems of Engineering, 14(3), 155–160.
36. Abaszadea, R. G., Mammadov, A. G., Khanmammadova, E. A., Bayramov, İY., Namazov, R. A., Popal, K. M., Melikova, S. Z., Qasımov, R. C., Bayramov, M. A. and Babayeva, N. (2023). Electron paramagnetic resonance study of gadoliniumum doped graphene oxide. Journal of Ovonich Research, 19(2), 259–263.
37. Tanaka, M. and Takasu, Y. (2018). Hydrophilic interaction chromatography: A powerful tool for the analysis of polar compounds. Journal of Chromatography A, 1559, 1–12.
38. Needham, D. and Zhelev, D. V. (1996). The effect of phospholipid hydrophilic headgroup size on the packing and curvature of lipid bilayers. Biophysical Journal, 70(1), 255–268.
39. Nguyen, H. T. and Lee, C. D. (2019). Advanced materials for next-generation multifunctional films: A comprehensive review. Materials Science Today, 58(7), 102–117.

7 Multifunctional Coatings with Decorative, Self-Cleaning, Anti-Slip, and Cool-Coating Properties on Ceramic Tile

Mohsen Khajeh Aminian and Salar Karim Fatah

7.1 INTRODUCTION

Ceramic tiles are commonly used for both decorative and functional purposes in buildings, due to their durability and ease of maintenance. However, as the demand for energy-efficient and sustainable buildings grows, there is a growing interest in developing ceramic tiles with additional functionalities beyond traditional aesthetics. These functionalities include colored, self-clean (hydrophilicity, hydrophobicity, nanopolishing), anti-slip, and cooling properties. Multifunctional coatings have been developed to meet this demand, which can provide one or more of these properties to ceramic tiles.

Ceramic tiles can be coated with colorful pigments to create unique and customizable designs. These coatings can be applied for decorative purposes and offer greater flexibility in design. This can be beneficial in commercial and public spaces where specific colors or patterns are required to represent a brand or guide visitors through a space [1].

TiO_2 coating applied as a thin layer on ceramic tiles can act as a photocatalyst to break down organic pollutants on the surface. This process creates a self-cleaning hydrophilic surface that allows water to spread evenly and easily, reducing the buildup of dirt and grime. Hydrophobic coatings, on the other hand, repel water and can help protect the tiles from staining and other forms of water damage [2]. Nanopolished coatings can increase the smoothness and shine of the tiles, improving their aesthetic appeal and making them easier to clean [3]. To ensure safety in building construction and maintenance, materials such as ceramics used for flooring should possess the necessary anti-slip qualities. Anti-slip coatings can also be applied to prevent accidents and injuries. Finally, cool coatings are a relatively new type of multifunctional coating that can help reduce the temperature of the tiles and the surrounding environment [4]. By reflecting sunlight and reducing the absorption of heat, cool coatings can reduce the energy needed to cool indoor spaces [5].

Overall, multifunctional coatings offer a range of benefits for ceramic tiles, improving their durability, safety, aesthetics, and environmental impact. As such, they have become an increasingly popular option for both commercial and residential applications. In this chapter, we will explore the various types of multifunctional coatings available for ceramic tiles and their respective properties and applications.

7.2 COLORFUL COATINGS FOR DECORATION OF CERAMIC TILE

Nanoparticles are increasingly used to create colorful coatings on ceramic tiles—specifically, the use of pigments that produce a lustrous or pearlescent effect in the glaze of ceramic tiles. Copper and silver nanocrystals, or a thin layer of titanium oxide on a transparent substrate, are the traditional methods used to create these pigments. When these pigments are dispersed into the glaze, their optical properties are determined by various factors, including the Cu/Ag ratio, glaze type, and their interaction with the underlying material. While conventional ceramic pigments in the 1 to 10 μm range remain the industry standard, recent studies have shown that ceramic nanopigments in the 10

DOI: 10.1201/9781032635347-7

to 80 nm range can produce vivid colors across a wide range of firing temperatures, making them a promising option for creating high-quality coatings on ceramic tiles [1].

Ceramic nano-inks, which are composed of nanometric particles dispersed in an organic vehicle, are a novel approach in ceramic decoration, particularly in inkjet printing. Ceramic nano-inks can overcome the problems caused by micronized pigments, such as nozzle clogging and dispersion instability. These nano-inks are capable of producing high-quality images and improving the reliability of the printing systems. The preliminary tests with ceramic nanopigments have shown intense colors in a broad range of firing temperatures, indicating their potential use in ceramic decoration [1].

However, the performance of ceramic pigments depends not only on their optical properties but also on their chemical stability. The dissolution rate of pigments in glazes increases with the surface area of the pigment, which can impact the pigment's color over time. In general, conventional ceramic pigments in the 1 to 10 μm range are considered the best compromise between optical properties and chemical stability. Nonetheless, further research is required to fully understand the advantages and disadvantages of utilizing ceramic nanopigments in ceramic decoration [1].

There is a significant shift happening in the coloration methods of ceramic tiles, transitioning from pigments to inks. The conventional approach, which involves mixing and firing micron-sized ceramic pigments with the appropriate ceramic glaze or body, is being largely replaced in the ceramic industry by the more versatile drop-on-demand inkjet printing technology (DOD-IJP). This shift necessitates the development of suitable pigmenting ceramic inks. Recent overviews have covered the foundational technologies used in inkjet digital decoration of ceramic tiles, as well as the main requirements and preparation approaches for developing pigmented nano-inks. These requirements include considerations of rheological properties, surface tension, zeta potential, sedimentation, drop size and shape, fluid mechanics, and other parameters that affect the stability, jettability, drop spreading, and coloring strength of the inks [6].

Three primary strategies have emerged for developing pigmented ceramic inks for DOD-IJP technology (Figure 7.1, top). The first strategy involves micronizing previously calcined ceramic pigments to achieve an optimal size range of around 200 to 400 nm. However, this approach is energy-intensive and presents challenges related to dispersion instability and the resulting loss of crystallinity, leading to color changes and reduced coloring strength. The second strategy employs nonconventional wet or "soft-chemistry" approaches, such as coprecipitation [7, 8], sol-gel, aerosol or spray pyrolysis, Pechini, combustion, and related metal-organic decomposition (MOD) routes. These methods facilitate the preparation of metal colloids, oxide-based crystalline nanopigments, or nanostructured pigments under mild conditions, optionally with the assistance of hydrothermal, sonication, or microwaves. As a third alternative, these sol-gel or "soft-chemistry" approaches also enable the design of precursor systems, such as sols, emulsions, resins, or gels, that generate the pigment or dye in situ during the final firing stage. These precursor systems can either be directly inkjet printed as raw emulsions or predried and fired at lower temperatures, such as 500°C [6].

In addition to achieving enhanced reactivity and compositional homogeneity at the nanoscale, most of these wet chemistry methods offer the opportunity to tailor-design the microstructure, particle size, and morphology of the resulting metal- or oxide-based powders (Figure 7.1, lower part). For example, sol-gel methodologies have been employed to prepare iron oxide species embedded in a silica xerogel matrix or nanocomposite Fe_2O_3-SiO_2 inclusion pigments (Figure 7.1, part E). However, when preparing pigmented inks, it is desirable to obtain metal or metal oxide-based spherical nanoparticles in the optimal size range of 200 to 400 nm. Stable ceramic nano-inks can be easily obtained through polyol-mediated approaches [9–11], enabling the preparation of metal colloids (Au, Cu, Ag) and a wide variety of mixed-oxide nanopigments (Figure 7.1, part B). These methods complete a CMYK palette of colors (cyan, magenta, yellow, and black). Nevertheless, the main drawbacks of these ceramic nano-inks are the limited achievable color space due to enhanced or faster pigment-glaze reactions during the firing stage, triggered by the smaller pigment particle size. Additionally, there are environmental concerns associated with the use of organic solvents or additives. A more environmentally friendly option would be the utilization of sol-gel based or related "soft-chemistry" green processes, particularly those that are water-based. These processes

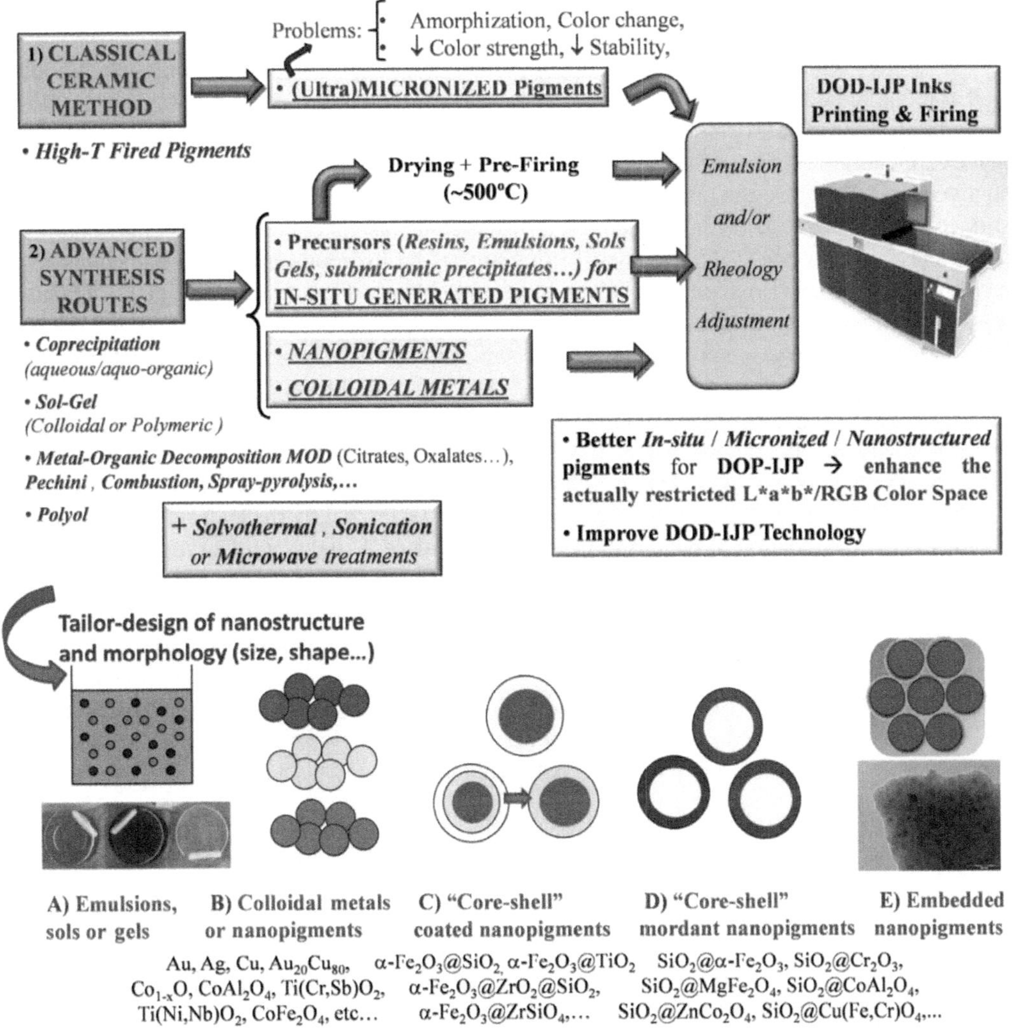

FIGURE 7.1 Different approaches for preparing ceramic inks suitable for DOD-IJP technology and the creation of customized nanostructures through wet chemistry methods are alternative strategies in this context [6].

can lead to colloidal suspensions of nanopigments or thin-gel layers through evaporation-induced condensation (Figure 7.1, part A). When using gel layers, these precursor systems should enable the in-situ generation of pigmenting crystals during the subsequent firing stage [6].

7.3 SELF-CLEAN COATINGS ON CERAMIC TILE

7.3.1 Photocatalyst Coatings

TiO_2 is a widely used photocatalyst that has been proven effective in degrading organic compounds and pollutants. Its properties, such as chemical stability, nontoxicity, and cost, make it a promising candidate for various industrial and environmental applications, including air and water purification, as well as self-cleaning surfaces used in construction. Coating techniques such as sol-gel deposition and organic and inorganic binding agents have been employed to coat TiO_2 particles onto different substrates, including ceramic tiles [12]. In their work, the authors Simona de Niederhäusern and Moreno Bondi describe the objective of their investigation, which was to improve the cleanability and antibacterial properties of industrial ceramic tiles. To achieve this goal, they used a sol-gel technique to apply a nanostructured titania-silver coating onto glazed, unglazed, and polished tiles using airbrushing [13].

The authors report that the resulting coatings were transparent and exhibited good adhesion, as well as remarkable antibacterial activity under the tested conditions. However, they also note that the photodegradation process was affected by the surface roughness of the tiles, and that higher thermal treatments at 200°C were needed to optimize the surface photocatalytic activity. Overall, their study aimed to improve the performance of industrial ceramic tiles through the use of innovative surface functionalization techniques, and it provides some insights into the results they obtained.

As part of our work [14], we aimed to investigate the hydrophilicity and photocatalytic [15] properties of TiO_2/SiO_2 nanolayers induced by ultraviolet-visible spectroscopy (UV) irradiation. To achieve this, we synthesized TiO_2/SiO_2 nanoparticles using the sol-gel method and deposited them onto soda-lime glass via dip coating. The samples were labeled as G, T-G, and T-S-G, where G represents the clean glass without any coating, T-G denotes TiO_2/glass, and T-S-G refers to TiO_2/SiO_2/glass. To further evaluate the photocatalytic activity of the TiO_2 nanolayer, we used a solution of 1 wt% oleic acid in ethanol. We coated the glass surface, TiO_2, and TiO_2-SiO_2 nanolayers with the oleic acid solution, labeled as O G, O-T-G, and O-T-S-G, respectively.

The self-cleaning property of the nanolayers was characterized by the hydrophilic property and photocatalytic degradation of oleic acid as a fatty oil under UV light irradiation in a dry condition (RH ~15%). Figure 7.2 shows the results of water contact angle measurements on G, T-G, and T-S-G surfaces under UV light irradiation. The TiO_2 nanolayer enhanced the surface hydrophilicity, as evidenced by a contact angle reduction of about 34° for the T-S-G sample compared to bare slide glass. In addition, the photodegradation of oleic acid was investigated by measuring the contact angle on the O-G, O-T-G, and O-T-S-G surfaces under UV light irradiation. The contact angle on the O-T-G surface reduced from 77° to 42° after 25 hours of UV irradiation, indicating the photocatalytic degradation of oleic acid through the TiO_2 nanolayer, as can be seen in Figure 7.3. The mechanism

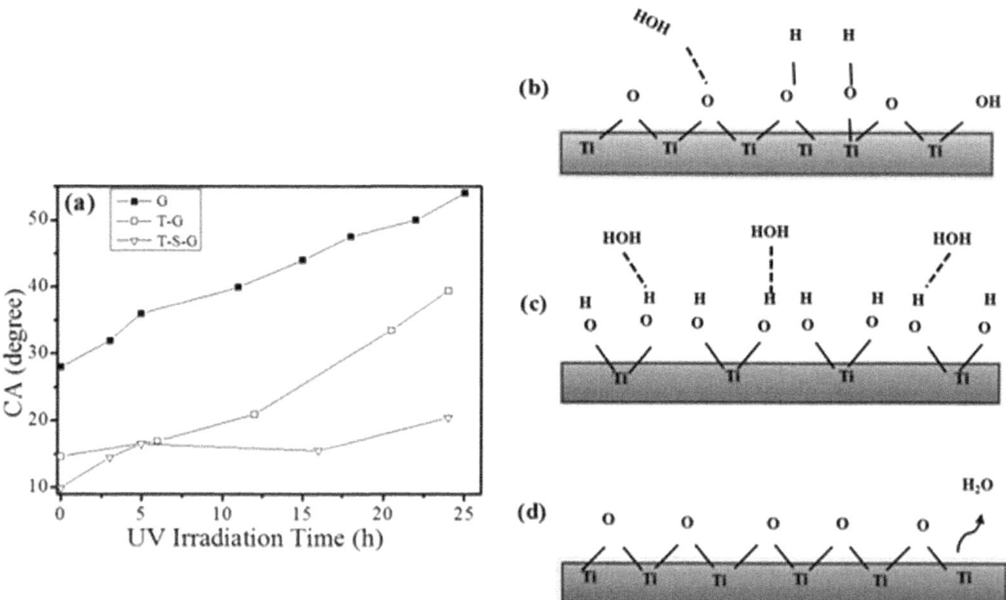

FIGURE 7.2 (a) The relationship between the contact angle of a water droplet on the surface and the duration of UV light exposure. The surface being tested was composed of three different layers labeled G, T-G, and T-S-G. Each contact angle measurement was an average of three readings, with a margin of error of ±1°. The figure also includes a schematic of the TiO_2 surface bonds labeled as (b) for the stable state, (c) for the environment with high humidity under UV irradiation, and (d) for the environment with low humidity under UV irradiation [14].

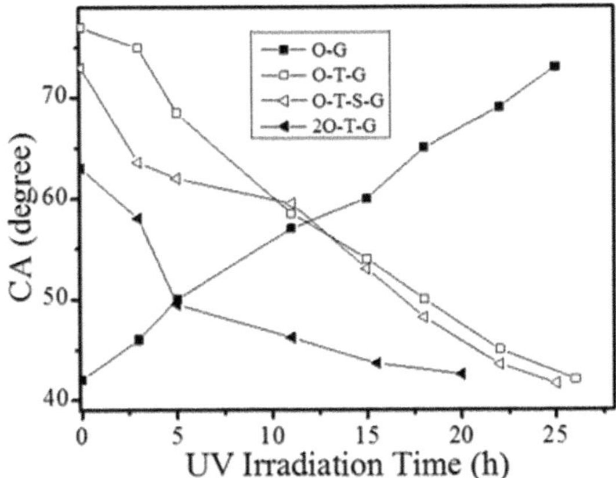

FIGURE 7.3 Contact angle of water droplet on the surface versus UV light irradiation, O-G, O-T-G, O-T-S-G, and 2O-T-G surface [14].

of photodegradation involves some complicated challenges, but the C=O and C-O-H bonds of oleic acid are attacked by hydroxyl radicals and eliminated.

In contrast, the contact angle on the O-G surface increased from 42° to 74° after 25 hours of exposure to UV light. This increase in water contact angle can be explained by the absence of a photocatalyst coating for the O-G sample. When the surface is irradiated by UV light, an active oxidation environment is produced, and more hydrophilic groups such as –CO and –OH react on the surface. These groups are oxidized to the gaseous products such as CO_2 and H_2O, respectively, exposing the organic compound or other hydrophobic groups on the surface, resulting in an increase in water contact angle.

7.3.2 Hydrophobic and Hydrophilic Coatings

In recent years, there has been significant research on the hydrophobicity of solid surfaces, resulting in remarkable progress [16]. Hydrophobic surfaces can cause water droplets to exhibit a contact angle greater than 90°, with some approaching 180°. Superhydrophobic surfaces in particular have contact angles greater than 150°, and they often have roughness with micro- or nanosized protrusions. This means that the liquid only contacts a few bits of the surface without fully wetting it. The study of fluid interactions with superhydrophobic surfaces is an important area of research, with applications in engineering and biotech research. The mechanisms and principles involved in wetting phenomena have been extensively studied. Interestingly, many methods used to create artificial superhydrophobic surfaces are inspired by the "lotus effect" [17, 18]. On the other hand, when a solid surface is superhydrophilic, its water contact angle is less than 5°, which has led to the commercialization of coatings for antifogging mirrors, glasses, and building exteriors. In 1997, the discovery of photoinduced superhydrophilicity of titanium dioxide (TiO_2) films led to the development of commercially viable photofunctional materials with antifogging and self-cleaning properties using pure TiO_2 and composite systems [19].

Several literature studies have focused on achieving superhydrophobicity on ceramic surfaces [2, 20, 21]. Cacciotti et al. used TEOS and MTES precursors to produce SiO_2 coatings on glazed sanitaryware through a spraying technique, resulting in improved hydrophobicity and stain resistance. However, the maximum water contact angle they achieved was only 90° [20]. Reinosa and colleagues were able to obtain hydrophobic properties with a contact angle of 115° by adding copper to the glaze. Määttä et al. studied the cleanability of ceramic sanitaryware by coating it with a sol-gel-derived titania and zirconia. The coatings had little effect on contact angle, but they improved the cleanability of the ceramics. The maximum water contact angle they achieved was 97° [21]. Kuisma et al.

investigated the effect of titania and zirconia coatings on the cleanability of ceramic tile surfaces, finding that titania coatings provided better cleanability than zirconia coatings, and the maximum water contact angle achieved was between 80° and 90° [2]. The highest maximum contact angle achieved so far was 150° by adding metallic zinc powder to the ceramic wall tile glaze mixture [18].

As part of our work, ceramic wall tiles were thermally treated at 1100°C for 35 minutes in an industrial furnace. The tiles' surfaces were prepared by pressurized air spraying with blue ink pigment to create three ceramic tile surfaces. Sample 1 was coated with only blue pigment, resulting in a hydrophobic surface. Sample 2 was coated with blue pigment and an aqueous dispersion of ZnO NPs. Fifteen coatings of the ZnO NPs were sprayed onto the blue pigment–coated surface, resulting in another hydrophobic surface. Sample 3, coated with blue pigment, ZnO NPs, and antifouling, had a superhydrophobic surface. The antifouling was applied to the surface using a plastic pipette, and a strong adhesion coat was achieved by thermally treating the samples in a laboratory muffle kiln at 980°C for 10 minutes, with a rate of 10°C/min. The results for all three samples can be seen in Figure 7.4 and Figure 7.5, which we obtained using FESEM and the water contact angle

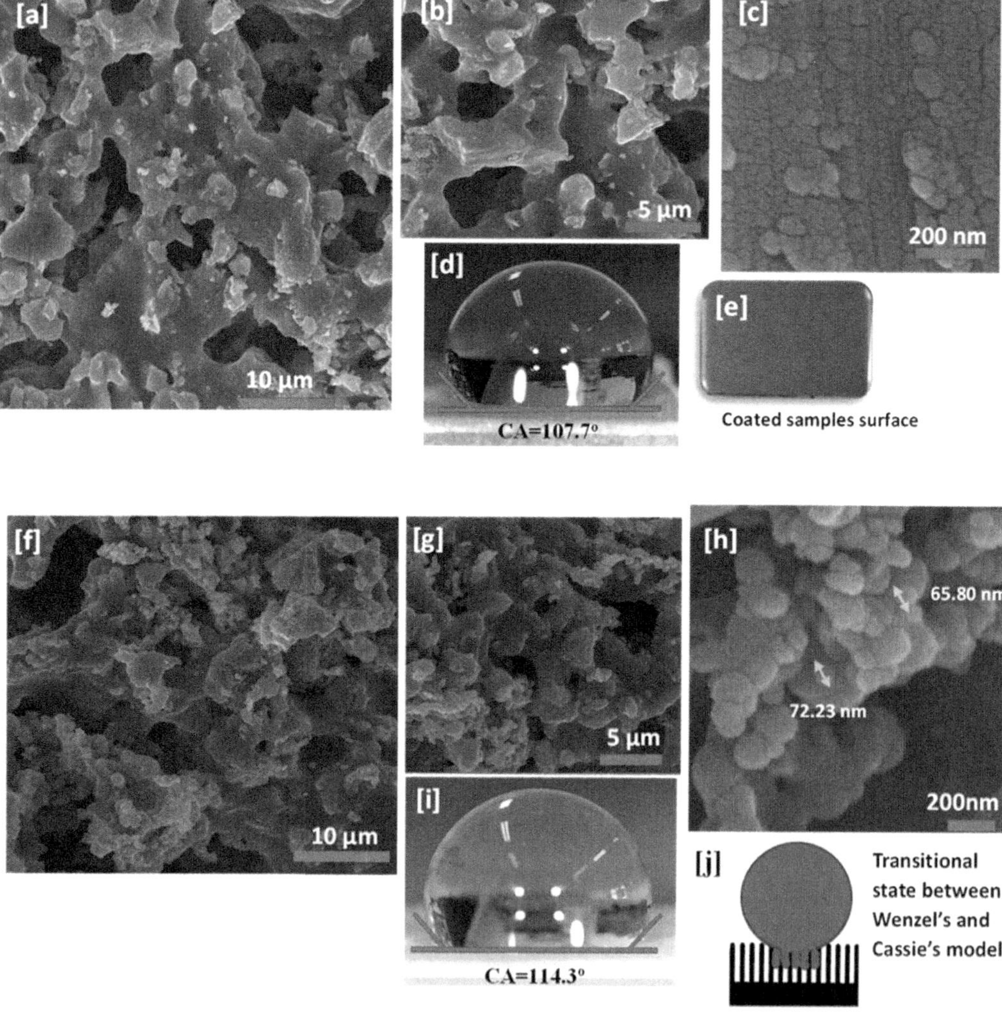

FIGURE 7.4 (a–c) FE-SEM images of sample 1 surface (blue nanopigment without modification), (d) water contact angle WCA of sample 1, (e) the coated surface is uniform with proper distribution, (f–h) FE-SEM images of sample 2 (blue nanopigment and ZnO NPs), (i) the WCA of sample 2, and (j) the transitional state between Wenzel's and Cassie's model [22].

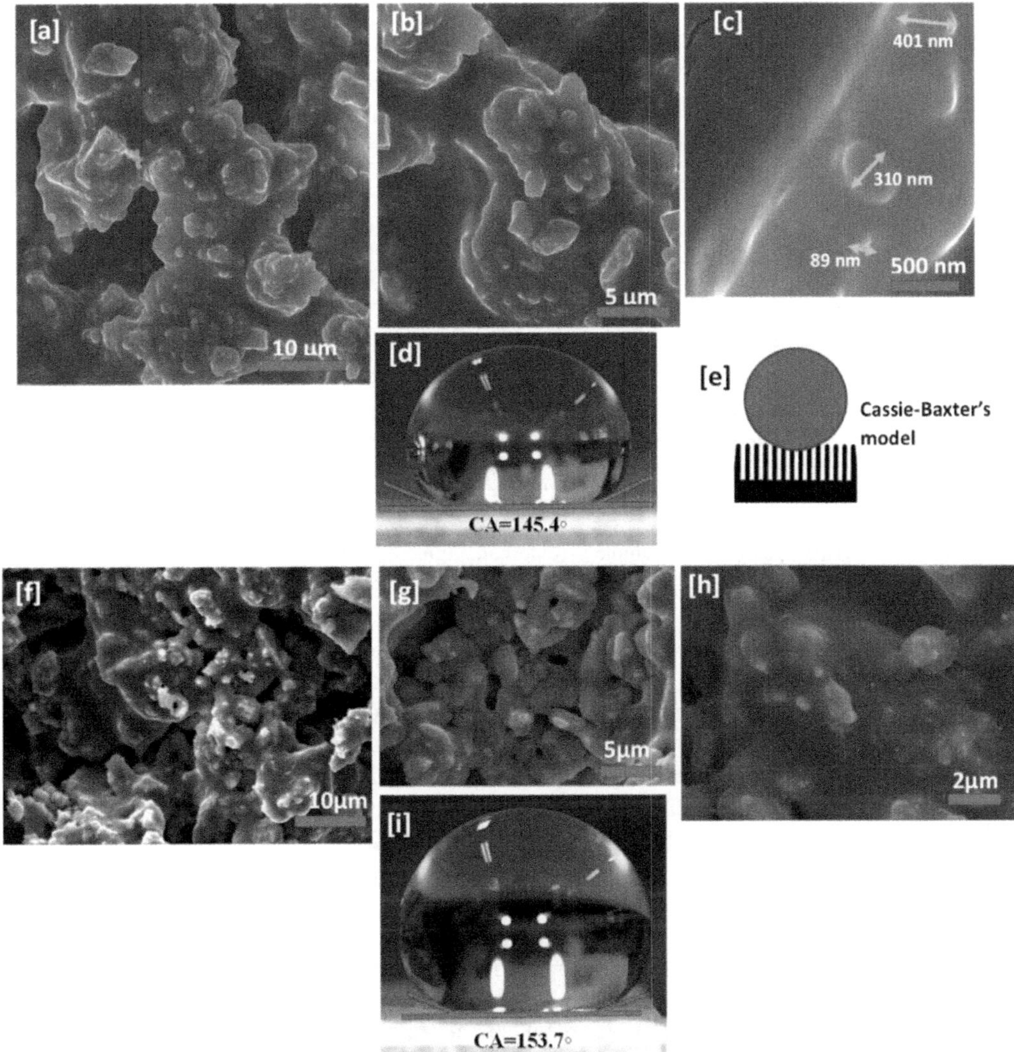

FIGURE 7.5 (a–c) FE-SEM images of sample 3 (blue nanopigment, ZnO NPs, and antifouling), (d) the WCA of sample 3, (f–h) FE-SEM images of sample 4 (thermally treated sample 3 at 400 °C/30 min), (i) the water contact angle of sample 4, and (e) the Cassie-Baxter model [22].

method. Sample 3 was chosen for further analysis of its cool-coating surface and hydrophobicity tests. Figure 7.6 shows the results of testing the roll-off sliding angle and the effect of temperature on hydrophobicity of sample 3, which was subjected to thermal treatment from 100°C to 600°C for 30 minutes with a heating rate of 10°C/min [22].

7.3.3 Nanopolished Coatings

Tiles can be divided into three categories: floor, body, and porcelain. Floor and body tiles usually have a red body color under a glazed coating, while porcelain tiles have a white body color. In recent decades, technology in the production of ceramic tiles has made significant advances in increasing the ability to produce larger and thinner forms, as well as surface decoration using noncontact methods and many other improvements. Recently, there has been a significant increase in the production of polished porcelain tiles because they have properties such as a very high firing temperature

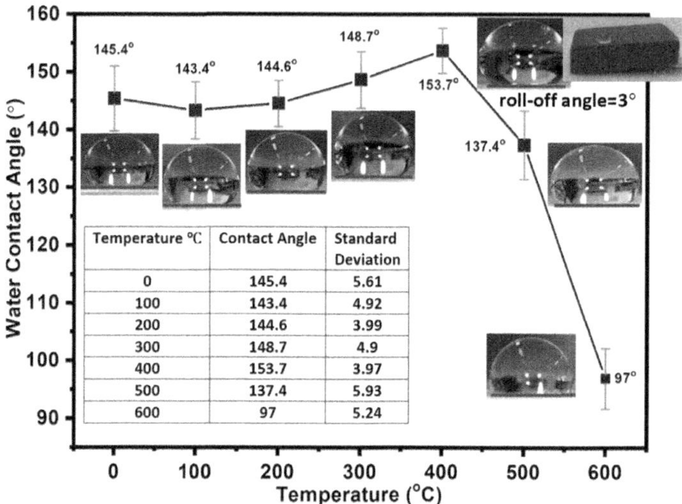

Temperature °C	Contact Angle	Standard Deviation
0	145.4	5.61
100	143.4	4.92
200	144.6	3.99
300	148.7	4.9
400	153.7	3.97
500	137.4	5.93
600	97	5.24

FIGURE 7.6 Thermal treatment of sample 3 (blue nanopigment, ZnO NPs, and antifouling) at 100°C to 600°C for 30 min. Also, the roll-off angle = 3° is shown for the contact angle of 153.7° [22].

of around 1170 to 1200°C, low water absorption, high hardness and strength, resistance to frost, stain resistance, and high chemical resistance. This process produces tiles and ceramics with good mechanical and chemical properties. The final stage of the process, the glazing stage, is necessary to produce a very smooth and glossy surface, which cannot be achieved without polishing. One of the main characteristics of this product is its almost zero water absorption (usually less than 0.5%). Therefore, tiles have good resistance to breakage, chemical resistance, stain resistance, and frost resistance, making them suitable for use in any environment [23].

In general, there are two ways to produce tiles and ceramics: ordinary production and polished production. For ordinary tiles, the process involves applying engobe, glaze, and printing in sequence on the biscuit before firing it in a kiln. On the other hand, for polished tiles, engobe is first applied to the biscuit, followed by printing and glaze application. The product is then fired in a kiln. The focus of this work is mainly on the final stage of polished tile and ceramic production, which involves making the surface of polished tiles water-repellent and stain-resistant.

Typically, in places where beauty and quality are of greater importance, nanopolished tiles are used, which have a higher surface glossiness compared to other types of tiles, as can be seen in Figure 7.7. It is well known and undeniable that mechanical operations after firing cause a decrease in some physical properties of ceramic surfaces. In the production process of tiles and ceramics, there are many factors that create fine and coarse porosity on the surface and body of the tile, and the presence of these pores affects the final quality of the tile. High porosity in tiles increases water absorption, tile weakness, and so on. Porosity in porcelain tiles can be the result of material processing conditions (crushing raw materials, compaction, and firing), the characteristics of the

FIGURE 7.7 Comparison of the appearance of regular tiles, polished tiles, and nanopolished tiles.

raw materials used in the formulation, glaze quality, firing temperature, and so on. The wax-coated pores on the surface of the tile in the production process are revealed after polishing and finishing and are the main cause of product staining [24]. Various opinions have been presented about the dimensions and characteristics of the pores that cause tile staining. Pores with a diameter less than 10 microns do not play a role in staining, and to minimize this phenomenon, pores with diameters of 5 to 20 microns should be eliminated [25]. Although polishing the surface of the tile increases surface glossiness to a great extent, if these pores are not filled, they will be filled with dust and debris that accumulates on the tile surface, reducing its beauty. Additionally, it makes the surface permeable to spilled liquids. If liquid penetration occurs, it will cause a decrease in the shine and beauty of these products [26]. Generally, polishing the surface of tiles will also change their resistance to wear [3].

To polish ceramic tiles, a polishing machine is used, which includes discs or pads that apply pressure and special speed to the surface of the tile. This process removes a thin layer of 0.4 to 0.8 mm from the porcelain tile surface, resulting in a polished finish [27, 28]. The polishing pads of the machine contain SiC, which is released onto the tile surface along with the waste generated during the polishing process. The waste produced from polishing ceramic tiles is mostly disposed of in landfill sites, which can contribute to serious environmental pollution. Therefore, recycling the waste from the polishing process can significantly benefit the environment [29, 30]. However, it is important to note that when these materials are recycled into the ceramic tile body, SiC particles are also present in the tile composition. During firing, these particles can cause the tile body to become spongelike and create different-sized pores in the body [31].

After the polishing process, the pores on the tile surface become open. To protect the polished surface from stains, the tile is covered with an antistain material. Over time, various types of coatings have been used to cover polished ceramic tile surfaces. Silane-based materials containing organosilane (-Si-C-) groups, silanol (-Si(OH)-), and silanolate (-Si-O-) groups have become important materials for the new generation of antistain technology due to their excellent water repellency and heat resistance. However, there are different types of silicon available, and the scientific selection of the appropriate type of silicon and the preparation of an antistain material with excellent performance and easy application through appropriate processes is a fundamental issue that requires a quick solution.

One example of an antistain material that can be used on polished porcelain tile surfaces is silicon oil containing methyl hydrogen, fluorosilicon oil, and amino-silicon oil. This material can penetrate the tile surface, create a stable chemical bond, and form a long-lasting, shiny, and resistant protective layer against washing. Overall, while the process of polishing ceramic tiles can have negative environmental impacts, there are steps that can be taken to mitigate them, such as recycling waste and using appropriate antistain materials.

It's important to note that silicones do not have water-repellent properties on their own because they have a high surface energy. However, they can be made water-repellent by bonding them with water-repellent groups such as ethylene or methyl. Another option is to use a connecting agent called TEOS, which can create cross-links between polymer chains and increase adhesion between the surface and the coating [32]. The way water droplets spread on a surface depends on the surface energy and roughness. There are many ways to change the surface structure of substrates to increase their water repellency. For example, studies on natural surfaces have shown that surface energy is primarily dependent on the chemical structure on the surface and that it decreases with increasing fluorine.

7.4 ANTI-SLIP COATINGS ON CERAMIC TILE

When it comes to building construction and maintenance, safety should always be a top priority. Basic requirements need to be met to ensure that buildings are not only protected against fire but also have the necessary load-bearing capacity and stability. Additionally, operational safety is crucial, which means that buildings should be able to withstand various loads during their construction and use, without collapsing or sustaining significant damage [33].

To ensure human safety, buildings should be designed and constructed in a way that prevents accidents or damage during use, such as slips, trips, or falls. Materials used in buildings, especially those on passageways, should be nonflammable and not prone to spreading fire. Furthermore, these routes should be made of materials that do not pose a risk of slipping, tripping, or falling. This will help prevent any accidents or injuries from occurring and promote safe usage of the building [33].

Currently, ceramics are commonly used in buildings not just for constructing walls but also for finishing floors. However, the primary concern with using ceramics for flooring is to ensure that the material has the necessary qualities to prevent accidents, such as slips, trips, or falls [34]. Such accidents can lead to physical injuries, which is why it is essential to choose a floor finishing material that provides the right amount of traction and slip resistance. Moreover, the performance of the building also depends on the quality of the floor finishing material. Therefore, it is crucial to select a ceramic material that not only meets safety requirements but also contributes to the overall performance of the building [33].

In their work, researchers G. Coskun and G. Sarıışık aimed to improve pedestrian safety on floor coverings by developing a new safety slip risk scale (SSRS). To accomplish this goal, they selected commercial floor coverings used in state institutions across six zones of flooring materials, as can be seen in Table 7.1. The researchers used a portable test device, the GMG 200-Tribometer shown in Figure 7.8, to measure the dynamic coefficient of friction (DCOF) values of floor coverings in both dry and wet conditions. DCOF is a measure of the resistance between two surfaces when in contact, which determines slip risk. The testing was conducted in accordance with DIN EN 51131 standards, a European standard for measuring slip resistance on floor coverings [35].

The DCOF data was analyzed by the researchers using the K average clustering (KMC) method, which is a statistical analysis technique that groups data into clusters based on similarity. Through this method, the DCOF values of the floor coverings were classified into four clusters. Based on this analysis, the researchers developed the safety slip risk scale (SSRS), a tool designed to assess slip risk on floor coverings using DCOF values. The scale was based on the mean DCOF values and enabled the researchers to classify the flooring materials into safe usage areas.

The study found that slip risk is particularly high in wet environments, as DCOF values decrease in such conditions (Table 7.2 and Figure 7.9). The study identified which floor coverings were conditionally safe in dry environments and which ones were in the insecure class in wet environments. The study recommends the use of nonslip floor coverings in both dry and wet conditions, particularly in areas with high foot traffic. The study also suggests that polished surface floor coverings should not be used in wet environments, as they increase slip risk. In contrast, natural stones are recommended for use in dry environments, as they provide safer movement. The study's findings have implications for the selection and use of floor coverings in various settings, including state institutions, where foot traffic can be high.

TABLE 7.1
Properties of Measured Areas

State Institution Code	State Institution Type	Floor coverings used in zones					
		Z1	Z2	Z3	Z4	Z5	Z6
ST1	State agency-1	Limestone	Ceramic	PVC flooring	Limestone	Ceramic	Ceramic
ST2	University	Ceramic	Ceramic	Laminate	Ceramic	Ceramic	Ceramic
ST3	Hospital-1	Limestone	PVC flooring	Laminate	Limestone	Ceramic	Ceramic

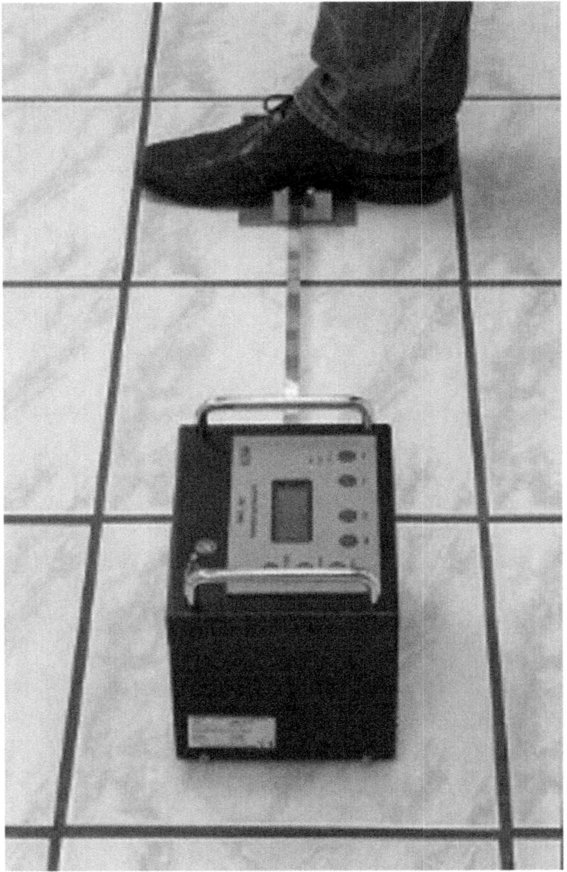

FIGURE 7.8 Measuring the dynamic friction coefficient (DCOF) values of floor coverings using the GMG 200 [35].

Overall, the study presents a new tool to improve pedestrian safety on floor coverings, and it provides valuable insights into the relationship between slip risk and DCOF values. By identifying safe usage areas and recommending nonslip floor coverings, the study can help prevent slip and fall accidents and promote safety in various settings [35].

In order to calculate the coefficient of sliding friction, it is necessary to measure the tensile force FR required to pull a test device with a known mass -m- and its normal force FN, which acts perpendicular to the direction of tension over a measuring section. The coefficient of sliding friction μ can then be computed using the formula

$$\mu = (FR)/(FN) \tag{7.1}$$

where FR represents the traction force and FN denotes the normal force or weight of the measuring device. It is noteworthy that the safety limit values are categorized into three distinct areas based on the coefficient of sliding friction, as specified in Table 7.3.

The results of the measurements taken on prewetted board surfaces, which had not undergone intentional cleaning, revealed that the cast stone slabs meet the necessary slip resistance requirements in the supermarket. The obtained μ values of 0.55 confirm this conclusion beyond doubt.

TABLE 7.2
Statistical Analysis of the DCOF According to Environmental Conditions and Zones of State Institutions

				95% CI
Slip Tests	**DV**	**M**	**SEM**	**LB, UB**
ST	ST1	0.35	0.021	[0.38, 0.39]
	ST2	0.37	0.021	[0.33, 0.41]
	ST3	0.39	0.021	[0.35, 0.43]
	ST4	0.43	0.021	[0.39, 0.47]
	ST5	0.43	0.021	[0.39, 0.47]
	ST6	0.47	0.021	[0.43, 0.51]
	ST7	0.48	0.021	[0.44, 0.53]
	ST8	0.43	0.021	[0.39, 0.47]
	ST9	0.41	0.021	[0.37, 0.45]
	ST10	0.53	0.021	[0.49, 0.57]
	ST11	0.50	0.021	[0.46, 0.54]
	ST12	0.27	0.021	[0.23, 0.31]
	ST13	0.20	0.021	[0.16, 0.25]
	ST14	0.24	0.021	[0.20, 0.28]
	ST15	0.19	0.021	[0.15, 0.23]
Z	Z1	0.40	0.013	[0.37, 0.43]
	Z2	0.38	0.013	[0.36, 0.41]
	Z3	0.38	0.013	[0.36, 0.41]
	Z4	0.38	0.013	[0.35, 0.40]
	Z5	0.37	0.013	[0.34, 0.40]
	Z6	0.36	0.013	[0.34, 0.39]
EC	DE	0.46	0.008	[0.44, 0.47]
	WE	0.30	0.008	[0.29, 0.32]

Note: DV = dependent variable, M = mean, LB = lower bound, UB = upper bound, ST = state institutions, SEM = standard error of the mean, Z = zones, EC = environmental conditions.

The process for measuring slip resistance in work areas and traffic routes is standardized according to the DIN 51130 standard. The method involves the use of an inclined plane and must be carried out by an independent testing institute.

The procedure consists of the following steps: two individuals wearing standardized shoes walk on the floor covering to be tested for three rounds, while motor lubricating oil 10W30 is applied as a lubricant. An inspector then walks continuously in small steps both forward and backward on the area, incrementally increasing the inclination angle of the plane until the individual can no longer walk safely. The resulting average acceptance angle derived from a series of measurements is used to classify the floor covering into the appropriate slip resistance class. It should be noted that the classification of the slip resistance is critical in ensuring that work areas and traffic routes meet the required safety standards. Therefore, precise and accurate measurement procedures are essential, which require the involvement of specialized testing institutes with the necessary expertise in carrying out the tests. The slip resistance classes (R groups) are shown in Table 7.4.

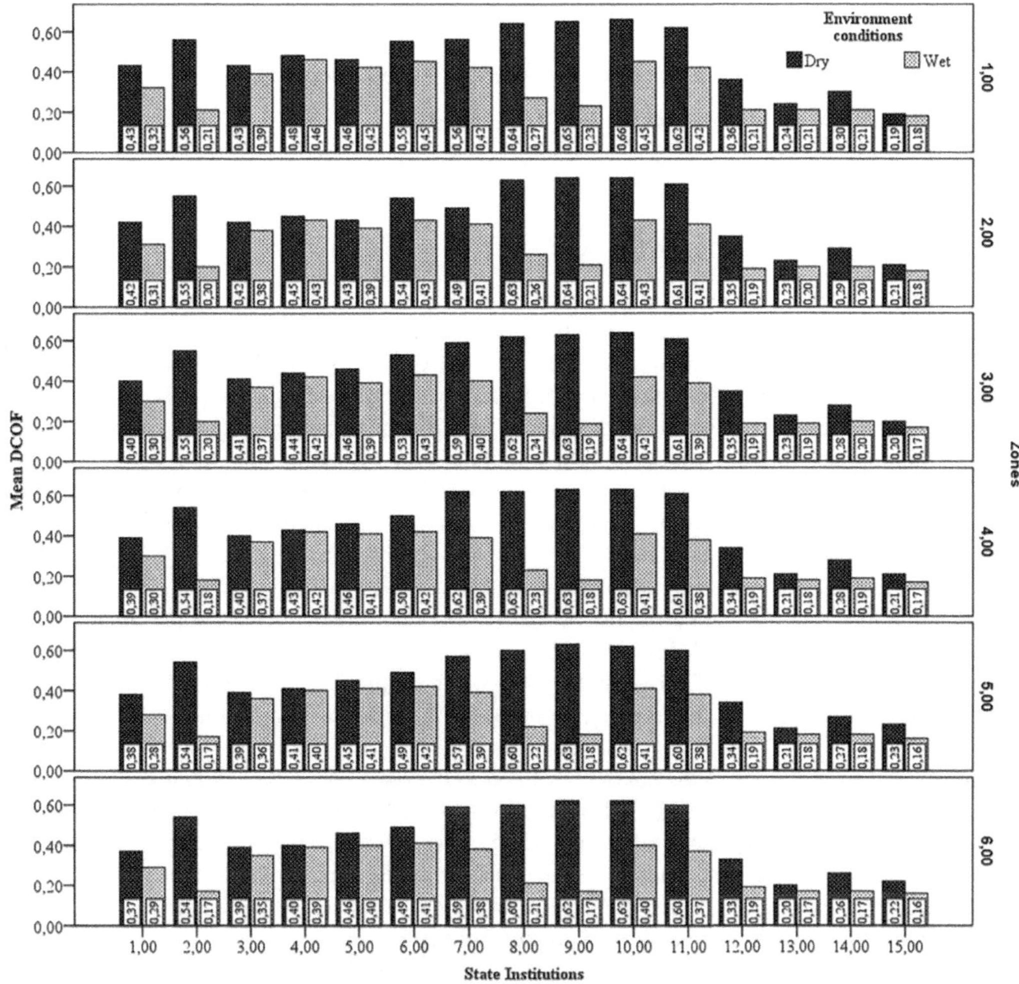

FIGURE 7.9 Mean DCOF (μ) values obtained in different state institutions and zones in dry and wet environments [35].

TABLE 7.3
Evaluation of the Measurement Results for the Sliding Friction Coefficient

Sliding Friction Coefficient (μ)	Slip Resistance Potential	Safety Area
$\mu < 0.30$	Even under ideal operating conditions, acute risk of slipping. The slip resistance potential is inadequate.	Unsafe (critical)
$0.30 \leq \mu < 0.45$	Sufficient for certain operating conditions. Regular control measurements are required.	Conditionally safe
$0.45 \leq \mu$	Sufficient slip resistance	Safe

TABLE 7.4
Slip Resistance Class in Terms of the Average Acceptance Angle

Average Overall Acceptance Angle	6°–10°	10°–19°	19°–27°	27°–35°	> 35°
Slip resistance class (R group)	R9	R10	R11	R12	R13

7.5 COOL COATINGS ON CERAMIC TILE

Besides aesthetic color characteristics, ceramic tile with a cool-coating surface is a fascinating topic academically and industrially. Solar radiation, particularly near-infrared radiation, accounts for around half of the solar spectrum. As a result, surfaces exposed to sunlight absorb radiation, which elevates their temperature. This causes a significant amount of heat to accumulate on the surface through conduction. To cool down building interiors, air conditioning is necessary. However, cool roofs are a more eco-friendly option for reducing the need for air conditioning. By reflecting most of the near-infrared radiation, cool roofs minimize heat absorption, thus cooling the surface during hot seasons. Pigments composed of Zn, Ti, and Al elements, which have high reflection properties, show great promise in this respect [36].

S. K. Fatah et al. [22] coated ceramic tiles using blue $CoAl_2O_4$ nanopigments, ZnO nanoparticles (NPs), and antifouling through air spray in their work. The coated surface was evaluated for its cooling performance using UV-Vis-NIR reflectance spectroscopy, which measures the reflectance of light across a broad range of wavelengths as shown in Figure 7.10. The results indicated a desirable coolant performance of the coating, as the total reflectance from 700 nm to 2500 nm was almost 64%.

However, the reflectance of the coating declined in the range of 1200 to 1700 nm, which can be attributed to the electronic transition of the Co ions in the blue phase nanopigments. Despite the demand for cool coating pigments other than white, other dark color pigments like blue, green, and brown tend to have a high absorption rather than reflection. But the developed blue phase nanopigment coating has a high reflectance of 64% due to its ZnO [37] content and surface modification, making it an attractive alternative for cool coating pigments.

The researchers observed that the blue nanopigments had a porous surface structure that enhanced their coolant property through diffuse reflection. This means that the porous structure in the micro-nano structure of the coating allowed for more light to be reflected, resulting in a higher

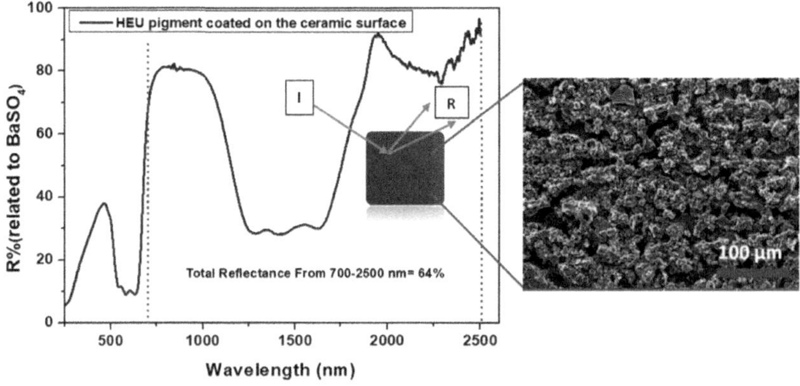

FIGURE 7.10 UV-visible and NIR reflectance of the coated ceramic tile surface encompassing blue $CoAl_2O_4$ nanopigments, ZnO NPs, and antifouling. The insert picture is the blue surface with a scheme of the incident and reflected lights and the FE-SEM image of the coated ceramic surface [22]. (*Note*: the HEU in the fig. stands for Heulandite zeolite that has been used in the nanopigments.)

reflectance. The researchers also used field-emission scanning electron microscopy (FE-SEM) to visualize the coated surface, which confirmed the presence of a porous surface structure.

Additionally, the blue nanopigment coating offered dual multifunctional properties, serving as a coolant and a superhydrophobic coating. The superhydrophobic properties make the coating repel water, which can reduce the risk of water damage and improve the durability of the ceramic tiles.

In our research [36], we have developed a blue pigment powder suitable for coating the surface of ceramic tiles, providing them with desirable cool-coating properties. The experimental procedure comprises a series of steps. Firstly, a precise mixture is created by combining 0.33 grams of zinc oxide and 1 gram of cobalt chloride with 1 gram of clinoptilolite zeolite in an aqueous solution adjusted to pH 5.5. Subsequently, an ion exchange and precipitation process is employed to facilitate the deposition of cobalt ions and zinc oxide within the pores and hollow channels of the zeolite structure. This process requires approximately 24 hours to complete. Following the ion exchange, the resulting powders are dried at a temperature of 90°C for a duration of 16 hours. The dried powders are then subjected to calcination at a temperature of 1000°C for a period of 3 hours. The resulting sample is subsequently applied as a coating on the ceramic tile surface using the third fire coating or decoration method [38]. This involves utilizing a sieve with a mesh size of 45 μm to ensure an even distribution of the prepared blue pigments over the ceramic surface, which measures 3 cm × 3 cm.

Upon the application of the coating, the ceramic tile is subjected to heating at a temperature of 1050°C for a duration of 40 minutes, followed by a soaking time of 30 minutes. To evaluate the performance of the coated surface, a double-beam spectrophotometer is employed to measure the reflectance across a range of wavelengths from 250 to 2500 nm. The obtained reflectance values are then compared to those of a reference white ceramic tile. The reflectance characteristics of the coated surface, as presented in Figure 7.11, demonstrate high reflectivity within the near-infrared (NIR) region, with values exceeding 0.6. However, a slight decrease in reflectance is observed within the 1200 to 1700 nm range, which can be attributed to the Co-electronic transition occurring within the sample. The results affirm the suitability of the coated pigment as an effective cool-coating material. The concept of cool coatings revolves around the utilization of highly reflective materials that minimize the absorption of solar radiation, thereby facilitating a cooler surface temperature [4]. This has led to an increased demand for cool-coating colors beyond traditional white coatings, with blue being a particularly sought-after choice due to its aesthetic appeal for wall and roof applications in residential settings. Conventional blue color pigments often lack the necessary cool-coating properties. However, our developed pigment, owing to the composite phases

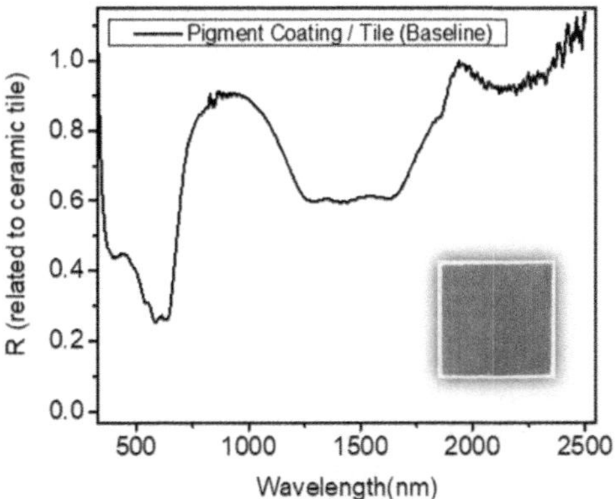

FIGURE 7.11 The UV-visible and near-infrared reflectance of a ceramic tile coated with prepared blue pigment in comparison to a white ceramic tile used as a reference. The reflected curve is adjusted to the baseline of the white ceramic tile surface that is not coated. The accompanying image shows the surface after the coating is applied [36].

of Co-willemite and spinel, along with the presence of sufficient Zn and Al in the sample, exhibits notable reflectivity within the visible and NIR regions.

In summary, our research endeavors have successfully yielded a blue pigment powder suitable for cool-coating applications on ceramic tiles. The comprehensive experimental procedure, coupled with the analysis of reflectance characteristics, substantiates the efficacy of the coated pigment as a cool-coating material, capable of providing the desired reflective properties for temperature regulation.

7.6 SUMMARY

Ceramic tiles serve both decorative and functional roles in buildings due to their durability and low maintenance. With the increasing demand for energy-efficient and sustainable buildings, the focus is on enhancing ceramic tiles beyond aesthetics by adding properties such as color variety, self-cleaning capabilities (hydrophilicity, hydrophobicity, nanopolishing), slip resistance, and cooling abilities. To meet these demands, multifunctional coatings have been developed to provide these features to ceramic tiles. Colorful pigments and TiO_2 coatings offer customizable designs and self-cleaning hydrophilic surfaces. Hydrophobic coatings prevent staining, while nanopolished coatings enhance aesthetics and ease of cleaning. Anti-slip coatings enhance safety, while cool coatings reflect sunlight, reducing tile and environmental temperatures and contributing to energy conservation.

In line with these advancements, researchers have turned to nanoparticles to create vibrant coatings with pearlescent effects, utilizing copper, silver nanocrystals, and titanium oxide layers. Ceramic nanopigments show potential for rich colors across firing temperatures, and ceramic nano-inks offer promise in inkjet printing for decoration, surpassing issues tied to micronized pigments. Photocatalyst coatings using TiO_2 have shown promise in degrading pollutants, and research into superhydrophobic and superhydrophilic surfaces is progressing, influencing tile applications. Nanopolished coatings enhance ceramic tile aesthetics and properties, including hydrophobic and superhydrophobic surfaces. Ensuring safety in building construction and usage is pivotal, and a safety slip risk scale (SSRS) aids slip risk assessment. Cool-coating surfaces using Zn, Ti, and Al elements aim to reduce heat absorption, with blue $CoAl_2O_4$ nanopigments showcasing cooling and superhydrophobic properties. A developed blue pigment powder offers highly reflective cool-coating applications for ceramic tiles.

As a culmination of these efforts, our own research has contributed to the development of multifunctional coatings that revolutionize ceramic tile properties. The integration of vibrant pigments, self-cleaning hydrophilic coatings, and advanced nanopolishing techniques has elevated both the aesthetic appeal and functional utility of ceramic tiles. Additionally, our exploration into photocatalyst coatings utilizing TiO_2 and the advancement of anti-slip solutions underscore the commitment to safety and practicality in building design. Moreover, our innovative approach to creating cool-coating surfaces using blue pigments has paved the way for enhanced temperature regulation and energy conservation. Through these multifaceted developments, we have not only summarized the current state of the field but also made strides toward shaping the future of ceramic tile technology.

ACKNOWLEDGMENTS

The authors wish to express their gratitude to Mr. Dehghan Banadaki, Mr. Abdi, and Mr. Vaselnia for their valuable contributions to the preparation of this chapter. Their assistance and support have been instrumental in the completion of this work.

REFERENCES

1. P.M.T. Cavalcante, M. Dondi, G. Guarini, M. Raimondo, G. Baldi, Colour performance of ceramic nano-pigments, Dye. Pigment. 80 (2009) 226–232. https://doi.org/10.1016/j.dyepig.2008.07.004.

2. J. Määttä, M. Piispanen, R. Kuisma, H.R. Kymäläinen, A. Uusi-Rauva, K.R. Hurme, S. Areva, A.M. Sjöberg, L. Hupa, Effect of coating on cleanability of glazed surfaces, J. Eur. Ceram. Soc. 27 (2007) 4555–4560. https://doi.org/10.1016/j.jeurceramsoc.2007.02.204.
3. A. Tucci, L. Esposito, L. Malmusi, A. Piccinini, Wear resistance and stain resistance of porcelain stoneware tiles, Key Eng. Mater. 213 (2001) 1759–1762. https://doi.org/10.4028/www.scientific.net/kem.206-213.1759.
4. L.M. Schabbach, D.L. Marinoski, S. Güths, A.M. Bernardin, M.C. Fredel, Pigmented glazed ceramic roof tiles in Brazil: Thermal and optical properties related to solar reflectance index, Sol. Energy. 159 (2018) 113–124. https://doi.org/10.1016/j.solener.2017.10.076.
5. Y. Chen, J. Mandal, W. Li, A. Smith-Washington, C.C. Tsai, W. Huang, S. Shrestha, N. Yu, R.P.S. Han, A. Cao, Y. Yang, Colored and paintable bilayer coatings with high solar-infrared reflectance for efficient cooling, Sci. Adv. 6 (2020) 1–9. https://doi.org/10.1126/sciadv.aaz5413.
6. M. Llusar, C. Gargori, S. Cerro, J.A. Badenes, G. Monrós, New ceramic pigments for the coloration of ceramic glazes, Adv. Sci. Technol. 92 (2014) 148–158. https://doi.org/10.4028/www.scientific.net/ast.92.148.
7. S.Y. Vaselnia, M. Khajeh Aminian, H. Motahari, Fe-doped titanite pigment: Synthesis, DFT/TDDFT calculations by Lanczos and Bethe-Salpeter equation methods and comparison of computational and experimental color properties, J. Phys. Chem. Solids. 138 (2020). https://doi.org/10.1016/j.jpcs.2019.109244.
8. S.Y. Vaselnia, M. Khajeh Aminian, R.D. Banadaki, Experimental and theoretical study on the structural, electronic, and optical properties within DFT+U, Fxc kernel for LRC model, and BSE approaches. Part I: CoTiO3 and Co2TiO4 pigments, Powder Technol. 390 (2021) 50–61. https://doi.org/10.1016/j.powtec.2021.05.070.
9. A. Babaei Darani, M. Khajeh Aminian, H. Zare, Synthesis and characterization of two green nanopigments based on chromium oxide, Prog, Color. Color. Coatings. 10 (2017) 141–148.
10. H. Heydari, M. Yousefpour, E. Emadoddin, M.H. Zori, M.K. Aminian, Effect of solvents and dispersants on polyol synthesis of V–ZrSiO4 nanopigment stable suspension for ink application, J. Coatings Technol. Res. 17 (2020) 1243–1253. https://doi.org/10.1007/s11998-020-00343-2.
11. H. Heydari, M. Yousefpour, E. Emadoddin, M.H. Zori, M.K. Aminian, Microwave-assisted polyol synthesis of V–ZrSiO4 nanoparticles and its use as a blue ceramic pigment, J. Coatings Technol. Res. 19 (2022) 1595–1607. https://doi.org/10.1007/s11998-022-00632-y.
12. A. Shakeri, D. Yip, M. Badv, S.M. Imani, M. Sanjari, T.F. Didar, Self-cleaning ceramic tiles produced via stable coating of TiO2 Nanoparticles, Materials (Basel). 11 (2018) 1–16. https://doi.org/10.3390/ma11061003.
13. S. De Niederhãusern, M. Bondi, F. Bondioli, Self-cleaning and antibacteric ceramic tile surface, Int. J. Appl. Ceram. Technol. 10 (2013) 949–956. https://doi.org/10.1111/j.1744-7402.2012.02801.x.
14. M. Khajeh Aminian, F. Sajadi, M.R. Mohammadizadeh, S. Fatah, Hydrophilic and photocatalytic properties of TiO2/SiO2 Nano-layers in dry weather, Prog, Color. Color. Coatings. 14 (2021) 221–232.
15. M. Khajeh Aminian, S.K. Fatah, Loading of alkaline hydroxide nanoparticles on the surface of Fe2O3 for the promotion of photocatalytic activity, J. Photochem. Photobiol. A Chem. 373 (2019) 87–93. https://doi.org/10.1016/j.jphotochem.2019.01.003.
16. Y. Cai, T.W. Coyle, G. Azimi, J. Mostaghimi, Superhydrophobic ceramic coatings by solution precursor plasma spray, Sci. Rep. 6 (2016) 1–7. https://doi.org/10.1038/srep24670.
17. Y.Y. Yan, N. Gao, W. Barthlott, Mimicking natural superhydrophobic surfaces and grasping the wetting process: A review on recent progress in preparing superhydrophobic surfaces, Adv. Colloid Interface Sci. 169 (2011) 80–105. https://doi.org/10.1016/j.cis.2011.08.005.
18. G. Acikbas, N. Calis Acikbas, The effect of sintering regime on superhydrophobicity of silicon nitride modified ceramic surfaces, J. Asian Ceram. Soc. 9 (2021) 734–744. https://doi.org/10.1080/21870764.2021.1915563.
19. T. Kamegawa, Y. Shimizu, H. Yamashita, Superhydrophobic surfaces with photocatalytic self-cleaning properties by nanocomposite coating of TiO_2 and polytetrafluoroethylene, Adv. Mater. 24 (2012) 3697–3700. https://doi.org/10.1002/adma.201201037.
20. I. Cacciotti, F. Nanni, V. Campaniello, F.R. Lamastra, Development of a transparent hydrorepellent modified SiO2 coatings for glazed sanitarywares, Mater. Chem. Phys. 146 (2014) 240–252. https://doi.org/10.1016/j.matchemphys.2014.03.005.
21. J.J. Reinosa, J.J. Romero, M.A. De La Rubia, A. Del Campo, J.F. Fernández, Inorganic hydrophobic coatings: Surfaces mimicking the nature, Ceram. Int. 39 (2013) 2489–2495. https://doi.org/10.1016/j.ceramint.2012.09.007.

22. S.K. Fatah, M. Khajeh Aminian, M. Bahamirian, Multifunctional superhydrophobic and cool coating surfaces of the blue ceramic nanopigments based on the heulandite zeolite, Ceram. Int. 48 (2022) 21954–21966. https://doi.org/10.1016/j.ceramint.2022.04.178.
23. I.M. Hutchings, Y. Xu, E. Sánchez, M.J. Ibáñez, M.F. Quereda, Porcelain tile microstructure: Implications for polishability, J. Eur. Ceram. Soc. 26 (2006) 1035–1042. https://doi.org/10.1016/j.jeurceramsoc.2004.12.019.
24. H.J. Alves, F.G. Melchiades, A.O. Boschi, Effect of spray-dried powder granulometry on the porous microstructure of polished porcelain tile, J. Eur. Ceram. Soc. 30 (2010) 1259–1265. https://doi.org/10.1016/j.jeurceramsoc.2009.11.018.
25. M. Romero, J.M. Pérez, Relation between the microstructure and technological properties of porcelain stoneware. A review, Mater. Constr. 65 (2015). https://doi.org/10.3989/mc.2015.05915.
26. H.J. Alves, M.R. Freitas, F.G. Melchiades, A.O. Boschi, Dependence of surface porosity on the polishing depth of porcelain stoneware tiles, J. Eur. Ceram. Soc. 31 (2011) 665–671. https://doi.org/10.1016/j.jeurceramsoc.2010.11.028.
27. I.M. Hutchings, Y. Xu, E. Sánchez, M.J. Ibáñez, M.F. Quereda, Development of surface finish during the polishing of porcelain ceramic tiles, J. Mater. Sci. 40 (2005) 37–42. https://doi.org/10.1007/s10853-005-5684-3.
28. I.M. Hutchings, K. Adachi, Y. Xu, E. Sánchez, M.J. Ibáñez, M.F. Quereda, Analysis and laboratory simulation of an industrial polishing process for porcelain ceramic tiles, J. Eur. Ceram. Soc. 25 (2005) 3151–3156. https://doi.org/10.1016/j.jeurceramsoc.2004.07.005.
29. R. De'Gennaro, S.F. Graziano, P. Cappelletti, A. Colella, M. Dondi, A. Langella, M. De'Gennaro, Structural concretes with waste-based lightweight aggregates: From landfill to engineered materials, Environ. Sci. Technol. 43 (2009) 7123–7129. https://doi.org/10.1021/es9012257.
30. R. de Gennaro, A. Langella, M. D'Amore, M. Dondi, A. Colella, P. Cappelletti, M. de' Gennaro, Use of zeolite-rich rocks and waste materials for the production of structural lightweight concretes, Appl. Clay Sci. 41 (2008) 61–72. https://doi.org/10.1016/j.clay.2007.09.008.
31. A. Shui, X. Xi, Y. Wang, X. Cheng, Effect of silicon carbide additive on microstructure and properties of porcelain ceramics, Ceram. Int. 37 (2011) 1557–1562. https://doi.org/10.1016/j.ceramint.2011.01.026.
32. S.J. Wang, X.D. Fan, Q.F. Si, J. Kong, Y.Y. Liu, W.Q. Qiao, G. Bin Zhang, Preparation and characterization of a hyperbranched polyethoxysiloxane based anti-fouling coating, J. Appl. Polym. Sci. 102 (2006) 5818–5824. https://doi.org/10.1002/app.24842.
33. E. Sudoł, What makes a floor slippery? A brief experimental study of ceramic tiles slip resistance depending on their properties and surface conditions, Materials (Basel). (2021). https://doi.org/10.3390/ma14227064.
34. A. Terjék, A. Dudás, Ceramic floor slipperiness classification – A new approach for assessing slip resistance of ceramic tiles, Constr. Build. Mater. 164 (2018) 809–819. https://doi.org/10.1016/j.conbuildmat.2017.12.242.
35. G. Coşkun, G. Sarıışık, Analysis of slip safety risk by portable floor slipperiness tester in state institutions, J. Build. Eng. 27 (2020). https://doi.org/10.1016/j.jobe.2019.100953.
36. S. Fatah, M. Khajeh Aminian, Cobalt-willemite and spinel as fractal blue nanopigments based on clinoptilolite zeolite: Synthesis, physical properties, and cool coating application, J. Solid State Chem. 307 (2022) 122753. https://doi.org/10.1016/j.jssc.2021.122753.
37. S.K. Fatah, Synthesis and characterisation of zinc oxide nanopowders prepared by precipitation method, Diyala J. Pure Sci. 14 (3) (2018) 40–47. http://dx.doi.org/10.24237/djps.1403.406A.
38. A. Campanile, B. Liguori, O. Marino, G. Cavaliere, V.L. De Bartolomeis, D. Caputo, Facile synthesis of nanostructured Cobalt pigments by Co- A zeolite thermal conversion and its application in porcelain manufacture, Sci. Rep. 10 (2020) 1–9. https://doi.org/10.1038/s41598-020-67282-1.

8 Antiviral Thin Surfaces and Surface Life

Elmira Khanmamadova, Rashad Abaszade, and Rasoul Moradi

8.1 INTRODUCTION

In recent years, the use of antiviral thin surfaces has gained significant importance in preventing the spread of infections and ensuring a hygienic environment. These thin surfaces aim to minimize the risk of infection by reducing the survival time of viruses on surfaces [1, 2]. Antiviral thin surfaces are defined as protective coatings or materials applied to surfaces to provide effective protection against viruses and to enhance hygiene standards. Viruses can survive for extended periods on various surfaces and can be contagious [3–5]. Especially in shared areas, the persistence of viruses on surfaces can increase the risk of infection. Therefore, the use of antiviral thin surfaces emerges as an effective strategy to prevent the spread of infections by rendering viruses ineffective when they come into contact with surfaces [6–9].

Antiviral thin surfaces have a wide range of applications in hospitals, clinics, laboratories, corporate environments, communal spaces, transportation hubs such as bus and train stations, airports, and commercial establishments like shopping malls [10–12]. Their utility is especially pronounced in places characterized by elevated rates of contact between people and a heightened susceptibility to infectious agents [13].

Antiviral thin surfaces typically contain components that are effective against viruses. These components prevent viral replication or render viruses ineffective, thereby reducing the risk of infection. Additionally, the use of antiviral thin surfaces yields more effective results when combined with regular cleaning and hygiene practices [14–17].

8.2 DIVERSITY OF ANTIVIRAL FORMULATIONS AND SCOPE OF EXPERIMENTS

Surface characterization: Various characterization methods can be used to evaluate the physical properties of antiviral thin surfaces (Figure 8.1).

These characterization methods help assess the effectiveness and quality of antiviral thin surfaces, contributing to the development and optimization of their formulations for various applications [18–20].

8.2.1 Scanning Electron Microscopy (SEM):

The methodology employed entails the investigation of surface morphology and topography. Scanning electron microscopy (SEM) serves as the primary means to acquire images of exceptional resolution, offering intricate insights into the attributes of surface roughness, structural composition, and various pertinent physical characteristics [21–25].

The working principle of SEM is to send a focused electron beam onto the surface. As a result of electron interactions on the surface, backscattered, scattered, and secondary electrons are produced. These electron-generated signals are detected by detectors, and the imaging process is performed. SEM visualizes the surface's roughness, porosity, perforations or cracks, surface layers, and other structural features in detail. It also allows analysis of parameters such as coating thickness, surface

DOI: 10.1201/9781032635347-8

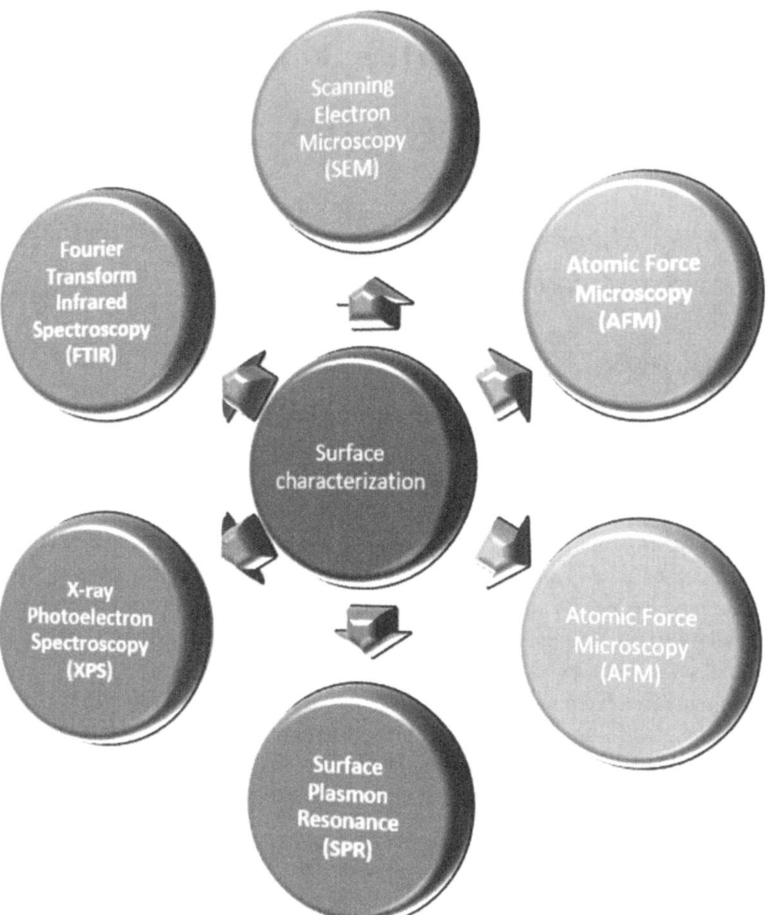

FIGURE 8.1 Methods to help evaluate the efficacy and quality of antiviral thin surfaces.

structure, particle distribution, and more. SEM is a commonly used method for characterizing antiviral thin surfaces. This technique assists in evaluating factors such as the adhesion of coatings to the surface, coating thickness, and determination of microstructures on the surface. The high-resolution images obtained from SEM provide valuable information in the design and development process of antiviral coatings [26–30].

8.2.2 Atomic Force Microscopy (AFM):

This is a method used to examine surface topography at the atomic scale. AFM measures the surface's roughness, height profile, and other nanoscale details by scanning a cantilever with a sharp tip over the surface. The working principle of AFM relies on the interaction between a scanning probe at the tip of the cantilever and the surface [31]. The probe at the tip interacts with the surface through atomic forces. These forces are detected by a detector connected to the tip, and an image representing the topographical features of the surface is generated. AFM visualizes the surface's roughness, perforations or cracks, surface layers, nanoparticle or nanoscale structure distributions, and other nanoscale details with high resolution [32]. It is also used to evaluate the surface's mechanical properties, surface reactivity, and coating thickness. AFM is a widely used method for characterizing antiviral thin surfaces. This technique helps in evaluating factors such as the adhesion of coatings to the surface, coating thickness, roughness, and determination of nanostructures on the surface [33, 34]. The high-resolution images obtained from AFM provide significant insights into the design and analysis of antiviral coatings' performance [35].

8.2.3 Surface Plasmon Resonance (SPR):

This technique is used to analyze the interactions of biomolecules on a surface [36]. This method enables the detection of chemical interactions on the surface using surface plasmon polaritons. The working principle of SPR is based on the optical properties of surface plasmon polaritons occurring at the interface between a metal layer and the surrounding environment. Plasmonic resonance occurs where electrons accumulate in the transition region between the metal layer and the environment [37–39]. This resonance is affected by changes caused by chemical or biochemical interactions on the surface. SPR is used to monitor the binding of biochemically or chemically labeled target molecules to a sensor chip placed in an optical setup. The sensor chip is coated with bioreceptors with which the target molecules interact. The plasmon resonance, along with changes caused by surface interactions, is detected by a detector in the sensor chip. SPR allows real-time monitoring of binding events on the surface, making it useful for evaluating the performance of antiviral coatings and detecting virus binding or interactions on the surface. It is also a useful tool for assessing ligand–receptor interactions, surface reactivity, and coating effectiveness [40–42]. SPR is a commonly used method for characterizing antiviral thin surfaces. This technique assists in analyzing chemical interactions on the surface, determining the performance of coatings, and evaluating ligand–receptor interactions. The real-time and label-free monitoring capability of SPR makes it an important tool in the design and development of antiviral coatings.

8.2.4 X-Ray Photoelectron Spectroscopy (XPS):

This method is used to analyze the chemical composition of a surface. It utilizes X-ray radiation to analyze the elements and chemical bonds on the surface. The working principle of XPS starts with sending X-rays to the surface. X-rays remove photoelectrons from the surface's atoms. These photoelectrons carry information about the energy levels and kinetic energies at which they were emitted. XPS determines the presence and quantity of elements and chemical bonds on the surface by analyzing these spectral data. XPS analyzes the chemical bond states of surface elements, surface composition, and ratios of chemical components [37]. It can also evaluate coating thickness, surface oxidation state, surface functionality, and layer structures. XPS is commonly used as a method for characterizing antiviral thin surfaces. This technique is used to determine the chemical composition of coatings, detect the presence of functional groups, control coating thickness, and analyze structural features. XPS is an important tool in the design and analysis of antiviral coatings' performance because it can determine the surface's chemical composition and molecular structure with high precision. Therefore, it is a commonly preferred technique for the analysis of antiviral thin surfaces.

8.2.5 Fourier Transform Infrared Spectroscopy (FTIR):

This technique is used to analyze the chemical composition and molecular structure of a surface. It measures the vibrations of chemical bonds on the surface using infrared radiation [11, 27].

The working principle of FTIR is based on the interaction of infrared radiation with the chemical components on the surface. Chemical bonds vibrate at specific frequencies, and these vibrations appear as characteristic peaks in the infrared spectrum [12]. FTIR determines the type and quantity of chemical bonds on the surface by analyzing this spectral information. FTIR is used to analyze functional groups, chemical bonds, polymer structures, oxidation states, and coating thickness on the surface. It can also assess the presence and interactions of biomolecules on the surface [13, 24].

FTIR is a commonly used method for characterizing antiviral thin surfaces. This technique is used to determine the chemical composition of coatings, detect the presence of functional groups, control coating thickness, and analyze structural features. FTIR is an important tool in the design and performance analysis of antiviral coatings. It is a characterization method that can determine the chemical composition and molecular structure of the surface rapidly and accurately, making it widely preferred for the analysis of antiviral thin surfaces.

8.3 IMPORTANCE OF VIRUS SURVIVAL TIME

The use of surfaces with antiviral properties can reduce the risk of infection by decreasing the survival time of viruses on surfaces. Numerous scientific studies have shown that surfaces with antiviral properties can affect the survival time of viruses on them. The application of antiviral components to surfaces or their use in surface coatings can prevent viral replication or render viruses ineffective on surfaces. Table 8.1 outlines the most commonly used methods for determining the antiviral properties and surface life of antiviral thin surfaces.

In evaluating the antiviral efficacy and longevity of antiviral coatings on surfaces, specific methodologies are employed to assess their capacity to impede viral replication, diminish viral titers, and obstruct viral transmission. Such methodologies furnish critical insights into the operational efficiency of antiviral materials or coatings in mitigating infection risk and curtailing viral dissemination. The selection of a particular methodology is contingent upon the intrinsic properties of the antiviral agent, the objectives of the assessment, and the envisaged application contexts.

Surfaces with antiviral properties typically interact with viruses that come into contact with the surfaces. The interaction has the potential to perturb the conformational integrity of viral entities, hinder their cellular ingress, or incapacitate the efficacy of their replication mechanisms. Consequently, this leads to a diminution in the duration of virus viability on external interfaces, thus mitigating the probability of viral transmission.

The antiviral potency of a given interface can be modulated by factors such as the viral genus, physicochemical attributes of the surface, and prevailing environmental conditions. Specific antiviral constituents might demonstrate heightened efficacy against particular viral strains. Moreover, the composition and topographical characteristics of the interfaces can significantly influence their antiviral capabilities. Ambient conditions including relative humidity, thermal fluctuations, and

TABLE 8.1
Methods Used to Determine the Duration of Action of Antiviral Thin Surfaces

Method	Description
Viral plaque assay	This method involves infecting cultured cells with a known amount of virus and measuring the formation of plaques (visible cell death).
Viral cytopathic effect (CPE) assay	It assesses the extent of virus-induced cell damage or death on a monolayer of susceptible cells.
Time-kill assay	It determines the rate at which the virus is inactivated or loses its infectivity on the surface.
Viral load reduction assay	It quantifies the reduction in the number of infectious viral particles after exposure to the antiviral surface.
Surface stability assay	This method evaluates the stability and integrity of the antiviral coating or material over time on the surface.
Contact transfer assay	It assesses the ability of the antiviral surface to prevent the transfer of infectious virus to another surface or host.
Antiviral efficacy testing	It involves exposing the antiviral surface to a specific virus and measuring its effectiveness in reducing viral activity or viability.

light intensity may further modulate this antiviral efficacy. Ensuring that surfaces possess antiviral attributes is paramount for efficacious infection mitigation.

1. Antiviral thin surfaces allow for application in various forms and on different types of surfaces. Each method has its own unique advantages and applications. The appropriate application method is important for the effectiveness and durability of antiviral thin surfaces. Surfaces with antiviral properties are typically applied in the form of sprays, coatings, or films. These types of antiviral products are used to prevent the adherence and spread of viruses on surfaces or to induce virus inactivation. They are commonly preferred in health care facilities, laboratories, public areas, and other settings where hygiene is crucial.

The spray application method involves the use of antiviral sprays containing chemicals designed to render viruses ineffective when sprayed onto surfaces. These sprays can typically contain alcohol-based disinfectants or solutions containing specific antiviral components. When applied to surfaces, they work by disrupting the structure of viruses or killing them.

The coating application method involves directly applying antiviral coatings to surfaces. These coatings can be applied by spreading or brushing them onto the surface (Figure 8.2). The coating method can be used to provide antiviral protection to the entire surface or specific areas.

Antiviral thin surfaces are commonly applied to surfaces in the form of sprays, coatings, or films. (Figure 8.2)

The film application method involves the application of antiviral films to surfaces. These films can be placed on the surface or adhered to the surface through their adhesive properties. Antiviral films are preferred in areas where surfaces need to be covered or protected [43].

Antiviral coatings or films, when applied to surfaces, create a protective layer and prevent viruses from adhering to the surface. These coatings are typically polymer or nanotechnology based. Antiviral coatings can shorten the lifespan of viruses on surfaces or completely eliminate them. Antiviral coatings usually require a drying or activation process after application. These processes enable the antiviral components to become effective and reveal the protective properties of the coatings against viruses.

The drying process occurs when the moisture on the surface where the antiviral coating is applied evaporates and the coating material fully hardens. During this time, the antiviral components form a film on the surface and start interacting with viruses. The drying time can vary depending on the type and thickness of the applied coating material and environmental factors. It is important to follow the instructions provided by the manufacturer.

Activation of the antiviral coating can be achieved through multiple modalities, including photoactivation, thermal induction, hydration, or specific chemical reactions. Once activated, the coating facilitates the interaction of its antiviral constituents with viruses present on the substrate,

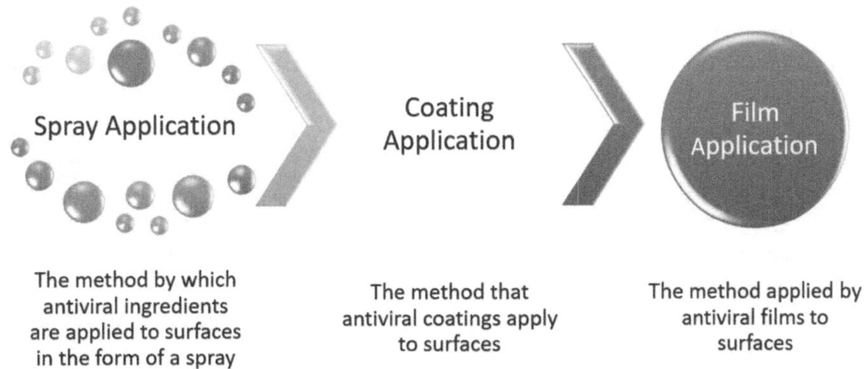

FIGURE 8.2 Antiviral thin surfaces are commonly applied in the form of sprays, coatings, or films.

TABLE 8.2
Antiviral Coating Drying or Activation Times, Application Methods, and Chemical Formulations

Duration	Method of Drying/ Activation Process	Material Type	Application Method	Chemical Formula
1 hour	Natural evaporation of moisture from the surface	Polymer-based coating	Spray, roller, brush, etc.	($C_{12}H_{22}O_3$)
2 hour	Natural evaporation of moisture from the surface	Ceramic-based coating	Dipping, spraying, brushing, etc.	(SiO_2)
6 hour	Natural evaporation of moisture from the surface	Nanocomposite coating	Dipping, spraying, brushing etc.	($TiO_2 + SiO_2$)
24 hour	Natural evaporation of moisture from the surface	Silicone-based coating	Spraying, dipping, roller, etc.	($CH_3[(CH_2)_3SiO]_3Si(CH_3)_3$)
1 month	Natural evaporation of moisture from the surface	Organic compound–based coating	Spraying, dipping, brushing, etc.	($C_{18}H_{38}NO_3$)
1 year	Natural evaporation of moisture from the surface	Metal-based coating	Spraying, dipping, brushing, etc.	(Ag- or Cu-based nanomaterials)
2 year	Natural evaporation of moisture from the surface	Glass-based coating	Spraying, dipping, roller, etc.	($SiO_2 + Na_2O + CaO$)
5 year	Natural evaporation of moisture from the surface	Polyurethane-based coating	Roll, spray, dip, etc.	($C_{12}H_{20}N_2O_2$)

subsequently neutralizing their activity [36]. These drying and activation processes enhance the effectiveness and durability of antiviral coatings. After these processes are completed, the coating forms a protective barrier on the surface and helps prevent the adherence or spread of viruses.

Table 8.2 provides information on antiviral coatings with different drying or activation times, the method of application, and the chemical formula used, typically activated through the evaporation of surface moisture.

In Table 8.2, various antiviral coatings with different drying or activation times are listed, indicating that they are generally activated through the natural evaporation of surface moisture. The table provides information on the material type, application method, and chemical formula used. It should be duly acknowledged that the specific composition of materials, methodologies of application, and chemical formulations may exhibit variations contingent upon the manufacturer and the precise product in question. Antiviral coatings find utility in diverse domains, including health care settings, public spaces, vehicular transport systems, and many other sectors.

Each coating product may have specific application guidelines and processes, so it is important to follow the manufacturer's instructions. Surfaces with antiviral properties can prevent the attachment and proliferation of microbes or viruses. These surfaces are typically achieved by using antiviral coatings or antimicrobial materials, which contain components that reduce or eliminate the effectiveness of viruses when applied to surfaces. These coatings typically include nanosized particles or chemical compounds. For example, silver nanoparticles or copper-based compounds can possess antiviral properties.

2. The process of making surfaces antiviral typically begins with surface cleaning and disinfection. Cleaning involves removing dirt and organic matter residues from the surface. Disinfection aims to kill or render microorganisms on the surface ineffective. Disinfectants, alcohol-based solutions, or various chemical agents are commonly used.

In the context of post-cleaning and disinfection procedures, it is possible to administer antiviral coatings to confer antiviral properties to various surfaces. Such coatings function by establishing a protective layer on the treated surfaces, inhibiting the adhesion and subsequent proliferation of viruses. Notably, these antiviral coatings often demonstrate prolonged efficacy even amidst repeated surface cleaning and disinfection.

Nevertheless, it is imperative to emphasize that the attainment of antiviral surfaces does not unequivocally ensure the comprehensive removal of viral entities or the absolute nullification of associated infection risks. The efficacy of antiviral coatings can be influenced by multiple factors including, but not limited to, the specific viral strain, the nature of the coating utilized, and the methodology of its application. Consequently, alongside the regular protocols of surface cleaning, disinfection, and overall hygiene, supplemental safety precautions and measures are recommended. The operational effectiveness of such antiviral coatings is contingent upon the choice of material, the precise technique of application, and inherent properties of the targeted surface.

Material Type

Silver Nanoparticles: Silver nanoparticles can possess antiviral properties. These nanoparticles can exhibit antiviral effects by breaking down the virus's membrane or disrupting its structure. They are typically applied to surfaces by being embedded in a polymer matrix.

Copper-Based Compounds: Copper is an element that has antimicrobial and antiviral properties. Copper-based compounds, when applied to surfaces, can prevent viruses from adhering and proliferating. These compounds may include copper oxide or copper ions.

There are various studies on the synthesis and application methods of silver nanoparticles. Here are some different methods and examples from the literature.

Chemical Reduction Method: Silver nanoparticles are frequently produced through the process of chemical reduction. In this methodological approach, a reductant is introduced to silver cations, leading to the nucleation and growth of nanoparticles. For instance, silver salts, with silver nitrate being the prototypical compound, undergo reduction in the presence of a reductant, yielding silver nanoparticles. Widely employed reductants include sodium borohydride, sodium citrate, and ascorbic acid. This methodology is characterized by its simplicity and expedited synthesis time [44].

Utilizing botanical extracts, silver nanoparticles were fabricated via a microwave-assisted approach. A systematic examination was conducted to discern the effects of these plant-derived extracts on the formation process of the silver nanoparticles [36].

In the Realm of Green Synthesis Techniques: The process of producing silver nanoparticles employing botanical extracts, microorganisms, or alternative biological entities is called green synthesis. A microemulsion is a system where water and oil phases mix with each other to form a stable system. Microwave heating is a method used for rapid and efficient synthesis of silver nanoparticles. Microwave energy enhances and controls the reaction rate, facilitating nanoparticle synthesis.

In one study, silver nanoparticles were synthesized using plant extracts through a microwave-assisted method. The influence of plant extracts on the synthesis of silver nanoparticles was investigated [45].

Green Synthesis Methods: The synthesis of silver nanoparticles using plant extracts, microorganisms, or other biological sources is referred to as green synthesis methods. These methods are considered environmentally friendly and less risky in terms of toxicity. A review study examined the synthesis of silver nanoparticles using plant extracts through green synthesis methods. The use of plant extracts is emphasized as an environmentally friendly and economical alternative [46].

Application Method: Spraying or Coating: Antiviral coatings can be applied to surfaces using spraying or coating methods. These methods ensure the even distribution of the coating on the surface and allow control over the thickness of the coating layer. The coating process may require specialized equipment or professional application.

These methods ensure the homogeneous distribution of the antiviral coating. The spraying method is preferred when the antiviral coating is in a liquid form. The coating is evenly distributed on the surface using a specialized spraying device. This method enables the rapid coverage of large and wide surfaces. The coating method, on the other hand, is used when the antiviral coating is in a thicker form. The coating material is applied to the surface and spread evenly. This method is typically preferred for smaller and detailed surfaces. The application of antiviral coatings through spraying or coating methods prevents the spread of viruses and other microorganisms on surfaces, ensuring hygiene. These coatings are commonly used in health care facilities, public transportation vehicles, hotels, restaurants, and other public areas.

Various antimicrobial surface has been described below:

1. **Silver:** Silver nanoparticles have antimicrobial and antiviral properties. Coatings containing silver nanoparticles can prevent viruses from adhering to surfaces and hinder their spread [47].
2. **Copper:** Copper surfaces can render viruses ineffective by damaging their membranes. Copper-based coatings can help reduce viral transmission in hospital environments [48].
3. **Polymers:** Some polymers possess antiviral properties that prevent viruses from adhering to cells or disrupt their membranes. These polymers can provide antiviral effects when integrated into surfaces or coatings [49].

In some cases, materials with antiviral effects can be implanted onto surfaces. For example, metal ions with antibacterial or antiviral properties can be embedded within a coating or layer formed on surfaces. Implants that continuously release antiviral drugs can be used to treat infections or prevent viral spread. These systems allow controlled delivery of the drug to the body [50]. Some implants can render viruses ineffective using electromagnetic fields or light. For example, an implant containing a material sensitive to ultraviolet light can be activated when exposed to light, effectively neutralizing viruses [51].

Surface Properties

Roughness: The roughness of a surface can influence the effectiveness of an antiviral coating. A smooth and flat surface allows better adhesion of the coating and reduces the chances of virus attachment and proliferation.

Chemical Composition: The chemical composition of a surface can affect the adhesion capability and effectiveness of an antiviral coating. The coating should be selected to be compatible with the surface's chemical composition.

Porosity: The porosity of a surface can affect the adhesion and effectiveness of an antiviral coating. A porous surface allows the coating to penetrate and provide better adhesion. This can result in longer-lasting coatings and hinder virus attachment. These studies investigate the effect of surface roughness on bacterial adhesion and corrosion resistance of stainless steel surfaces, focusing on the relationship between surface roughness and porosity.

3. Nanotube and graphene-based antiviral thin surfaces are a strategy used to prevent the spread or to inactivate viruses. In recent scientific and technological advancements, carbon-based materials, namely nanotubes and graphene, have garnered considerable interest due to their distinctive physical and chemical attributes [29–31].

Nanotubes can be characterized as cylindrical constructs formed by the systematic alignment of carbon atoms in a planar configuration. These configurations confer notable properties such as remarkable tensile strength, flexibility, and high thermal conductivity. Conversely, graphene is constituted by a monoatomic layer of carbon, manifesting properties analogous to those of nanotubes.

The rationale behind the incorporation of nanotubes and graphene in antiviral applications hinges on their inherent properties. Primarily, these materials present an expansive surface area, which augments antiviral efficacy. This extensive surface area furnishes increased loci for viral interaction and adherence. Secondarily, they promote the optimal binding of antiviral entities—be they antiviral pharmaceuticals, antibodies, or other agents—and subsequent viral neutralization.

Furthermore, both nanotubes and graphene are distinguished by their exceptional electrical conductivity. Such a trait harbors the potential to pioneer electronic-driven antiviral modalities.

For instance, sensors based on nanotube and graphene structures have demonstrated potential in the molecular detection of viral pathogens [37]. The sensitivity of these sensors at the molecular scale establishes them as pivotal instruments for early viral diagnosis and infection management. Furthermore, both nanotubes and graphene exhibit remarkable mechanical robustness. This intrinsic mechanical strength facilitates the fabrication of resilient materials suitable for antiviral coatings and filtration applications. As an illustration, coatings derived from nanotube and graphene structures can be strategically applied to various surfaces, impeding viral adhesion and subsequently mitigating viral transmission [22, 25].

In addition to their mechanical properties, nanotubes and graphene exhibit biocompatibility, signifying their synergistic operation within biological matrices. Such compatibility augments the efficacy of antiviral interventions and diminishes potential adverse effects. For instance, utilizing nanotubes and graphene as vehicular agents for drug delivery can target specific loci of viral infection, yielding precise therapeutic outcomes.

Consequently, the incorporation of nanomaterials, specifically nanotubes and graphene, into antiviral applications is becoming increasingly prevalent in contemporary research. The following list presents the advantages and disadvantages of nanotube and graphene-based antiviral thin surfaces.

Advantages:

- Large surface area, increasing antiviral effectiveness
- Excellent electrical conductivity, enabling electronic-based antiviral systems
- High mechanical strength, allowing for the production of durable antiviral materials
- Biocompatibility, enhancing antiviral treatments and reducing side effects

Disadvantages:

- Cost of production may be higher compared to traditional materials
- Complex synthesis methods for nanotubes and graphene
- Potential concerns regarding the environmental impact of nanomaterials

It's important to note that while nanotube and graphene-based antiviral surfaces show promise, further research is needed to fully understand their effectiveness, long-term stability, and potential limitations in practical applications. However, the advantages and disadvantages may differ under certain conditions and may vary depending on the product or technology.

8.4 A STRATEGY FOR USING ANTIVIRAL NANOPARTICLES AND SELF-STERILIZING SURFACES

This section covers the integration of antiviral nanoparticles into surfaces, application methods, surface sterilization processes, and evaluation of effectiveness [7, 10].

Antiviral nanoparticles have emerged as a promising solution to improve surface disinfection and prevent the spread of pathogens. Their small size and high surface area/volume ratio allow them to interact effectively with viruses and inhibit their activity. By integrating these nanoparticles into surface materials or coatings, self-sterilizing surfaces can be created that can consistently combat viral contamination.

Different application are examined,such as spray coating, dip coating, and electrostatic deposition, while addressing the process of integrating antiviral nanoparticles into surfaces. It also evaluates surface sterilization processes such as exposure to UV light, heating, or chemical treatments, and considers solutions to increase the antiviral effectiveness of nanoparticles [6].

This strategy illustrates how antiviral nanoparticles can be used and how surfaces can be self-sterilized (see Figure 8.3).

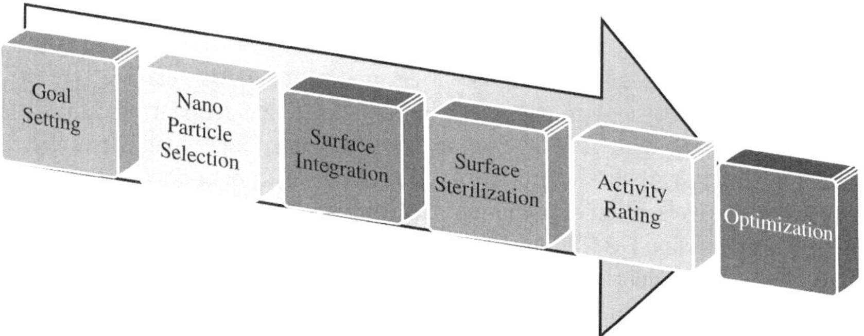

FIGURE 8.3 A strategy for using antiviral nanoparticles and self-sterilize surfaces

Goal Setting

a. Identify areas and applications where antiviral nanoparticles will be used.
b. Detection of risky areas where surfaces are in contact with viruses.

Nanoparticle Selection

a. Selection of nanoparticles with antiviral properties.
b. Integration of nanoparticles on surfaces and evaluation of their adhesion abilities.

Surface Integration

a. Identification of suitable methods for integrating antiviral nanoparticles into surfaces (spray coating, dip coating, electrostatic deposition, etc.).
b. Optimizing the integration methods in accordance with the properties of the surfaces and the intended application area.

Surface Sterilization

a. Determination of surface sterilization processes (UV light, heating, chemical sterilization, etc.).
b. Use of optimized sterilization processes to maximize the interaction of antiviral nanoparticles with surface viruses.

Activity Rating

a. Evaluation of the effectiveness of antiviral nanoparticles on surfaces (virus inactivation tests, antiviral performance analyses, etc.).
b. Evaluation of the ability of surfaces to self-sterilize (long-term effectiveness, durability, etc.).

Optimization

a. Optimizing processes and components to increase the antiviral effectiveness and self-sterilizing abilities of surfaces.
b. Integration of nanoparticles into surfaces and sterilization processes and continuous improvement of efficacy evaluations.

This strategy provides a roadmap for the use of antiviral nanoparticles and improving the self-sterilizing ability of surfaces [8, 9]. Each step plays an important role in the development of antiviral fines and supports a continuous improvement cycle to increase efficacy.

8.5 ENHANCING THE EFFECTIVENESS AND LIFETIME OF ANTIVIRAL THIN SURFACES WITH GRAPHENE-BASED MATERIALS

Research is ongoing regarding the durability of nanotube and graphene-based antiviral surfaces in terms of surface lifetime. These studies aim to evaluate the durability of surfaces under repeated cleaning and disinfection [18, 20].

Nanomaterials such as nanotubes and graphene generally exhibit physical and chemical durability. However, they need to be optimized to maintain their antiviral properties and ensure the long-term lifespan of surfaces. Therefore, research can assess the following aspects of nanotube and graphene-based antiviral surfaces:

- **Surface Cleanliness**: It is important for surfaces to maintain their antiviral effectiveness and physical durability under repeated cleanings and disinfections.
- **Resistance to Chemical and Physical Effects**: Surfaces should be resistant to environmental factors, chemical substances, corrosion, and mechanical stress. This is important to extend the surface lifespan and preserve its effectiveness [16, 19].
- **Surface Stability**: Surfaces should not lose their antiviral properties over time and should work consistently. It is important for surfaces to effectively retain and release antiviral compounds.

In recent investigations, empirical evidence has been presented to substantiate the advancement of enduring and resilient antiviral surfaces derived from nanotubes and graphene. Graphene is a two-dimensional lattice of carbon atoms configured in a consistent planar arrangement. Such a configuration impairs the propensity of viral entities to bind and propagate on the surface, concurrently enhancing its mechanical robustness. Additionally, graphene's high conductivity can provide electrical stimulability in antiviral surfaces and support rapid heat transfer for surface sterilization.

Electrochemical deposition is a technique in which an electrical current facilitates chemical reactions, resulting in the deposition of a material onto a substrate. Electrical current is used to facilitate the attachment of graphene layers to the surface. An electrochemical cell contains an electrolyte solution and two electrodes. The electrolyte solution typically consists of a salt solution or a specialized chemical solution.

During the electrochemical deposition process, electrical current is applied, initiating chemical reactions within the electrolyte solution [2]. These reactions enable the pulling of graphene layers onto the surface. Electrochemical interactions allow for the attachment and uniform distribution of graphene layers on the surface. In the realm of material science, the method of electrochemical deposition offers a proficient approach for the incorporation of graphene sheets onto nanoscale substrates. The surface's electrostatic potential or charging attributes promote the affinity of graphene layers to adhere to the substrate. Nevertheless, the electrochemical deposition technique is not devoid of inherent challenges. The quality and homogeneity of graphene layers are critical for the success of the method.

By combining electrical current and chemical reactions, this method enables the attachment of graphene layers to the surface and facilitates the achievement of a homogeneous coating [4].

Techniques such as chemical vapor deposition (CVD) enable the integration of graphene layers onto surfaces. Additionally, the addition of antimicrobial agents or antiviral compounds to graphene can enhance the antiviral effectiveness of the surface [52, 53].

Research aims to investigate the potential of using graphene-based materials to enhance the effectiveness and surface lifespan of antiviral thin surfaces. In light of its distinct physicochemical properties, graphene holds potential as a pivotal material in the advancement of antiviral thin film coatings. The objective of this research is to elucidate novel methodologies, leveraging graphene's attributes, to foster enhanced safety and hygiene standards within various environments.

8.6 CONCLUSION

The research conducted on the effectiveness and surface lifespan of antiviral thin surfaces highlights the potential of nanomaterials such as nanotubes and graphene in developing surfaces that can inhibit the spread of viruses. These studies evaluate the integration of specific antiviral compounds or nanoparticles onto nanotube or graphene surfaces to assess their effectiveness and long-term performance. The effectiveness of antiviral thin surfaces can be determined through experiments that examine virus attachment, proliferation, and viability on the surfaces. These experiments demonstrate antiviral efficacy by either inactivating the viruses or facilitating their easy removal from the surface. Notably, studies on viruses like SARS-CoV-2 have shown the effectiveness of nanotube and graphene-based antiviral surfaces in preventing virus transmission. The surface lifespan of nanotube and graphene-based antiviral surfaces can be assessed through investigations into their durability. These studies evaluate the surfaces' ability to withstand long-term use and repeated cleaning. While nanomaterials like nanotubes and graphene generally exhibit physical and chemical resilience, the surface lifespan can vary and should be optimized for specific application conditions. The research on the effectiveness and surface lifespan of nanotube and graphene-based antiviral thin surfaces contributes to the development of surfaces that provide antiviral protection and aid in preventing the spread of viruses. However, further research and development are necessary as it is crucial to establish specific outcomes and standards for each material and application.

REFERENCES

1. Kern, W., & Puotinen, D. A. (1970). Cleaning solutions based on hydrogen peroxide for use in silicon semiconductor technology. RCA Review, 31(2), 187–205.
2. Wallace, R. M., & Seabaugh, A. (2008). High-k dielectrics: Past, present, and future. Journal of Applied Physics, 104(11), 111101.
3. Lide, D. R. (Ed.). (2003). CRC handbook of chemistry and physics, CRC press, 2475.
4. Li, X., Cai, W., Colombo, L., & Ruoff, R. S. (2009). Evolution of graphene growth on Ni and Cu by carbon isotope labeling. Nano Letters, 9(12), 4268–4272.
5. Chhowalla, M., Shin, H. S., Eda, G., Li, L. J., Loh, K. P., & Zhang, H. (2013). The chemistry of two-dimensional layered transition metal dichalcogenide nanosheets. Nature Chemistry, 5(4), 263–275.
6. Grass, G., Rensing, C., & Solioz, M. (2011). Metallic copper as an antimicrobial surface. Applied and Environmental Microbiology, 77(5), 1541–1547.
7. Elechiguerra, J. L., Burt, J. L., Morones, J. R., Camacho-Bragado, A., Gao, X., Lara, H. H., & Yacaman, M. J. (2005). Interaction of silver nanoparticles with HIV-1. Journal of Nanobiotechnology, 3(1), 6.
8. Dizaj, S. M., Lotfipour, F., Barzegar-Jalali, M., Zarrintan, M. H., & Adibkia, K. (2014). Antimicrobial activity of the metals and metal oxide nanoparticles. Materials Science and Engineering: C, 44, 278–284.
9. Li, J., Peng, X., Wei, G., He, P., Zhang, B., & Chen, G. (2019). Self-sterilizing properties and mechanical performance of bacterial nanocellulose membrane incorporated with ZnO nanoparticles for potential wound dressing application. Journal of Applied Polymer Science, 136(23), 47631.
10. Foster, H. A., Ditta, I. B., Varghese, S., & Steele, A. (2011). Photocatalytic disinfection using titanium dioxide: spectrum and mechanism of antimicrobial activity. Applied Microbiology and Biotechnology, 90(6), 1847–1868.
11. Harper, J. D., Fedorov, A. G., & Shelnutt, J. A. (2003). Fourier transform infrared spectrometry analysis of nanodiamond surfaces. Analytical Chemistry, 75(4), 771–777.
12. Kazarian, S. G., & Chan, K. L. A. (2013). Applications of ATR-FTIR spectroscopic imaging to biomedical samples. Biochimica et Biophysica Acta (BBA) - Biomembranes, 1828(10), 2332–2343.
13. Marigheto, N. A., Snowden, M. J., Mitchell, J. C., & Davis, A. L. (2000). Application of FTIR and Raman spectroscopy to the study of calcite and gypsum at high pressure. Spectrochimica Acta Part A: Molecular and Biomolecular Spectroscopy, 56(3), 667–673.
14. Siddiqui, M. S. et al. (2020). Antiviral coatings: An effective tool to combat COVID-19. Journal of Materials Chemistry B, 8(40), 9013–9037.
15. Banerjee, D. et al. (2021). Recent advances in antiviral coatings for combating COVID-19. ACS Biomaterials Science & Engineering, 7(1), 12–36.

16. Galdiero, S. et al. (2020). Nanotechnology-based antiviral agents. Journal of Nanobiotechnology, 18(1), 1–26.
17. Nguyen, K. T. et al. (2020). Nanotechnology solutions for COVID-19. ACS Nano, 14(7), 8219–8246.
18. Li, Y., Leung, P., Yao, L., Song, Q. W., & Newton, E. (2017). Antiviral effect of graphene oxide: How graphene oxide affects the attachment and entry of herpes simplex virus 1. Journal of the American Chemical Society, 139(21), 7629–7632.
19. Joshi, R. K., Alwarappan, S., Yoshimura, M., & Sahajwalla, V. (2017). Graphene-based antiviral composite materials. Methods in Molecular Biology, 1539, 293–302.
20. Zhang, Y., Ali, S. F., Dervishi, E., Xu, Y., Li, Z., Casciano, D., & Biris, A. S. (2012). Cytotoxicity effects of graphene and single-wall carbon nanotubes in neural phaeochromocytoma-derived PC12 cells. ACS Nano, 6(1), 680–690.
21. Abaszade, R. G., Kapush, O. A., Mamedova, S. A., Nabiyev, A. M., Melikova, S. Z., & Budzulyak, S. I. (2020). Gadolinium doping influence on the properties of carbon nanotubes. Physics and Chemistry of Solid State, 21(3), 404–408.
22. Abaszade, R. G., Kapush, O. A., & Nabiyev, A. M. (2020). Properties of carbon nanotubes doped with gadolinium. Journal of Optoelectronic and Biomedical Materials, 12(3), 61–65.
23. Abaszade, R. G., Mamedov, A. G., Bayramov, I. Y., Khanmamadova, E. A., Kotsyubynsky, V. O., Kapush, O. A., Boychuk, V. M., & Gur, E. Y. (2022). Structural and electrical properties of sulfur-doped graphene oxide/graphite oxide composite. Physics and Chemistry of Solid State, 23(2), 256–260.
24. Abaszade, R. G. (2022). Synthesis and analysis of flakes graphene oxide. Journal of Optoelectronic and Biomedical Materials, 14(3), 107–114.
25. Abaszade, R. G., Babanli, M. B., Kotsyubynsky, V. A., Mammadov, A. G., Gür, E., Kapush, O. A., Stetsenko, M. O., & Zapukhlyak, R. I. (2022). Influence of gadolinium doping on structural properties of carbon nanotubes. Physics and Chemistry of Solid State, 24(1), 153–158.
26. Li, Y., Lin, Z., Zhao, M., Xu, T., Wang, C., & Zhang, Y. (2019). Antiviral activity of carbon nanomaterials. Nanoscale, 11(41), 19140–19163.
27. Arun, K. S., Rudra, S., Thamizharasan, G., Pradhan, M., Rani, B., Sahu, N. K., & Nayak, A. K. (2022). Crystal structure controlled synthesis of tin oxide nanoparticles for enhanced energy storage activity under neutral electrolyte. Journal of Materials Science: Materials in Electronics, 33(17), 13668–13683.
28. Boychuk, V. M., Zapukhlyak, R. I., Abaszade, R. G., Kotsyubynsky, V. O., Hodlevsky, M. A., Rachiy, B. I., Turovska, L. V., Dmytriv, A. M., & Fedorchenko, S. V. (2022). Solution combustion synthesized NiFe2O4/reduced graphene oxide composite nanomaterials: Morphology and electrical conductivity. Physics and Chemistry of Solid State, 23(4), 815–824.
29. Stetsenko, M. O., & Abaszade, R. G. (2023). X-ray phase analysis of carbon nanotubes obtained by the arc discharge method. UNEC Journal of Engineering and Applied Sciences, 3(1), 15–20.
30. Figarova, S. R., Aliyev, E. M., Abaszade, R. G., & Figarov, V. R. (2023). Negative thermal expansion of sulphur-doped graphene oxide. Advanced Materials Research, 1175, 55–62.
31. Abaszade, R. G., Mammadov, A. G., Kotsyubynsky, V. O., Gur, E. Y., Bayramov, I. Y., Khanmamadova, E. A., & Kapush, O. A. (2022). Photoconductivity of carbon nanotubes. International Journal on Technical and Physical Problems of Engineering, 14(3), 155–160.
32. Figarova, S. R., Aliyev, E. M., Abaszade, R. G., Alekberov, R. I., & Figarov, V. R. (2021). Negative differential resistance of graphene oxide/sulphur compound. Journal of Nano Research Submitted, 67, 25–31.
33. Abaszade, R. G., Mamedova, S. A., Agayev, F. H., Budzulyak, S. I., Kapush, O. A., Mamedova, M. A., Nabiyev, A. M., & Kotsyubynsky, V. O. (2021). Synthesis and characterization of graphene oxide flakes for transparent thin films. Physics and Chemistry of Solid State, 22(3), 595–601.
34. Abaszade, R. G., Mammadov, A. G., Kotsyubynsky, V. O., Gur, E. Y., Bayramov, I. Y., Khanmamadova, E. A., & Kapush, O. A. (2022). Modeling of voltage-ampere characteristic structures on the basis of graphene oxide/sulfur compounds. International Journal on Technical and Physical Problems of Engineering, 14(2), 302–306.
35. Abaszade, R. G., Mammadov, A. G., Khanmammadova, E. A., Bayramov, İ. Y., Namazov, R. A., Popal Kh, M., Melikova, S. Z., Qasımov, R. C., Bayramov, M. A., & Babayeva, N. (2023). Electron paramagnetic resonance study of gadoliniumum doped graphene oxide. Journal of Ovonich Research, 19(2), 259–263.
36. Khanmamedova, E. A. (2023) Thermal processing analysis of graphene oxide, Seoul, South Korea: International scientific and practical conference, Theoretical and Practical aspects of Modern Scientific research, 151–153
37. Nayak, A. K., & Swain, A. K. (2019). Facile room temperature synthesis of reduced graphene oxide as efficient metal-free electrocatalyst for oxygen reduction reaction. In: Sahoo, S., Tiwari, S., Nayak, G. (eds) Surface Engineering of Graphene, pp. 259–271. Springer.

38. Liu, J., Fu, S., Yuan, B., Deng, Z., & Shen, M. (2010). Facile synthesis of graphene oxide and its reduction. Journal of Materials Chemistry, 20(35), 7491–7496.
39. Binnig, G., Quate, C. F., & Gerber, C. (1986). Atomic force microscope. Physical Review Letters, 56(9), 930.
40. Müller, D. J., & Dufrene, Y. F. (2008). Atomic force microscopy as a multifunctional molecular toolbox in nanobiotechnology. Nature Nanotechnology, 3(5), 261–269.
41. Radmacher, M. (1997). Measuring the elastic properties of biological samples with the AFM. IEEE Engineering in Medicine and Biology Magazine, 16(2), 47–57.
42. Homola, J. (2008). Surface plasmon resonance sensors for detection of chemical and biological species. Chemical Reviews, 108(2), 462–493.
43. Rodríguez-Padilla C, Rodríguez-González JA. (2021). Strategies to Design Antiviral Surfaces Based on Metal and Metal Oxide Nanoparticles: A Review. J Mater Sci Technol, 71:65–78. doi: 10.1016/j.jmst.2020.06.038. Epub 2020 Jun 24. PMID: 34249072; PMCID: PMC8255593.
44. Balamurugan, M., Saravanan, S. Green Synthesis of Silver Nanoparticles by using Eucalyptus Globulus Leaf Extract. J. Inst. Eng. India Ser. A 98, 461–467 (2017). https://doi.org/10.1007/s40030-017-0236-9
45. Parveen, M., Ahmad, F., Malla, A.M. et al. Microwave-assisted green synthesis of silver nanoparticles from Fraxinus excelsior leaf extract and its antioxidant assay. Appl Nanosci 6, 267–276 (2016). https://doi.org/10.1007/s13204-015-0433-7
46. Jadoun, S., Arif, R., Jangid, N.K. et al. Green synthesis of nanoparticles using plant extracts: a review. Environ Chem Lett 19, 355–374 (2021). https://doi.org/10.1007/s10311-020-01074-x
47. Roy, K. et al. (2017). Antiviral applications of silver nanoparticles in the treatment of respiratory viruses. Nano-Micro Letters, 9(3), 1–12.
48. Warnes, S. L. et al. (2015). Antimicrobial copper alloys for touch surfaces: a systematic review. Applied and Environmental Microbiology, 81(6), 1915–1925.
49. Zhou, J. et al. (2018). Antiviral strategies for emerging viral infections. ACS Infectious Diseases, 4(6), 708–717.
50. Siepmann, J. et al. (2012). Drug. Advanced Drug Delivery Reviews, 64, 1590–1600. delivery devices: Issues and challenges
51. Buonanno, M. et al. (2020). UV-C and UV-C plus violet filter lamps for SARS-CoV-2 inactivation: Effectiveness and hazards prevention. Physics of Fluids, 32(6), 061704.
52. Rich, R. L., & Myszka, D. G. (2011). Grading the commercial optical biosensor literature-Class of 2008: The Mighty Binders. Journal of Molecular Recognition, 24(6), 892–914.
53. Schasfoort, R. B., & Tudos, A. J. Eds (2008). Handbook of Surface Plasmon Resonance, Royal Society of Chemistry, 554

9 Abradable Coatings and Their Application

Amlan Prabhujyoti Sahu and Ajit Behera

9.1 INTRODUCTION

Abradable coatings are advanced materials used in aerospace and turbomachinery applications to minimize friction and wear between moving components. These coatings exhibit controlled erosion, allowing them to conform to the shape of mating surfaces while maintaining efficient clearance. Scientific research in various journals highlights the development and optimization of abradable coatings. Studies explore materials such as polymer composites, metallic powders, and ceramic coatings, focusing on their mechanical properties, erosion resistance, and manufacturing techniques [1–3]. The abradable coating is applied to the stationary component, creating a sealing interface with the rotating part. During the operation of the engine, the rotor makes interaction with the abradable coatings [4]. Due to the relative motion and the properties of the coating, it is selectively worn away, creating a tailored gap between the rotating and stationary components. This gap allows for minimal gas leakage while avoiding detrimental contact and damage to the parts. Various materials can be used to manufacture abradable coatings, including metals, ceramics, polymers, and composites. Commonly used metals include aluminum and nickel-based alloys, while ceramics like alumina and zirconia are also utilized. Composites can be formed by combining metallic and ceramic components to achieve desired properties such as improved wear resistance or thermal stability [5].

Abradable coatings are used in various industries and applications where there is a need for controlled clearance and reduced wear between moving parts. Here are some target applications of abradable coatings:

Gas turbines: Abradable coatings are widely employed in gas turbine engines, particularly in the turbine and compressor segments. They are applied to stator vanes, rotor blades, and shroud segments. The coating allows for a closer tolerance between the rotor and stator, improving engine efficiency and reducing the risk of blade tip rubbing [6].

Aircraft engines: Abradable coatings are also applied to components in aircraft engines, including compressor blades, turbine seals, and fan blades. The coatings help maintain the desired clearance between rotating and stationary parts, reducing the chances of blade rubbing and improving overall engine performance [7].

Steam turbines: In steam turbines, abradable coatings are used in the casing and blade tips to minimize clearance and prevent contact between the rotor and stator. This improves efficiency and reduces energy losses [8].

Automotive industry: Abradable coatings find applications in the automotive industry, particularly in components such as piston rings and cylinder liners. The coatings help reduce friction and improve engine performance by maintaining optimal clearances between moving parts [5].

Industrial machinery: Abradable coatings are used in various industrial machinery, such as pumps, compressors, and turbines. They help control clearances and reduce wear between components, resulting in improved efficiency and extended component life [4].

DOI: 10.1201/9781032635347-9

9.2 DEMANDS ON ABRADABLE COATINGS

Abradable coating is an essential component in various industrial applications, particularly in the aerospace and gas turbine industries. It is a specialized coating that possesses unique properties, making it crucial for efficient and reliable operation of rotating machinery, such as turbine blades and compressor systems. This coating is designed to reduce the contact and friction between moving components, thereby enhancing performance, minimizing wear, and increasing overall efficiency [3, 9]. One of the primary functions of abradable coating is to create a tight sealing interface between stationary and rotating parts in gas turbines. Gas turbines operate at high temperatures and speeds, and the clearance between the rotating blades and the surrounding casing needs to be carefully controlled to optimize efficiency. Abradable coatings are applied to the casing or shroud surfaces, allowing the blades to pass through without direct contact, while maintaining a minimal and controlled clearance. This ensures efficient operation and prevents undesirable effects, such as vibration, loss of performance, or damage to the turbine components [3].

The main advantage of abradable coatings lies in their ability to self-adjust and conform to the shape of the rotating parts. When the blades pass through the coated surface, the abradable material undergoes controlled erosion or wear, accommodating the blade's profile and reducing the clearance. This conformability helps maintain a tight seal, improves aerodynamic efficiency, and minimizes leakage losses. Consequently, abradable coatings contribute to higher turbine efficiency, lower fuel consumption, and reduced emissions [9].

Furthermore, abradable coatings help protect critical components from damage caused by foreign object ingestion (FOI). In gas turbine engines, the ingestion of debris, such as sand, dust, or ice, can lead to blade erosion and damage. Abradable coatings act as sacrificial layers, absorbing the impact of foreign objects and preventing direct contact with the blades. This feature significantly improves the durability and lifespan of turbine blades, reducing maintenance and replacement costs [10].

Abradable coatings are essential in various engineering applications where there is a need for close-clearance interfaces between rotating and stationary components. These coatings are designed to undergo controlled wear, providing an effective sealing mechanism and reducing the risk of undesirable contact between surfaces. The primary purpose of abradable coatings is to create a self-machining interface that allows for tight tolerances while minimizing the risk of damage or performance degradation. The problem that necessitates the use of abradable coatings arises from the inherent clearance gap between these surfaces [1]. In many engineering systems, such as gas turbine engines, compressors, or turbochargers, there is a requirement for components to operate in close proximity while maintaining a minimal clearance gap. However, due to manufacturing tolerances, thermal expansion, and other factors, it is challenging to achieve a perfect fit between rotating and stationary parts. Consequently, there is a risk of contact or interference, leading to friction, wear, and ultimately, system failure. Abradable coatings address this problem by providing a sacrificial material that can be easily worn away by the rotating component, allowing for a controlled and predictable clearance. Typically, these coatings are composed of soft and compressible materials, such as thermoplastic composites, ceramics, or metallic powders, which can be applied as a coating or a lining on the stationary component. When the rotating part comes into contact with the abradable coating, it gradually wears away, ensuring sufficient clearance is maintained and preventing undesirable contact [4]. The selection and design of abradable coatings involve careful consideration of factors such as material properties, operating conditions, and performance requirements. The coating must possess suitable mechanical properties to withstand the operational stresses and temperatures while exhibiting the desired wear characteristics. The coating should also be compatible with the mating materials and exhibit good adhesion to the substrate.

9.3 TYPES OF ABRADABLE COATING

There are several types of abradable coatings that serve different purposes and are used in various industries. Let's explore some of the common types of abradable coatings and their composition.

1. Metallic abradable coatings
2. Polymer abradable coatings
3. Ceramic abradable coatings
4. Composite abradable coatings

9.3.1 Metallic Abradable Coatings

Metallic abradable coatings are typically composed of a metallic matrix material mixed with a solid lubricant or filler material. The primary purpose of these coatings is to create a sacrificial layer that can be easily abraded by an opposing surface, such as a rotor or impeller. The coating's ability to be worn away in a controlled manner reduces contact forces, minimizes friction, and prevents damage to the mating components [11]. The application of metallic abradable coatings involves a multistep process. It typically begins with the preparation of the substrate surface, which may involve cleaning, roughening, or applying a bond coat. The abradable material is then applied using techniques such as plasma spraying, high-velocity oxygen fuel (HVOF) spraying, or physical vapor deposition (PVD). After deposition, the coating may undergo additional treatments, such as heat treatment or machining, to achieve the desired properties and dimensions [12]. The development and optimization of metallic abradable coatings involve extensive research and testing. Factors such as coating composition, microstructure, thickness, and deposition technique significantly affect the performance and durability of the coating. Researchers and engineers employ a range of characterization techniques, including microscopy, spectroscopy, and mechanical testing, to evaluate the coating's properties and behavior under various operating conditions. By using abradable coatings, engine designers can achieve tighter clearances between rotating and stationary components, which improves engine efficiency, reduces fuel consumption, and enhances engine performance.

The benefits of metallic abradable coatings are

Improved efficiency: By reducing clearances, metallic abradable coatings minimize leakage flows and improve the efficiency of rotating machinery. This leads to enhanced overall system performance and reduced energy consumption.

Damage prevention: Abradable coatings act as sacrificial layers, protecting contacting components from damage by allowing controlled wear to occur on the coating surface. This prevents contact between the rotating and stationary parts and reduces the risk of catastrophic failure.

Vibration reduction: Metallic abradable coatings help dampen vibrations that can occur between rotating and stationary components. By minimizing vibration, these coatings contribute to smoother operation and increased system reliability.

Sealing effect: The compliant nature of abradable coatings allows them to conform to the shape of the opposing surface, providing an effective sealing mechanism. This sealing effect helps prevent gas or fluid leakage, improving system performance and efficiency.

9.3.2 Polymer Abradable Coatings

Polymer abradable coatings, in particular, offer several advantages over metallic abradable coatings, including lighter weight, better thermal stability, and improved erosion resistance [13]. Polymer abradable coatings are typically composed of a polymeric matrix reinforced with solid lubricants or fillers.

The polymeric matrix is often made of thermosetting resins such as epoxy or phenolic, which provides the coating with excellent mechanical properties and thermal stability. The solid lubricants or fillers, such as PTFE (polytetrafluoroethylene) or graphite, are incorporated to enhance the coating's abradability and reduce friction. The manufacturing process involves spraying or applying the coating onto the substrate using techniques like plasma spray, thermal spray, or resin impregnation [14, 15].

Several types of polymer abradable coatings are available. Here they are divided on the basis of their characteristics and applications.

Epoxy-based coatings. Epoxy resins are frequently used as the base material for abradable coatings due to their excellent adhesion, wear resistance, and thermal stability.

Polyimide-based coatings. Polyimide coatings offer high-temperature stability, excellent dimensional stability, and good wear resistance, making them suitable for demanding applications.

Composites. Polymer matrix composites, reinforced with materials such as fibers or particles, can enhance the mechanical properties and wear resistance of abradable coatings [16].

The unique characteristic of polymer abradable coatings lies in their ability to selectively abrade or wear away in a controlled manner. During operation, the rotating component, such as a turbine blade, comes into contact with the stationary abradable coating. The coating material is designed to have a lower hardness than the blade material. As a result, the blade's leading edge erodes or abrades the coating, creating a tailored clearance that minimizes contact and reduces friction between the components. The abraded particles are carried away by the gas flow in the engine [3, 17]. There are several key characteristics that make polymer abradable coatings desirable for their applications:

Compressibility: Polymer abradable coatings are typically formulated with elastomers or thermoplastics that exhibit a high degree of compressibility. This property allows the coating to deform under pressure, reducing the contact forces and preventing damage to the mating component.

Low friction: The coatings are designed to have low coefficients of friction, which further minimizes the wear and heat generation during contact between the moving and stationary parts. This low-friction behavior helps to reduce energy losses and improve the overall efficiency of the machinery.

Wear resistance: While the coating is designed to be easily abraded, it should still exhibit sufficient wear resistance to ensure an acceptable service life. The balance between wear resistance and abradability is a critical factor in the formulation and selection of the coating material [18].

Thermal stability: Polymer abradable coatings must be able to withstand the high temperatures generated in the operating conditions of gas turbines and other machinery. These coatings are typically formulated with high-temperature-resistant polymers or reinforced with ceramic fillers to enhance their thermal stability.

Polymer abradable coatings are typically applied using various techniques, including plasma spraying, thermal spraying, or by applying a precured composite tape. The choice of application method depends on the specific requirements of the application and the properties of the coating material. Polymer abradable coatings find extensive applications in various industries. Some notable examples include the following:

Gas turbine engines: Polymer coatings are commonly employed in gas turbine engines, particularly in compressor systems and clearance control rings. These coatings help reduce clearance gaps and improve engine efficiency.

Automotive industry: Polymer abradable coatings are utilized in piston rings and cylinder liners to reduce friction and improve fuel efficiency in internal combustion engines.

Power generation: They are employed in steam and gas turbines to minimize leakage and improve efficiency by maintaining precise clearances.

Polymer abradable coatings offer several advantages over other sealing solutions.

Improved efficiency: By reducing clearances, abradable coatings minimize air leakage in the engine, which improves overall efficiency and performance.

Enhanced component life: The controlled clearance provided by abradable coatings reduces contact forces, wear, and stress on components, extending their operational life.

Weight reduction: Polymer-based coatings are lighter than metallic alternatives, resulting in weight savings and improved fuel efficiency.

Erosion resistance: The incorporation of solid lubricants or fillers in the coating enhances its resistance to erosion caused by high-speed gas flows.

Ease of repair: Polymer coatings can be repaired or reapplied easily, reducing maintenance time and costs [16, 19].

9.3.3 Ceramic Abradable Coatings

The primary function of ceramic abradable coatings is to provide a sacrificial layer that can accommodate dimensional changes, vibrations, and thermal expansion within a rotating machinery system. By having an abradable coating, a small and controlled gap can be maintained between the rotating and stationary components during normal operation. This clearance helps to reduce friction and improve efficiency by minimizing contact and reducing the risk of blade tip rubs or other detrimental interactions [20]. Ceramic abradable coatings are typically composed of a blend of ceramic powders, binders, and additives. The specific composition may vary depending on the application and the desired performance requirements. Common ceramic materials used in abradable coatings include aluminum oxide (Al_2O_3), silicon carbide (SiC), and boron nitride (BN). These materials are chosen for their thermal stability, wear resistance, and low coefficient of friction. The binders used in abradable coatings are typically organic compounds that provide the necessary adhesion and flexibility to the ceramic particles. These binders are carefully selected to ensure proper coating deposition and subsequent performance under operational conditions. Additives such as liquid, powdered, or fluid lubricants or solid lubricants may also be incorporated to further enhance the coating's tribological properties [21].

There are several key properties and characteristics that make ceramic abradable coatings suitable for their intended applications:

Abradability: The coating should have controlled wear characteristics, allowing it to abrade selectively and predictably when in contact with the mating surface. This controlled wear ensures the maintenance of a desired gap while minimizing the risk of excessive wear or damage [22].

Thermal stability: Ceramic abradable coatings must be able to withstand the high temperatures encountered in gas turbines and jet engines. They should have high-temperature stability to avoid degradation or failure during operation.

Mechanical integrity: The coating should possess sufficient strength and toughness to withstand the mechanical stresses and vibrations experienced in the operating environment.

Chemical resistance: Ceramic abradable coatings should be resistant to the corrosive and oxidative environments typically found in gas turbines and jet engines to ensure long-term performance.

Bond strength: The coating must have good adhesion to the underlying substrate to prevent delamination or spalling during operation [23].

Low friction: Ceramic abradable coatings typically have a low coefficient of friction, reducing the risk of damage or excessive wear between the coating and the mating surface.

Wear resistance: While the coating is designed to wear away, it still needs to maintain sufficient wear resistance to ensure its longevity and prevent premature failure. The selection of ceramic materials and appropriate formulation are pivotal factors in attaining the intended level of durability against wear [18].

Several ceramic materials are commonly used as abradable coatings, including

Aluminum-based materials: Aluminum-based coatings, such as aluminum oxide (Al_2O_3) or aluminum titanate (Al_2TiO_5), are frequently used due to their excellent thermal stability and abradability. They can be applied by various methods, including plasma spraying, chemical vapor deposition (CVD), or physical vapor deposition (PVD).

Ceramics with polymer binders: Composite coatings, consisting of ceramic particles incorporated within a polymer matrix, offer improved abradability and can be tailored for specific applications. These coatings typically use polymers like epoxy, phenolic resins, or polyimides as binders.

Other ceramic materials: Certain ceramics, such as zirconia-based materials (e.g., partially stabilized zirconia), can also be used as abradable coatings due to their excellent high-temperature properties and abradability.

The application process for ceramic abradable coatings involves various techniques, including plasma spraying, CVD, PVD, or slurry-based methods. Plasma spraying is one of the most commonly used methods, where the process involves the heating of ceramic powder material followed by its acceleration onto a base through the use of a plasma spray. The sprayed coating is then typically machined or ground to achieve the desired clearance or abradability.

Ceramic abradable coatings find extensive use in various industrial applications, including

Gas turbine engines: Abradable coatings are applied to stationary components, such as shrouds and casings, in gas turbine engines. They allow for tighter clearances between rotating and stationary parts, reducing leakage and improving overall engine efficiency.

Compressors: In centrifugal and axial compressors, abradable coatings can be applied to improve aerodynamic performance and reduce gas leakage.

Sealing applications: Abradable coatings are used in sealing applications, such as piston rings in internal combustion engines. The controlled wear characteristics of these coatings allow for better sealing efficiency and reduced oil consumption.

9.3.4 Composite Abradable Coatings

Composite abradable coatings are advanced materials used in various industries, particularly in the aerospace sector, to minimize the wear and damage caused by contact between rotating and stationary components. These coatings are designed to provide a sacrificial layer that can be easily abraded by the counterpart surface, thereby reducing friction and improving overall system efficiency [24]. Composite abradable coatings are typically composed of two main constituents: a matrix material and a filler material. The matrix material is usually a polymer or a polymer-based composite, while the filler material is typically a solid particulate material. The selection of these materials depends on the specific application requirements and operating conditions. The matrix material provides the structural integrity and adhesion properties to the coating, while the filler material enhances its abradability. Common matrix materials used in composite abradable coatings include thermosetting polymers like epoxy, polyimides, and phenolics, as well as thermoplastic polymers like poly ether ether ketone (PEEK). These polymers offer excellent mechanical properties, high-temperature stability, and good adhesion to the substrate. The filler materials used in composite abradable coatings can vary depending on the desired abradability, thermal conductivity, and wear resistance. Some commonly used fillers include soft metals like aluminum and bronze powders, ceramic materials like graphite, mica, and boron nitride, as well as organic materials like polytetrafluoroethylene (PTFE) or other fluoropolymers. These fillers provide the necessary low-friction and easy-abrasion characteristics to the coating [25].

Composite abradable coatings possess several key properties that make them suitable for various applications. These properties include

Abradability: The primary characteristic of composite abradable coatings is their ability to be easily abraded or worn down by contact with the counterpart surface. This property allows the coating to accommodate dimensional changes, reduce contact forces, and prevent damage to the rotating or stationary components.

Low friction: Composite abradable coatings offer low-friction properties, which minimize the energy losses associated with friction and improve overall system efficiency. The low-friction behavior is mainly attributed to the presence of solid lubricant fillers within the coating.

Thermal stability: These coatings are designed to withstand high temperatures, as they are often used in hot sections of engines and turbines. The selection of appropriate matrix materials ensures thermal stability and resistance to degradation under elevated operating conditions.

Mechanical strength: Composite abradable coatings exhibit sufficient mechanical strength to withstand the operational stresses and loads experienced during use. The choice of matrix materials and reinforcement techniques can enhance the mechanical properties of the coatings.

Chemical resistance: The matrix materials used in composite abradable coatings offer good chemical resistance, protecting the underlying components from corrosive environments or chemical attack [26].

Composite abradable coatings find wide applications in various industries, particularly in aerospace and gas turbine engines. Some specific applications include

Gas turbine engines: Composite abradable coatings are widely employed in engines for gas turbines, specifically in regions where clearance control is necessary within stationary components such as shrouds or casings and revolving compressor blades. These coatings allow for tighter clearances during operation, reducing leakage and improving engine efficiency.

Aircraft engines: In aircraft engines, composite abradable coatings are applied to components such as fan blades, stators, and vanes to minimize clearance gaps and improve aerodynamic performance. These coatings also help reduce noise and vibration levels in the engine.

Seals and bearings: Composite abradable coatings are employed in various sealing applications, such as labyrinth seals and bearing clearances. They provide effective sealing while minimizing wear and improving operational efficiency [27].

9.4 CLASSIFICATION BASED UPON THE BASE METAL

9.4.1 Co-Based Abradable Coating

Co-based abradable coatings are a type of thermal spray coating used in various engineering applications, particularly in gas turbine engines. Co-based abradable coatings are specifically formulated to provide excellent abradability, which refers to the ability of a material to wear or be abraded by a mating component while maintaining its integrity [28]. Co-based abradable coatings are primarily composed of cobalt (Co) as the main constituent, along with various other elements and additives.

The specific composition may vary depending on the desired properties and application requirements. Nickel (Ni) is often added to enhance the ductility and toughness of the coating. Other elements such as chromium (Cr) and aluminum (Al) may be included to improve oxidation resistance and adhesion to the substrate [29]. Co-based abradable coatings are typically applied using thermal spray techniques, such as high-velocity oxygen fuel (HVOF) spraying or air plasma spraying (APS). In the HVOF process, the coating material is fed into a combustion chamber along with a fuel gas and oxygen mixture. The resulting high-velocity flame propels molten particles onto the substrate, creating a dense and well-bonded coating. APS involves the use of an electric arc to heat and melt the coating material, which is then accelerated by a compressed air stream onto the substrate [30, 31].

Co-based abradable coatings exhibit several key properties that make them suitable for gas turbine applications, such as

Abradability: The primary characteristic of these coatings is their ability to abrade easily under contact with mating components. This allows for the creation of tighter clearances, reducing gas leakage and improving overall engine efficiency [28].

Ductility and toughness: Co-based coatings are typically formulated to have high ductility and toughness, enabling them to withstand the high rotational and vibrational stresses experienced in gas turbine engines [31].

Thermal resistance: The addition of elements like chromium and aluminum enhances the oxidation resistance of the coating, enabling it to withstand high-temperature operating conditions.

Bond strength: Co-based coatings have excellent bond strength with the substrate, ensuring good adhesion and minimizing the risk of delamination or spalling.

Wear resistance: While the coatings are designed to be abradable, they also exhibit a certain level of wear resistance to prevent excessive material loss during operation [28].

9.4.2 YSZ-Based Abradable Coating

YSZ (yttria-stabilized zirconia) is a widely used material in various industrial applications. Its distinct integration of properties makes it suitable for abradable coatings. It is a ceramic material that exhibits high hardness, excellent thermal stability, and good resistance to wear and corrosion, excellent mechanical properties, high melting point, and resistance to thermal shock. Additionally, it has low thermal conductivity and high fracture toughness, which further enhance its suitability for abradable coatings. The key characteristics of YSZ, including its low coefficient of thermal expansion, high hardness, wear resistance, and toughness enable YSZ to withstand the high temperatures and mechanical stresses experienced in gas turbines and jet engines. The addition of yttria (Y_2O_3) to zirconia (ZrO_2) stabilizes the cubic phase of zirconia at room temperature, resulting in YSZ. The yttria content in YSZ can vary, typically ranging from 6 to 8 mol%. This stabilization mechanism prevents the transformation of zirconia into its monoclinic phase, which is associated with a significant volume expansion and reduced mechanical properties. The addition of yttrium oxide enhances the thermal and chemical stability of the zirconia, making it suitable for high-temperature applications [32]. Plasma spraying is one of the most commonly used techniques for applying YSZ coatings. In this process, the powder is injected into a plasma jet, where it is rapidly heated and accelerated toward the substrate. Upon collision, the liquefied or partially liquefied particles adhere to the surface of the substrate, resulting in the formation of a layer of coating. The plasma spraying process allows for the formation of a highly porous structure, which is desirable for abradable coatings as it facilitates the controlled wear of the material. The porosity in YSZ-based abradable coatings is typically achieved by controlling the spray parameters, such as particle size, spray distance, and spray angle. The resulting coating has a cellular or honeycomb-like microstructure, with interconnected voids that enable the coating to be easily abraded by the mating component during operation [33].

The performance of YSZ-based abradable coatings is influenced by several factors, including the coating thickness, porosity, and surface roughness. Thicker coatings generally provide better wear resistance, while higher porosity allows for more significant material removal. Surface roughness plays a crucial role in achieving the desired tribological behavior because it affects the frictional contact and the ability of the coating to conform to the mating component. YSZ-based abradable coatings offer several advantages in gas turbine and compressor applications. They provide effective clearance control between rotating and stationary components, reducing the risk of blade tip rubbing and associated damage. This clearance control contributes to improved engine efficiency, reduced fuel consumption, and increased overall performance. Additionally, YSZ coatings can provide thermal barrier properties, protecting the underlying substrate from high-temperature exposure [34, 35].

9.4.3 Silica-Based Abradable Coating

Silica-based abradable coatings are a type of coating used in high-temperature applications, particularly in gas turbines and jet engines. These coatings are designed to provide a sacrificial layer that can easily wear away or abrade when in contact with a mating surface, such as a rotor or stator blade. The controlled wear of the coating minimizes the risk of damage, improves engine efficiency, and reduces clearance gaps between components. Silica, also known as silicon dioxide (SiO_2), is a widely available ceramic material with excellent thermal and chemical stability. Silica-based abradable coatings offer several advantages, including low thermal conductivity, high hardness, and good wear resistance, making them suitable for demanding engine environments [36].

9.4.4 Ni-Based Abradable Coating

Ni-based abradable coatings are a specialized type of coating used in high-temperature applications, particularly in gas turbines and jet engines.

Nickel-based alloys are commonly used as the primary material for abradable coatings due to their excellent mechanical properties, resistance to high temperatures, and compatibility with turbine engine environments. These abradable coatings typically consist of a metallic matrix, primarily composed of nickel, and embedded particles of a softer material, such as ceramic or polymer [37]. The softer particles are intentionally chosen to be less resistant to wear than the mating surfaces, ensuring controlled material removal during contact. The design and optimization of nickel-based abradable coatings involve several factors, including the selection of appropriate materials, particle size and distribution, coating thickness, and deposition methods. Nickel is chosen as the primary matrix material due to its excellent mechanical properties, including high strength, ductility, and resistance to oxidation and corrosion [38]. To achieve the desired abradability, softer particles are added to the nickel matrix. These particles can be ceramic materials like alumina, zirconia, or glass, or polymeric materials such as PTFE (polytetrafluoroethylene) or epoxy. The choice of the softer material depends on various factors, including its wear resistance, thermal stability, and compatibility with the nickel matrix [39].

9.5 DEMAND FOR ABRADABLE COATINGS

Abradable coatings play a critical role in various engineering applications, particularly in the aerospace and gas turbine industries. These specialized coatings are designed to provide a controlled wear mechanism, enabling clearance control and minimizing damage in contact situations. Abradable coatings exhibit controlled wear characteristics, allowing them to accommodate close tolerances in rotating machinery. Typically, they are applied as a thin layer on a substrate, such as turbine shrouds, casings, and seal components. The coating material is engineered to be softer than the opposing surface, allowing it to wear away gradually during contact. This controlled wear helps maintain proper clearances, reduce vibrations, and increase efficiency in high-speed rotating systems.

Factors driving demand for abradable coatings include the following:

Efficiency and performance: As the demand for more efficient gas turbine engines and aerospace systems increases, the need for abradable coatings grows. These coatings allow engines to operate at higher temperatures and speeds, resulting in improved performance and reduced fuel consumption.

Maintenance and cost savings: By utilizing abradable coatings, manufacturers can reduce maintenance costs associated with engine wear. Instead of replacing expensive components, the worn abradable coating can be easily reapplied or repaired, extending the service life of the underlying parts.

Environmental regulations: The aerospace and gas turbine industries are subject to stringent environmental regulations. Abradable coatings help reduce air leakage, improving the overall efficiency of engines and reducing emissions. This makes them an attractive solution for meeting regulatory requirements.

9.6 MARKET SCENARIOS FOR ABRADABLE COATINGS

Abradable coatings play a significant role in various industries, particularly in aerospace and gas turbine applications. They are designed to create a controlled frictional interface between moving components, such as rotor blades and casing, to improve efficiency and reduce wear. According to a market research report by Mordor Intelligence, the global abradable coatings market was valued at USD $1.38 billion in 2020 and is projected to reach USD $1.79 billion by 2026, growing at a compound annual growth rate (CAGR) of 4.4% during the forecast period (2021–2026) [40]. Another market research report by Research Dive estimates that the global abradable coatings market will experience substantial growth due to increasing demand from the aerospace industry. It predicts a CAGR of 5.6% from 2020 to 2027 [41].

9.7 VARIOUS PROCESSING ROUTES OF ABRADABLE COATINGS

The processing routes of abradable coatings can vary depending on the specific requirements of the application. Here are some commonly used processing routes.

9.7.1 Thermal Spray Coating

Thermal spray techniques, such as high-velocity oxygen fuel (HVOF) or atmospheric plasma spraying (APS), are frequently employed for abradable coating deposition [42]. These methods involve the deposition of molten or semimolten particles onto the substrate, resulting in a porous coating with good mechanical properties. Several materials, including metals, ceramics, and composites, have been investigated for thermal spray abradable coatings [43].

9.7.2 Sol-Gel Coating

Sol-gel processing is another route for fabricating abradable coatings. It involves the synthesis of a sol, which is a colloidal suspension of nanoparticles, followed by a controlled deposition onto the substrate. The sol-gel technique offers advantages such as low-temperature processing and the ability to tailor the coating's composition and porosity [44, 45].

9.7.3 Polymer Infiltration

In this method, a preform or substrate is first prepared using a porous material, such as carbon or ceramics. The preform is then infiltrated with a polymer resin, which can be cured to form a solid, abradable coating. Polymer infiltration techniques allow for the creation of highly porous coatings with excellent abradability [46, 47].

9.7.4 Plasma-Sprayed Polymer Coating

Plasma spraying can also be used to deposit polymer-based abradable coatings. In this approach, polymer powders are heated up to the melting point and propelled onto the surface of the substrate using a plasma jet. The resulting coating exhibits good abradability due to the high porosity and controlled particle bonding [48].

9.8 SUMMARY

An abradable coating is a specialized type of coating used in various mechanical systems, particularly in turbomachinery such as gas turbines and jet engines. The purpose of an abradable coating is to create a controlled and adjustable clearance between rotating and stationary components, such as blades and casings, to prevent contact and minimize frictional losses. The coating is typically applied to the stationary component, such as the inner surface of the casing or shroud, and is designed to wear away gradually as it comes into contact with the rotating component, such as the turbine blades. This controlled wearing process helps maintain the necessary clearance between the components while minimizing unwanted rubbing or scraping. Abradable coatings are usually composed of a soft and porous material, such as composite polymers or ceramics, that can be easily abraded or worn down by contact with the rotating component. The coating's composition and structure are engineered to optimize its abradability and provide the desired clearance control. Upon contact between the rotor and the abradable coating, the soft material easily deforms or wears away, allowing the rotating component to pass through without causing damage or significant friction. This controlled wear process helps improve the overall efficiency and performance of the mechanical system by reducing losses due to rubbing or scraping. Abradable coatings are widely used in various applications, including gas turbine engines, aircraft engines, and industrial compressors. They play a crucial role in optimizing the performance, reliability, and longevity of these systems by minimizing contact and friction between rotating and stationary components.

REFERENCES

1. Dorfman, M., Erning, U., & Mallon, J. (2002). Gas turbines use 'abradable' coatings for clearance-control seals. Sealing Technology, 2002(1), 7–8. https://doi.org/10.1016/S1350-4789(02)80002-2.
2. Yi, M., He, J., Huang, B., & Zhou, H. (1999). Friction and wear behaviour and abradability of abradable seal coating. Wear, 231(1), 47–53. https://doi.org/10.1016/S0043-1648(99)00093-9.
3. Rhys-Jones, T. N. (1990). Thermally sprayed coating systems for surface protection and clearance control applications in aero engines. Surface and Coatings Technology, 43–44(1), 402–415. https://doi.org/10.1016/0257-8972(90)90092-Q.
4. Jacquet-Richardet, G., Torkhani, M., Cartraud, P., Thouverez, F., Nouri Baranger, T., Herran, M., Gibert, C., Baguet, S., Almeida, P., & Peletan, L. (2013). Rotor to stator contacts in turbomachines. Review and application. Mechanical Systems and Signal Processing, 40(2), 401–420. https://doi.org/10.1016/j.ymssp.2013.05.010.
5. DeMasi-Marcin, J. T., & Gupta, D. K. (1994). Protective coatings in the gas turbine engine. Surface and Coatings Technology, 68–69, 1–9. https://doi.org/10.1016/0257-8972(94)90129-5.
6. Zhao, M., Zhang, L. X., & Pan, W. (2012). Properties of yttria-stabilized-zirconia based ceramic composite abradable coatings. Key Engineering Materials, 512–515, 1551–1554. doi:10.4028/www.scientific.net/kem.512-515.1551
7. Nyssen, F., & Batailly, A. (2019). Thermo-mechanical modeling of abradable coating wear in aircraft engines. Journal of Engineering for Gas Turbines and Power, 141(2) 1–11.
8. Sporer, D., Wilson, S., Fiala, P., & Schuelein, R. (2010). Thermally sprayed abradable coatings in steam turbines: Design integration and functionality testing. Turbo Expo: Power for Land, Sea, and Air, 44021, 2309–2317.
9. Sporer, D., Refke, A., Dratwinski, M., Dorfman, M., Metco, S., Giovannetti, I., Giannozzi, M., & Bigi, M. (2008). New high-temperature seal system for increased efficiency of gas turbines. Sealing Technology, 2008(10), 9–11. https://doi.org/10.1016/S1350-4789(08)70517-8.

10. Hajmrle, K., Fiala, P., Chilkowich, A. P., & Shiembob, L. T. "Abradable Seals for Gas Turbines and Other Rotary Equipment." Proceedings of the ASME Turbo Expo 2004: Power for Land, Sea, and Air. Volume 4: Turbo Expo 2004. Vienna, Austria. June 14–17, 2004. pp. 673–682. ASME. https://doi.org/10.1115/GT2004-53865.
11. Liu, A., Marshall, M., Rahimov, E., & Panizo, J. (2022). Investigation of wear mechanics and behaviour of NiCr metallic foam abradables. Proceedings of the Institution of Mechanical Engineers, Part C: Journal of Mechanical Engineering Science, 236(9), 4962–4972.
12. Qi, Y., Ma, W., & Zhuang, X. et al. (2021). Thermal shock failure behavior of TiZrNiCuBe metallic Glass/NiCrAl-bentonite abradable flame-retardant composite coatings. Journal of Thermal Spray Technology, 30, 2155–2160. https://doi.org/10.1007/s11666-021-01273-0.
13. Zhang, B., Marshall, M., Lewis, R. (2022). Investigating Al-Si base abradable material removal mechanism with axial movement in labyrinth seal system. Wear, 510–511, 204496. https://doi.org/10.1016/j.wear.2022.204496.
14. Zhang, B., & Marshall, M. (2019). Investigating material removal mechanism of Al-Si base abradable coating in labyrinth seal system. Wear, 426–427(Part A), 239–249. https://doi.org/10.1016/j.wear.2019.01.034.
15. Grimenstein, L. F. "Polymers in Thermal Spray." Proceedings of the ITSC2011. Thermal Spray 2011: Proceedings from the International Thermal Spray Conference. Hamburg, Germany. September 27–29, 2011. pp. 782–784. ASM. https://doi.org/10.31399/asm.cp.itsc2011p0782.
16. Rajendran, R. (2012). Gas turbine coatings – An overview. Engineering Failure Analysis, 26, 355–369. https://doi.org/10.1016/j.engfailanal.2012.07.007.
17. Mohammad, F., Mudasar, M., & Kashif, A. (2021). Criteria for abradable coatings to enhance the performance of gas turbine engines. Journal of Material Science and Metallurgy, 2, 1–7.
18. Batailly, A., Legrand, M., & Pierre, C. "Influence of Abradable Coating Wear Mechanical Properties on Rotor Stator Interaction." Proceedings of the ASME 2011 Turbo Expo: Turbine Technical Conference and Exposition. Volume 6: Structures and Dynamics, Parts A and B. Vancouver, British Columbia, Canada. June 6–10, 2011. pp. 941–950. ASME. https://doi.org/10.1115/GT2011-45189.
19. Novinski, E. R. (1989). The selection and performance of thermal sprayed abradable seal coatings for gas turbine engines. SAE Transactions, 98, 60–63. http://www.jstor.org/stable/44471607.
20. Sporer, D. R., Taeck, U., Dorfman, M. R., Nicoll, A. R., Giannozzi, M., & Giovannetti, I. "Novel Ceramic Abradable Coatings with Enhanced Performance." Proceedings of the ASME Turbo Expo 2006: Power for Land, Sea, and Air. Volume 4: Cycle Innovations; Electric Power; Industrial and Cogeneration; Manufacturing Materials and Metallurgy. Barcelona, Spain. May 8–11, 2006. pp. 1017–1023. ASME. https://doi.org/10.1115/GT2006-90993.
21. Nava, Y., Mutasim, Z., & Coe, M. (2001). Ceramic Abradable Coatings for Applications up to 1100° C. Thermal Spray 2001: New Surfaces for a New Millenium, (pp. 119–126) ASM International.
22. Clegg, M. A., & Mehta, M. H. (1988). NiCrAl/bentonite thermal spray powder for high temperature abradable seals. Surface and Coatings Technology, 34(1), 69–77. https://doi.org/10.1016/0257-8972(88)90090-4.
23. Wei, X., Mallon, J. R., Correa, L. F., Dorfman, M. R., & Ghasripoor, F. "Microstructure and Property Control of CoNiCrAlY Based Abradable Coatings for Optimal Performance." Proceedings of the ITSC 2000. Thermal Spray 2000: Proceedings from the International Thermal Spray Conference. Montreal, Quebec, Canada. May 8–11, 2000. pp. 407–412. ASM. https://doi.org/10.31399/asm.cp.itsc2000p0407.
24. Fois, N., Watson, M., & Marshall, M. B. (2017). The influence of material properties on the wear of abradable materials. Proceedings of the Institution of Mechanical Engineers, Part J: Journal of Engineering Tribology, 231(2), 240–253.
25. Bardi, U., Giolli, C., & Scrivani, A. et al. (2008). Development and investigation on new composite and ceramic coatings as possible abradable seals. Journal of Thermal Spray Technology, 17, 805–811. https://doi.org/10.1007/s11666-008-9246-5.
26. Guo, M., Cui, Y., Wang, C., Jiao, J., Bi, X., & Tao, C. (2023). Design and characterization of BSAS-polyester abradable environmental barrier coatings (A/EBCs) on SiC/SiC composites. Surface and Coatings Technology, 465, 129617. https://doi.org/10.1016/j.surfcoat.2023.129617.
27. Duramou, Y., Bolot, R., Seichepine, J. L., Danlos, Y., Bertrand, P., Montavon, G., & Selezneff, S. (2014). Relationships between Microstructural and Mechanical Properties of Plasma Sprayed AlSi-Polyester Composite Coatings: Application to Abradable Materials. In Key Engineering Materials (Vol. 606, pp. 155–158). Trans Tech Publications, Ltd. https://doi.org/10.4028/www.scientific.net/kem.606.155.
28. Xie, Z., Zhang, C., Wang, R., Li, D., Zhang, Y., Li, G., & Lu, X. (2021). Microstructure and wear resistance of WC/Co-based coating on copper by plasma cladding, Journal of Materials Research and Technology, 15, 821–833. https://doi.org/10.1016/j.jmrt.2021.08.114.

29. Xu, J., Hu, Z., Wang, S., Tan, W., & Zhou, J. (2022). Laser cladding co-based coating coupled with electromagnetic/ultrasonic compound energy field. Materials Science and Technology, 39, 803–814. https://doi.org/10.1080/02670836.2022.2142742.
30. https://www.samaterials.com/thermal-spraying-coatings/1919-cobalt-based-alloy-powder-for-thermal-spray.html.
31. Meng, L., Hu, M., & Jia, K. (2022). Fabrications and microstructure analysis of cobalt-based coatings by an easy-coating and sintering process. Science and Engineering of Composite Materials, 29, 529–534. https://doi.org/10.1515/secm-2022-0178.
32. Nayak, A. K. (2022). Bismuth series photocatalytic materials for the treatment of environmental pollutants. In A. K. Nayak & N. K. Sahu (Eds.), Nanostructured Materials for Visible Light Photocatalysis (pp. 135–151). Elsevier.
33. Wang, Z., Du, L., Lan, H., Huang, C., & Zhang, W. (2019). Preparation and characterization of YSZ abradable sealing coating through mixed solution precursor plasma spraying. Ceramics International, 45(9), 11802–11811. https://doi.org/10.1016/j.ceramint.2019.03.058.
34. Sun, X., Liu, Z., & Du, L. et al. (2021). A study on YSZ abradable seal coatings prepared by atmospheric plasma spray and mixed solution precursor plasma spray. Journal of Thermal Spray Technology, 30, 1199–1212. https://doi.org/10.1007/s11666-021-01178-y.
35. Lamuta, C., Di Girolamo, G., & Pagnotta, L. (2015). Microstructural, mechanical and tribological properties of nanostructured YSZ coatings produced with different APS process parameters. Ceramics International, 41(7), 8904–8914. https://doi.org/10.1016/j.ceramint.2015.03.148.
36. Morrell, P., Bettridge, D. F., Greaves, M. D., Dorfman, M. R., Russo, L. D., Britton, C., & Harrison, K.. (1998) "A New Aluminum-Silicon/Boron Nitride Powder for Clearance Control Applications." Proceedings of the ITSC 1998. Thermal Spray 1998: Proceedings from the International Thermal Spray Conference. Nice, France. May 25–29, 1998. (pp. 1187–1192). ASM. https://doi.org/10.31399/asm.cp.itsc1998p1187.
37. Marx, R., Faramarzi, R., Jungwirth, F., Kleffner, B. V., Mumme, T., Weber, M., & Wirtz, D. C., Silicate Coating of Cemented Titanium-Based Shafts in Hip Prosthetics Reduces High Aseptic Loosening. Z Orthop Unfall. 2009 Mar–Apr;147(2):175-82. German. ttps://doi.org/10.1055/s-0029-1185456. Epub 2009 Apr 8. PMID: 19358071
38. Batra, U. (2009) Thermal spray coating of abradable Ni based composite, Surface Engineering, 25(4), 284–286. DOI: 10.1179/174329407X215087
39. Xu, C., Du, L., Yang, B., & Zhang, W. (2011). The effect of Al content on the galvanic corrosion behaviour of coupled Ni/graphite and Ni–Al coatings. Corrosion Science, 53(6), 2066–2074. https://doi.org/10.1016/j.corsci.2011.02.019.
40. Intelligence. "Abradable Coatings Market - Growth, Trends, COVID-19 Impact, and Forecasts (2021-2026)." June 2021. https://www.globenewswire.com/news-release/2022/01/25/2372581/0/en/Protective-Coatings-Market-Growth-Trends-COVID-19-Impact-and-Forecasts-2021-2026.html.
41. Research Dive. "Abradable Coatings Market Size, Share, Trends, Industry Forecast 2020-2027." September 2020. https://www.researchdive.com/covid-19-insights/316/global-abradable-coatings-market#impact-analysis.
42. Davis, J. R. (Ed.). (2004). Handbook of Thermal Spray Technology. ASM International.
43. Faraoun, H. I., Grosdidier, T., Seichepine, J.-L., Goran, D., Aourag, H., Coddet, C., Zwick, J., & Hopkins, N. (2006). Improvement of thermally sprayed abradable coating by microstructure control. Surface and Coatings Technology, 201(6), 2303–2312. https://doi.org/10.1016/j.surfcoat.2006.03.047.
44. Rudra, S., Janani, K., Thamizharasan, G., Pradhan, M., Rani, B., Sahu, N. K., & Nayak, A. K. (2022). Fabrication of Mn3O4-WO3 nanoparticles based nanocomposites symmetric supercapacitor device for enhanced energy storage performance under neutral electrolyte. Electrochimica Acta, 406, 139870.
45. Rak, Z. S. (2001). A process for Cf/SiC composites using liquid polymer infiltration. Journal of the American Ceramic Society, 84, 2235–2239. https://doi.org/10.1111/j.1151-2916.2001.tb00994.x.
46. Berman, D., & Shevchenko, E. (2020). Design of functional composite and all-inorganic nanostructured materials via infiltration of polymer templates with inorganic precursors, Journal of Materials Chemistry C, 8, 10604–10627. https://doi.org/10.1039/D0TC00483A.
47. Heimann, R. B. (2008). Plasma-Spray Coating: Principles and Applications. John Wiley & Sons.
48. Gill, B. J., & Tucker, R. C. (1986). Plasma spray coating processes. Materials Science and Technology, 2(3), 207–213. https://doi.org/10.1179/mst.1986.2.3.207.

10 Multifunctional Thin Film for Energy Applications

Ftema W. Aldbea and A. Sharma

10.1 INTRODUCTION

Multifunctional thin films could profoundly affect the energy industry with enhanced efficiency and versatility [1–6]. Solar energy collection, energy storage, and energy conversion are all possible with slight adjustments to these films; see Figure 10.1 and Table 10.1. Perovskite solar cells are one application of thin films [7]. The crystalline substance perovskite is stacked into atomically thin layers to create perovskite solar cells. The ability of this material to convert solar energy into electricity makes it a good choice for use in solar cells.

Perovskite solar cells have shown promise for low-cost mass production thanks to their excellent power conversion efficiencies. Energy storage is another application for multifunctional thin films. Thin films of transition metal oxides are functional in many applications, including producing the electrodes used in rechargeable batteries [8]. Electric vehicles and portable gadgets could also benefit from the films' ability to store and release energy effectively. Research is ongoing to create novel materials and improve the performance of multifunctional thin films because of their excellent uses in the energy sector [9, 10]. Understanding multifunctional thin films for energy applications is the focus of this work.

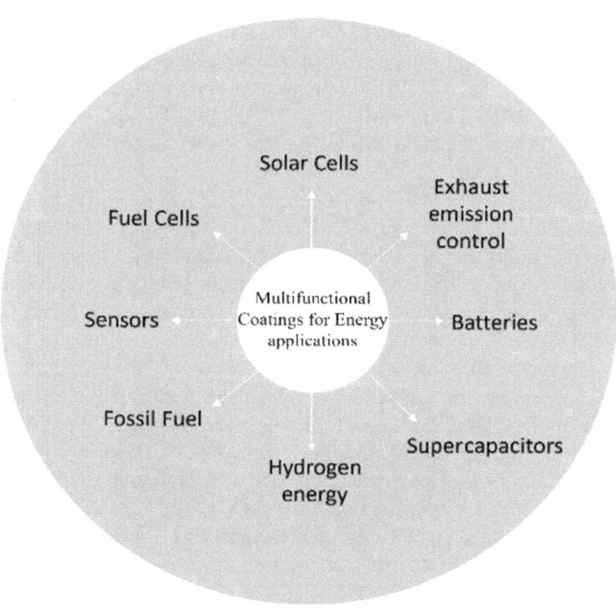

FIGURE 10.1 Exploration of the diverse possibilities where coatings can be applied to enhance energy-related functions.

DOI: 10.1201/9781032635347-10

TABLE 10.1
Different Multifunctional Thin-Film Coatings for Various Energy Applications and Subsequent Performance Evaluations

S. No.	Type of Material	Coating Material	Improvement/Efficiency/ Performance	Reference
	A range of coatings tailored for various thin-film solar cell technologies			
1	Organic solar cells	Graphene	2.07% with superior stability	[58]
2	CdTe thin-film solar cells	CdTe	12.78%	[59]
3	Quantum dot-sensitized solar cells Dye-sensitized solar cell	CdS on carbon nanotube Metal sulfide/graphene	PCE = 7.5%-5.25% / 6.01%	[30, 34, 60]
4	Amorphous silicon	Hydrogenated amorphous silicon thin films	11.9%	[61]
5	Copper indium gallium selenide thin film	Boron-doped ZnO	18.3%	[62]
6	Solar cells with single or double junctions based on silicon	Silicon in both amorphous and microcrystalline forms	Stabilized efficiency of 11.5%	[63]
7	Photovoltaic cells utilizing metal-insulator semiconductor transistor technology	Oxide of indium and tin	10.85%	[64]
	A range of coatings used on various catalysts for the purpose of exhaust emission management			
8	Catalyst including a supporting platinum group metal			
8 (i)	$Pt_n/CeCuO_2$	Pt	CO oxidation reaction	[65]
8 (ii)	Palladium–tin oxide on hexagonal boron nitride (h-BN) substrate	Pd/SnO_2	CO oxidation reaction	[66]
8 (iii)	$Pt/LaFeO_3/MgAl_2O_4$	$Pt/LaFeO_3$	CO oxidation reaction	[67]
9	Encapsulated Pt group metal catalyst			
	Manganese oxide (MnOx) over platinum (Pt) on alumina (Al_2O_3) support	MnO_x/Pt	CO oxidation reaction	[68]
10	Supported metal or oxide catalyst			
10 (i)	Fe/SiO_2	Fe	CO oxidation reaction	[69]
10 (ii)	Vanadium oxide (VO_x) supported on titanium dioxide (TiO_2) within a SBA-15 matrix	VO_x/TiO_2	Selective catalytic reduction of nitrogen oxides	[70]
10 (iii)	Cu-SSZ-13@SiO_2	SiO_2	Selective catalytic reduction of nitrogen oxides	[71]
10 (iv)	Fe_2O_3@CeO_2/TiO_2	Fe_2O_3	Selective catalytic reduction of nitrogen oxides	[72]
	Cathode materials coated by ALD for lithium-ion battery			
11	$LiFePO_4$	ZrO_2	Lessened the undesired secondary processes at the cathode–electrolyte junction	[73]
12	$LiNi_xCo_yMn_zO_2$ (x = 0.5, y = 0.2, z = 0.3)	Aluminum oxide	Suppressed the transition metal dissolution during cycling	[74]
13	$LiNi_xMn_yCo_zO_2$ (x = 0.6, y = 0.2, z = 0.2)	Aluminum oxide	Enhanced durability during cyclic operation at elevated cutoff voltages	[75]
14	$LiNi_xMn_yCo_zO_2$ (x = 0.6, y = 0.2, z = 0.2)	Aluminum oxide	Capacity retention 92.2% at 1 C rate after 300 cycles	[76]

(Continued)

TABLE 10.1 *(Continued)*
Different Multifunctional Thin-Film Coatings for Various Energy Applications and Subsequent Performance Evaluations

S. No.	Type of Material	Coating Material	Improvement/Efficiency/ Performance	Reference
15	$LiNi_xMn_yCo_zO_2$ (x = 0.83, y = 0.05, z = 0.12)	Aluminum oxide	Minimized direct current (DC) resistance, resulting in improved cycling performance	[77]
16	$LiNi_xMn_yCo_zO_2$ (x = 0.8, y = 0.1, z = 0.1)	Aluminum oxide	Lowered cathode alkalinity and managed the extent of dissolution of transition metal	[78]
17	$LiNi_xCo_yMn_zO_2$ (x = 0.8, y = 0.1, z = 0.1)	Titanium nitride	Maintained 70.9% at a current density of 100 mA • g^{-1} to the initial capacity after 200 cycles within the voltage range spanning 2.8 to 4.5 V.	[79]
18	$LiNi_xCo_yMn_zO_2$ (x = 0.8, y = 0.1, z = 0.1)	$Li_xZr_yPO_z$	Retained 80.5% of its original capacity at a cycle rate of 1 C after undergoing 100 cycles.	[80]
19	$LiNi_xMn_yO_4$ (x = 0.5, y = 1.5)	Aluminum oxide	Retained 93.7% of its original capacity at a rate of 0.5 C at a temperature of 55°C after undergoing 150 cycles	[81]
20	$LiNi_xMn_yO_4$ (x = 0.5, y = 1.5)	Titanium dioxide	Noted better capacity retention and an increase in coulombic efficiency across the full cell configuration	[82]
21	$LiMn_yNi_xO_4$ (x = 0.5, y = 1.5)	FeO_x	93.5% and 86.1% capacity retention under room temperature and 55°C after 500 cycles at 1 C rate	[83]
22	$LiMn_yNi_xO_4$ (x = 0.5, y = 1.5)	LiF	Observed enhanced mitigation of Mn dissolution	[84]
		Catalysts coating for hydrogen fuel cell		
23	Platinum /CNT	Platinum	1.4 times higher mass activity than the commercial 20% Pt/C catalyst for ORR	[85]
24	Platinum 1/ZIF-NC	Platinum	Electrochemical surface area: 229 $m^2 \cdot g_{Pt}^{-1}$; specific activity at 0.9 V: 0.51 mA • cm_{Pt}^{-2}; mass activity at 0.9 V: 1.17 A • mg_{Pt}^{-1}	[86]
25	Platinum /TiN	Platinum	Half-wave potential is 20 mV higher than that of commercial Pt/C	[87]
26	Platinum /Sb-doped SnO_2	Platinum	Electrochemical surface area: 74 $m^2 \cdot g_{Pt}^{-1}$; mass activity at 0.9 V: 102 mA • g_{Pt}^{-1}	[88]
27	Platinum / Cerium dioxide (CeO_2)/ Carbon nanotubes (CNTs)	Platinum	Electrochemical surface area: 74.13 $m^2 \cdot g_{Pt}^{-1}$; half-wave potential: 0.865 V	[89]

(Continued)

TABLE 10.1 ***(Continued)***
Different Multifunctional Thin-Film Coatings for Various Energy Applications and Subsequent Performance Evaluations

S. No.	Type of Material	Coating Material	Improvement/Efficiency/ Performance	Reference
		Coating effects on initial charge and long-term capacity in anode LIBs		
28	C/MnO	Al_2O_3	After 100 cycles, a discharge capacity of 1165 mAh • g^{-1} was obtained at a current rate of 150 mA • g^{-1}, showcasing an impressive performance. Moreover, the material demonstrated a significant specific capacity of around 600 mAh • g^{-1} when subjected to a higher current rate of 1000 mA • g^{-1}.	[90]
29	Si	MgO	At a current density of 0.1 A • g^{-1}, the material demonstrated an initial capacity (IC) of 2218.3 mAh • g^{-1}, emphasizing its impressive charge storage capabilities. Following an extensive cycling test spanning 200 cycles, a retained capacity (RC) of 765.1 mAh • g^{-1} was sustained at a more demanding current density of 0.5 A • g^{-1}.	[91]
30	$Li_4Ti_5O_{12}$	$ZnO + Al_2O_3$	At T = 55°C and a 1 C rate, the initial capacity (IC) measured 224 mAh • g^{-1}. Following 250 cycles, the material exhibited a remarkable retained capacity (RC) of 216 mAh • g^{-1}.	[92]

10.2 APPLICATION OF THIN FILMS IN SENSORS

Thin-film technology finds various applications in sensor systems, offering several advantages in terms of performance and versatility [11, 12]. One such application is in thermoelectric sensors, which are capable of measuring environmental temperatures by utilizing the thermoelectric effect. This effect generates a voltage when a temperature gradient is present at a material junction. Thermocouples, composed of two different metals bonded together, are commonly used in thermoelectric sensors [9]. By applying a temperature gradient across the junction, a voltage is produced that is directly proportional to the temperature difference, allowing for accurate temperature readings. Another type of thermoelectric sensor is the thermopile. Thermoelectric sensors offer several benefits compared to other types of temperature sensors. They are highly reliable and can operate effectively even under extremely low temperatures, making them suitable for arctic environments. Additionally, these sensors are compact and lightweight, making them ideal for mobile applications. One significant advantage of thermoelectric sensors is that they do not require an external power source since they directly monitor the voltage generated by the temperature gradient.

Another innovative application of thin-film technology is found in a fiber-coupled photodetector. This photodetector incorporates an optical sensor head employing thermal detection techniques (a

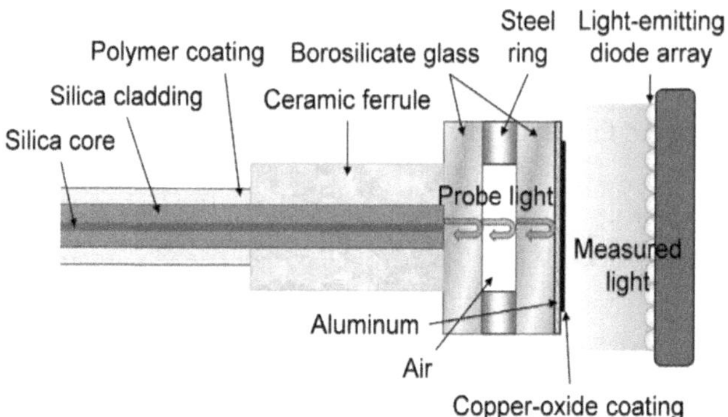

FIGURE 10.2 Diagram illustrating the head of optical sensor configuration of the photodetector revealing polymer coating and copper-oxide coatings [12]. (Reprinted with permission from ref 12. Copyright © 2017, The Author(s), open access article distributed under the terms of the Creative Commons CC BY license, which permits unrestricted use.)

schematic design is shown in Figure 10.2), enhancing its ability to counteract strong electromagnetic influences and maintain functionality in contexts with electrical delicacy. Three Fabry-Perot interferometers are connected one after the other to create the optical sensor head, and a hydrogel layer containing copper-oxide microparticles is used to coat the surface located at its end face [11].

This unique photothermal coating can be easily applied to various surfaces. When the photothermal coating is exposed to irradiation, it absorbs photons and converts them into heat. The probing light's optical path length is modified by the heat's presence, resulting in a shift in wavelength of the resonant. The photodetector demonstrates impressive power sensitivity and detection limits when subjected to white-light irradiation. It exhibits excellent response and recovery times, making it highly efficient in detecting optical signals. Furthermore, by integrating a multicore optical fiber, this photodetector enables spatially resolved measurements. Thin-film technology also plays a crucial role in radiation detectors, which are used to identify and quantify the presence of radiation in various fields such as medical imaging, radiation therapy, and nuclear power plants [13]. Different types of radiation detectors, including Geiger-Muller counters, scintillation detectors, and ionization chambers, are available. These detectors effectively convert radiation energy into an electrical signal, enabling the measurement and analysis of radiation levels. Chemical sensors, another application of thin-film technology, are used to identify and quantify the concentration of individual chemicals [14]. They find application in environmental monitoring, food and beverage processing, and medical diagnostics. These sensors can be based on electrochemical, optical, or gas-based principles. By reacting with the target chemical in a way that generates a detectable signal, such as a shift in electrical conductivity or light absorption, chemical sensors provide valuable information about the presence and concentration of specific substances. Flow sensors, also utilizing thin-film technology, are instrumental in determining the velocity of liquids or gases. These sensors have diverse applications ranging from HVAC systems to car engines and medical devices. Examples of flow sensors include differential pressure sensors, magnetic flowmeters, and ultrasonic flowmeters [15–17]. By monitoring fluid or gas pressure, velocity, or displacement, flow sensors enable accurate measurement and control of flow rates in various systems. Lastly, vacuum sensors gauge the level of atmospheric pressure within a vacuum environment. They are used in vacuum packaging, deposition processes, and semiconductor fabrication, among other applications. Vacuum sensors come in different types, such as ionization gauges, thermal conductivity gauges, and capacitance manometers [18]. These sensors analyze the characteristics of gas molecules to determine the pressure within a vacuum chamber.

Overall, thin-film technology plays a crucial role in various sensor applications, offering robust sensing capabilities and opening up new possibilities in industries and scientific research [12].

10.3 USE OF THIN-FILM ELECTROLYTES IN THE FRAMEWORK OF SOLID OXIDE FUEL CELLS

Critical significance is attributed to thin-film electrolytes in enhancing the efficiency and durability of solid oxide fuel cells (SOFCs), with the goal of transforming the fuel's chemical energy into electricity while maintaining high efficiency and minimal emissions. By depositing thin coatings on the electrode surface, the electrode polarization resistance can be reduced, and catalytic activity can be increased [19–21]. Traditionally, achieving densification of the conventional yttria-stabilized zirconia (YSZ) electrolyte layer in SOFCs requires high sintering temperatures exceeding 1400°C. Even so, this impairs the structural integrity of other functional layers within the cell. To overcome this challenge, researchers have utilized a 3 wt% titanium dioxide-doped YSZ electrolyte material (3TiYSZ) to fabricate thin electrolyte films. These films, with a thickness of 300 nm, are deposited on nickel oxide (NiO) and yttria-stabilized zirconia anode composites and fused quartz substrates using electron beam evaporation. By optimizing the sintering temperature during post-deposition within the range of 950°C to 1150°C, researchers have achieved improved structural, topographical, and electrochemical performance. X-ray diffraction analysis confirms the absence of impurity phases, while revealing the cubic YSZ to have prominent reflections. Upon sintering at 1150°C, electrolyte films display elevated crystallite size, lowered microstrain and grain boundary density, and improved surface texture. Impedance plots indicate proper response of the electrolyte, with the ionic conductivity being thermally activated and enhanced. The experimental setup depicted in Figure 10.3 showcases the step-by-step methodology for making NiO/YSZ anode composites and depositing electrolytes of thin 3TiYSZ through the utilization of e-beam evaporation. The figure also highlights the subsequent characterizations carried out to analyze the properties and performance of the fabricated materials [22]. Zirconia, also known as zirconium dioxide (ZrO_2), is another promising material used in thin-film electrolytes for SOFCs due to its pronounced ability to conduct oxygen ions and maintain chemical stability under elevated temperatures [19, 23–25]. Yttria-stabilized zirconia (YSZ), a ceramic material composed of zirconium oxide and yttrium oxide as a stabilizer, has gained widespread application in this context, thanks to its unique properties and ability to maintain the integrity of the zirconia crystal lattice without flaws or crazing [26].

Furthermore, cerium dioxide (ceria) thin-film electrolytes have also been employed in SOFCs and analogous systems employing electrochemistry, benefiting from their elevated oxygen ion conductivity at elevated temperatures. Ceria (cerium oxide) shows promise as an electrolyte material for oxide fuel cells, offering potential advantages for enhancing cell performance [27, 28].

In SOFCs, thin-film electrolytes typically consist of multiple layers of heterogeneous materials. The performance of these electrolytes can be optimized by adjusting the properties of the various alternating layers that comprise them. This approach allows for fine-tuning the electrolyte's characteristics and improving its overall functionality within the specific device application; see Table 10.1 for more specific examples.

10.4 ADVANCED COATING MATERIALS FOR SOLAR CELLS

The use of advanced coating materials has been widely acknowledged as a means to enhance the efficiency and longevity of energy applications such as solar cells. These coatings possess various properties such as thermal barrier, antireflective, hydrophobic, corrosion resistance, and nanocomposite characteristics, which have proven beneficial in numerous applications in solar cells [29]. Coatings play a vital role in solar cells, safeguarding them against elemental damage, improving efficiency, and minimizing energy loss. Antireflective, conductive, passivation, back reflector, and antisoiling coatings are commonly employed in solar cell technology. The choice of coating depends on factors such as the type of solar cell and the operating environment, as these factors influence the optimal coating selection. Solar cells, also referred to as photovoltaic (PV)

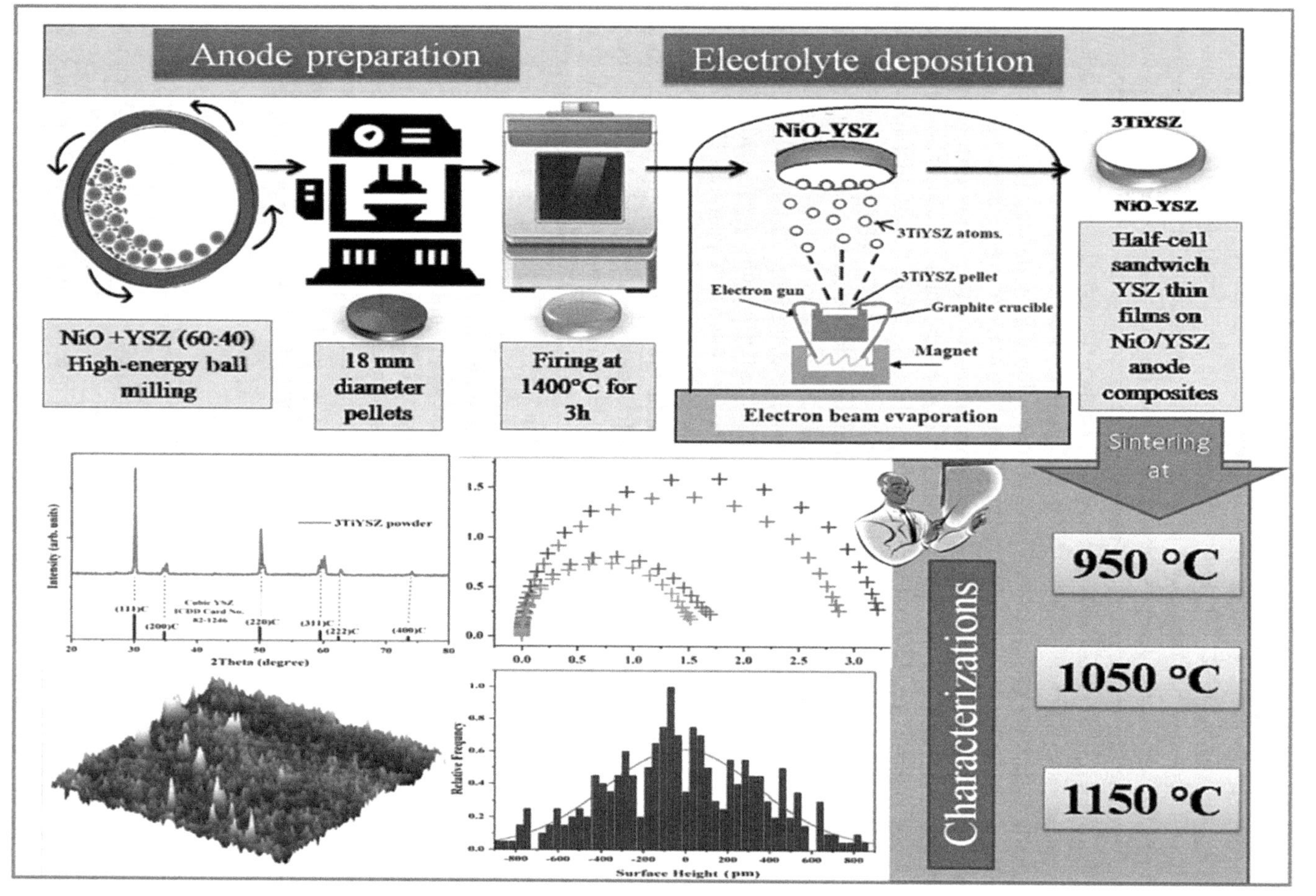

FIGURE 10.3 The sequence of steps demonstrating how the nickel oxide/yttria-stabilized zirconia anode composites are prepared and followed by the deposition process of 3TiYSZ electrolytes with reduced thickness using electron-beam deposition. The figure further illustrates the characterization methods employed to evaluate and understand the characteristics and behavior of the produced materials [22]. (Reprinted with permission from ref 22. Copyright © 2023 Elsevier B.V. All rights reserved.)

cells, hold great promise in addressing the global energy crisis and mitigating environmental pollution by directly converting solar energy into electrical power [1, 30]. However, the challenge lies in developing efficient and cost-effective solar cells. First-generation silicon-based solar cells exhibit high efficiency but are costly and challenging to produce [31, 32]. To address production costs, silicon-based thin film-solar cells of the second generation were developed. Although more affordable, they generally perform less efficiently than their first-generation counterparts. One significant issue affecting solar cell efficiency is the high reflection of incident light at the front interface, resulting in significant loss. Antireflection coatings, such as selenium (Se) vapor-based selenization coatings, have been developed to reduce reflection losses and enhance the overall efficiency of solar cells, specifically copper indium gallium selenide (CIGS) solar cells. The application of these coatings has shown promising results in improving the efficiency of CIGS solar cells [33]. Dye-sensitized solar cells (DSSCs) utilize an electrolyte between a working electrode (anode) and a counter electrode (cathode). Commonly, platinum and oxide of titanium nanostructure are employed in DSSCs. However, finding alternative substances for Pt coatings on DSSC counter electrodes is an ongoing research area. Nickel sulfide (NiS) and graphene composite films have demonstrated potential as alternatives to Pt coatings due to their high current density and low charge transfer resistance [34, 35]. Furthermore, graphene with vertically erected walls has been found to enhance the photovoltaic effectiveness of DSSCs while providing superhydrophobic properties. Superhydrophobicity is a desired property for real-world applications due to its exceptional, cost-effective, environmentally friendly, and self-cleaning qualities. The technique of deposition of glancing angle, a type of electron-beam deposition, is employed to create porous patterns in Pt counter electrodes for DSSCs. The development of porous Pt counter electrodes through GLAD holds promise for achieving greater DSSC efficiency [36]. Research by Hsu et al. revealed that employing GLAD to coat a porous Pt counter electrode led to a significant increase in DSSC efficiency, up to 10.71%. Despite significant breakthroughs, the high resistance of cell components remains a challenge in increasing the steady efficiency of solar cells. Researchers are diligently working to overcome this limitation and further enhance the light conversion efficiency of DSSCs [37, 38].

What's more, in Figure 10.4, the layers of a solar cell with a configuration of superstrate are depicted. The first layer is a shockproof glass that makes way for light to enter. This glass is coated with a transparent conducting oxide (TCO) film, which maintains electrical contact while allowing light transmission. The TCO coating is divided into multiple cells using laser scribing, enabling series connection and efficient charge carrier separation. A semiconductor layer is then deposited to generate charges, followed by the deposition of a back contact to collect the carriers. The solar cell is encapsulated with an ethylene vinyl acetate (EVA) polymer layer and covered with a backsheet. Finally, a junction box is connected to the back of the solar module for electrical connections. The thin-film amorphous silicon PV modules contain a significant amount of recyclable glass, including the TCO layer, providing the added advantage of glass extraction without damage [39].

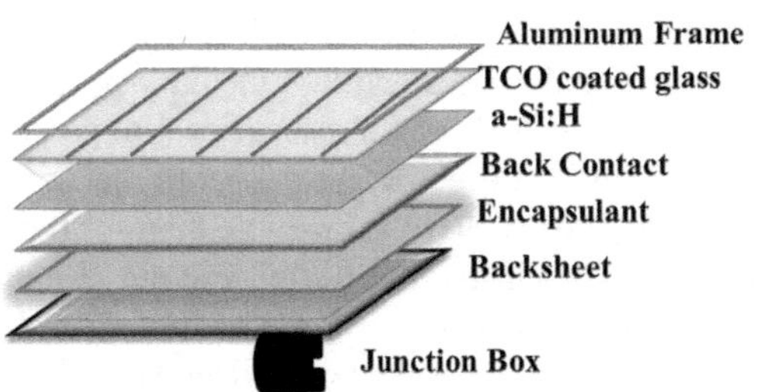

FIGURE 10.4 A schematic representation of a solar cell with a configuration of superstrate, depicting the distinct layers involved in its structure [39]. Reprinted with permission from ref 39. Copyright © 2023 Elsevier B.V. All rights reserved.

In short, the development of efficient solar cell coatings is crucial for improving efficiency and reducing production costs. Alternative semiconductor materials, nanotextured surfaces, antireflection coatings, and advanced techniques like GLAD hold promise in enhancing solar cell performance. Ongoing studies aim to address current challenges and increase the stable efficiency of solar cells; see Table 10.1 for more specific examples [35, 36].

10.5 THIN-FILM COATINGS FOR FUEL CELLS, BATTERIES AND EXHAUST EMISSION CONTROL, AND OTHER

The utilization of the chemical interplay between hydrogen and oxygen allows fuel cells to create electricity. However, one drawback of fuel cells is their susceptibility to corrosion and degradation over time. To overcome this challenge, the application of coatings has shown promise in preventing corrosion and enhancing the efficiency of fuel cells. Various materials, including ceramics, polymers, and metals, can be used to coat fuel cells and provide protection. Coatings also play a crucial role in improving the performance, safety, and lifespan of batteries. Graphene, metals, polymers, and solid electrolytes are examples of commonly used coatings for batteries. Implementing coatings in batteries can significantly enhance their efficiency, safety, and longevity. A notable development in Li-ion battery technology is the use of ZnO quantum dot coatings, as reported by Sun et al. Through the atomic layer deposition technique, ZnO quantum dots of varying sizes are uniformly deposited on graphene sheets [40, 41]. Enhanced rate performance is observed in smaller particles, resulting in a significant upsurge in reversible specific capacity. Additionally, ZnO nanostructures undergo notable volume fluctuations during the charge and discharge processes. Qin et al. have detailed the fabrication of an improved anode for lithium batteries, comprising a three-dimensional porous graphene network hosting Sn nanoparticles encased in graphene shells (Sn@G-PGNWs) [42]. This unique structure, fabricated through a single-step chemical vapor deposition technique, enhances the overall structural integrity and electrical conductivity of the electrode. This approach demonstrates an exceptional rate performance and extended cycling stability. In the case of solid electrolytes, their ionic conductivity is generally lower compared to liquid electrolytes. However, by incorporating carbon nanotubes (CNT) as fillers, Tang et al. have successfully improved the ionic flow through polyethylene oxide (PEO) coatings. To mitigate the risk of battery shorting, the CNTs are strategically placed within insulating clay layers, resulting in enhanced Li salt dissociation [43].

The modification of cathode surfaces is achieved by creating thin protective films using techniques such as atomic layer deposition (ALD). ALD coatings, including metal oxides, nitrides, phosphates, fluorides, and alloy coatings, offer several benefits, such as improved electronic and ion flow, adjustment of surface chemical properties, suppression of metal dissolution, and physical protection [45–47]. ALD, as a manufacturing technique operating at the atomic and nearly atomic scale, supplies superior catalysts and electrodes with increased accuracy when compared to standard wet-chemistry methodologies; see Figure 10.5. Developed by Tuomo Suntola during the 1970s, a method for producing thin films by atomic layer deposition centered around self-limiting reactions that take place solely on the surface of the substrate [48]. It offers precise control over film growth through alternating pulses of precursor gases and intermediary purging stages. Compared to conventional chemical vapor deposition, ALD provides superior film properties.

In the context of batteries using solid electrolytes for lithium ions, single-crystalline Ni-rich cathodes, $LiNi_{0.5}Co_{0.2}Mn_{0.3}O_2$ (SC-NCM) have gained attention for their higher capacity retention. However, they suffer from interfacial instability during cycling. To address this issue, Li_3PO_4 is introduced as a coating for SC-NCM electrodes using ALD [49–53]. Through in-situ AFM experiments and subsequent characterizations, it has been established that this coating suppresses undesirable side reactions and enhances interfacial stability; see Figure 10.6.

The utilization of coating demonstrates efficacy in surface alteration, particularly for enhancing the capabilities of materials used in energy storage applications. This strategy aims to enhance specific capacity and cycling stability. Recent progress has been made in carbon coating techniques

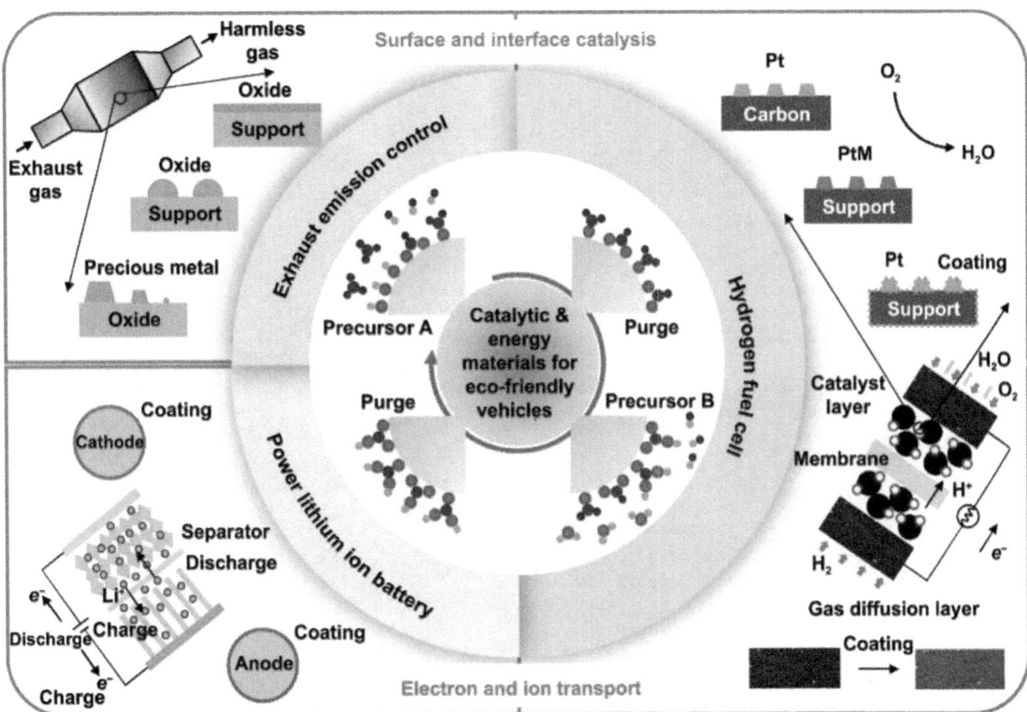

FIGURE 10.5 Schematic diagram illustrating the enhancement of surfaces for state-of-the-art catalytic and energy materials, intended for use in sustainable vehicle technologies [44]. (Copyright https://creativecommons.org/licenses/by/4.0/#.)

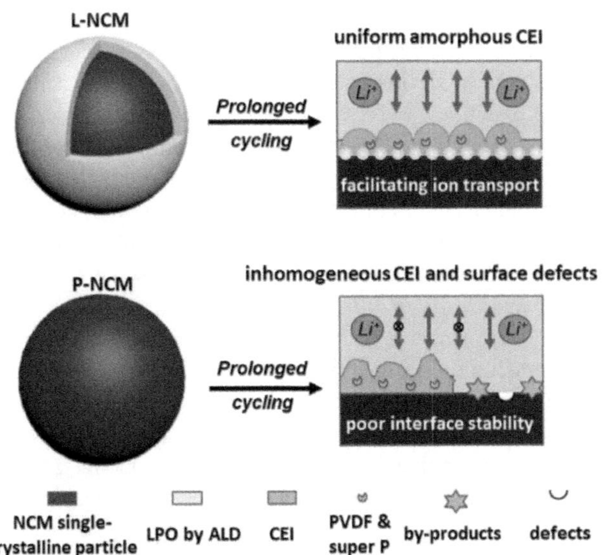

FIGURE 10.6 Study of the electrochemical behaviors exhibited by P-NCM and L-NCM in solid-state lithium batteries, supported by a visual representation illustrating the mechanism governing the degradation of electrode surfaces and the regulation of interfaces for both P-NCM and L-NCM [49]. (Permission from Wiley© 2023 Copyright - All Rights Reserved.)

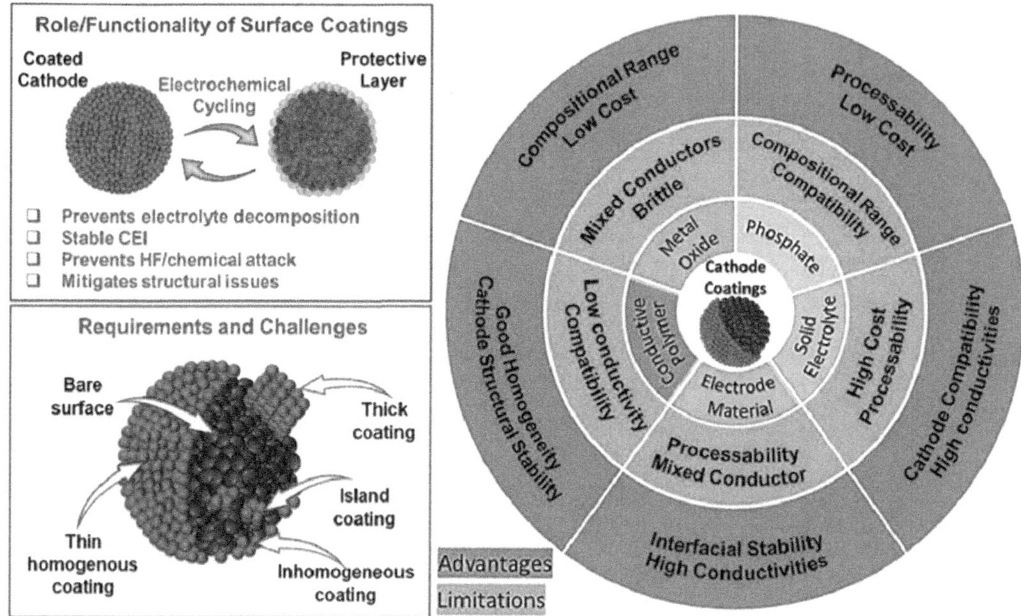

FIGURE 10.7 Different types of surface coatings on and effect on properties of high-energy-density lithium-ion battery cathode materials [56]. (Copyright © 2021 Elsevier B.V. All rights reserved.)

for cathode materials such as $LiFePO_4$, $LiMn_2O_4$, $LiCoO_2$, NCA ($LiNiCoAlO_2$), and NCM ($LiNiMnCoO_2$); see Figure 10.7 [54, 55].

In addition, coatings also play a crucial role in enhancing the performance of supercapacitors. Supercapacitors, also known as ultracapacitors or electrochemical capacitors, can benefit from coatings that offer increased energy density, improved stability, and reduced internal resistance. Commonly used coatings in supercapacitors include carbon, conductive polymer, metal oxides, and graphene. The application of these coatings has shown significant potential in enhancing the overall performance of supercapacitors. In the realm of fossil fuel-based energy systems, coatings find applications in various components such as oil rigs, pipelines, and power stations. These coatings serve specific purposes based on their intended function and environmental factors. Antifouling coatings help mitigate the accumulation of unwanted substances, thermal barrier coatings offer protection against high temperatures, abrasion-resistant coatings enhance durability, and corrosion-resistant coatings prevent deterioration caused by chemical reactions [57]. The selection of coatings depends on the specific requirements and operating conditions of each component within the fossil fuel-based energy infrastructure.

In short, the strategic application of thin-film coatings empowers energy storage systems with enhanced electrochemical performance, stability, and functionality. These coatings offer valuable solutions for supercapacitors, improving energy density, stability, and internal resistance. Additionally, they serve as vital protective measures for fossil fuel-based energy systems, safeguarding against fouling, high temperatures, abrasion, and corrosion. By embracing thin–film coatings, the path to advancements in various energy applications is illuminated, unlocking greater efficiency, reliability, and progress in the field of energy storage. See Table 10.1 for more specific examples.

10.6 FUTURE TRENDS IN COATINGS

Coatings play an indispensable role in various energy sectors, and their significance will continue to grow in the future. While thin film coatings have found use in both renewable and nonrenewable energy domains, relying solely on their implementation will not adequately address the escalating demand for energy. To ensure the prospects of human development, substantial focus must be channeled toward sophisticated coatings integrating features like mesoporous architectures, carbon allotropes, nanowires, alloys, nanotubes, chiral nanostructures, nanocomposites, superlattices, and hybrid nanostructures. Moreover, addressing prevailing energy challenges like carbon emissions, energy harvesting, energy storage, and energy conservation, advanced coating materials and techniques must be combined to tackle these issues effectively. We can expect the rise of energy-related coatings, including solar reflective coatings for improved energy efficiency, anticorrosion coatings to protect renewable energy infrastructure, and hydrophobic coatings for enhanced energy storage. Moreover, coatings with intelligent energy efficiency and harvesting functionalities will become increasingly prominent. Looking ahead, the coatings sector is poised for numerous advancements in the coming years, paving the way for a transition to a more sustainable energy future. One exciting development is the thin-film thermoelectric device, which leverages heat to generate electricity. These flexible devices can be coated with thermoelectric materials, allowing them to conform to various surfaces and enable versatile energy conversion technology. Among the promising applications of multifunctional thin films for energy applications, the use of thin La_2O_3 films with porous honeycomb structures holds great potential. These films can serve as electrodes in supercapacitors and gas sensors, facilitating the detection of specific gases in the environment and offering diverse applications [10]. The research focus has also extended to multifunctional coatings for solar panels, aiming to reduce costs without compromising durability. Recent findings include coatings that reduce glare, self-clean, change color based on temperature, and provide transparent conductivity, among other functionalities. Furthermore, the utilization of amorphous $Ge_2Sb_2Te_5$ (GST) thin films as a multifunctional platform has garnered significant attention. These films possess unique characteristics such as phase transition behavior and high electrical conductivity, making them strong contenders for applications in photo/thermosensing and nonvolatile memory [9]. By combining these features, a versatile platform can be built, opening up novel possibilities in sensing, computing, and data storage. In short, the future of multifunctional thin films for energy applications, including solar, sensors, batteries, and more, looks promising. Ongoing research and development efforts will drive advancements in coating technologies, enabling improved energy efficiency, enhanced functionality, and expanded applications across various energy sectors. These advancements will contribute to a more sustainable and innovative energy landscape.

10.7 CONCLUSION

The development of multifunctional thin-film coatings for energy production is a key objective in modern clean energy technologies. These coatings offer high efficiency, low emissions, and adaptability to various fuel sources, making them a highly promising energy solution. However, several scientific challenges need to be addressed in order to optimize their performance and facilitate device fabrication. Throughout this chapter, we have explored the diverse applications of multifunctional thin films in the realm of energy production. We have also delved into their significance as anode and cathode materials, as well as their relevance in other energy-related areas. The extensive discussion highlights the wide-ranging potential of thin-film coatings in various energy applications, including solar panels, sensors, batteries, and more. Moving forward, it is crucial to continue research and development efforts aimed at overcoming scientific hurdles and refining the fabrication processes of multifunctional thin films. By doing so, we can unlock their full potential and pave the way for more efficient and sustainable energy systems. In summary, multifunctional thin film-coatings hold immense promise for revolutionizing energy production. Their versatility and adaptability make them a vital technology in the pursuit of clean energy solutions. As we strive for a more sustainable future, the advancements made in the field of multifunctional thin films will undoubtedly play a crucial role in driving innovation and shaping the energy landscape of tomorrow.

REFERENCES

1. L. El Chaar, L.A. Lamont, N. El Zein, Review of photovoltaic technologies, Renew. Sustain. Energy Rev. 15 (2011) 2165–2175. https://doi.org/10.1016/j.rser.2011.01.004.
2. A. Evans, A. Bieberle-Hütter, J.L.M. Rupp, L.J. Gauckler, Review on microfabricated micro-solid oxide fuel cell membranes, J. Power Sources. 194 (2009) 119–129. https://doi.org/10.1016/j.jpowsour.2009.03.048.
3. B. Wu, C. Chen, D.L. Danilov, R.A. Eichel, P.H.L. Notten, All-solid-state thin film Li-ion batteries: new challenges, new materials, and new designs, Batteries. 9 (2023) 186. https://doi.org/10.3390/batteries 9030186.
4. T. Osaka, H. Nara, T. Momma, T. Yokoshima, New Si-O-C composite film anode materials for LIB by electrodeposition, J. Mater. Chem. A. 2 (2014) 883–896. https://doi.org/10.1039/c3ta13080k.
5. D. Go, J.W. Shin, S. Lee, J. Lee, B.C. Yang, Y. Won, M. Motoyama, J. An, Atomic layer deposition for thin film solid-state battery and capacitor, Int. J. Precis. Eng. Manuf. - Green Technol. 10 (2022) 851–873. https://doi.org/10.1007/s40684-022-00419-x.
6. J. Zhou, P. Cui, Z. Wang, Thin film oxide solid electrolytes towards high energy density batteries: Progress of preparation methods and interface optimization, J. Mater. Chem. A. 11 (2023) 15122–15139. https://doi.org/10.1039/d3ta02448b.
7. H.S. Jung, N.G. Park, Perovskite solar cells: From materials to devices, Small. 11 (2015) 10–25. https://doi.org/10.1002/smll.201402767.
8. E. Rauwel, P. Rauwel, Metal Oxide Thin Films, Synthesis, Characterization, and Application. Materials. *14* (2021) 1834. https://doi.org/10.3390/ma14081834.
9. M.M. Alkhamisi, S.Y. Marzouk, A.R. Wassel, A.M. El-Mahalawy, R.A. Almotiri, Multi-functional platform based on amorphous Ge2Sb2Te5 thin films for photo/thermodetection and non-volatile memory applications, Mater. Sci. Semicond. Process. 149 (2022) 106856. https://doi.org/10.1016/j.mssp.2022.106856.
10. A.A. Yadav, A.C. Lokhande, R.B. Pujari, J.H. Kim, C.D. Lokhande, The synthesis of multifunctional porous honey comb-like La2O3 thin film for supercapacitor and gas sensor applications, J. Colloid Interface Sci. 484 (2016) 51–59. https://doi.org/10.1016/j.jcis.2016.08.056.
11. Z. Liu, B. Tian, B. Zhang, J. Liu, Z. Zhang, S. Wang, Y. Luo, L. Zhao, P. Shi, Q. Lin, Z. Jiang, A thin-film temperature sensor based on a flexible electrode and substrate, Microsystems Nanoeng. 7 (2021). https://doi.org/10.1038/s41378-021-00271-0.
12. G.Y. Chen, X. Wu, X. Liu, D.G. Lancaster, T.M. Monro, H. Xu, Photodetector based on Vernier-enhanced fabry-perot interferometers with a photo-thermal coating, Sci. Rep. 7 (2017) 1–9. https://doi.org/10.1038/srep41895.
13. K. Arshak, O. Korostynska, Thin- and thick-film real-time gamma radiation detectors, IEEE Sens. J. 5 (2005) 574–580. https://doi.org/10.1109/JSEN.2005.850992.
14. P.K. Sekhar, E.L. Brosha, R. Mukundan, F.H. Garzon, Chemical sensors for environmental monitoring and homeland security, Electrochem. Soc. Interface. 19 (2010) 35–40. https://doi.org/10.1149/2.F04104if.
15. B.R. Bracio, R.J. Fasching, F. Kohl, J. Krocza, A smart thin-film flow sensors for biomedical applications, Annu. Int. Conf. IEEE Eng. Med. Biol. - Proc. 4 (2000) 2800. https://doi.org/10.1109/iembs.2000.901445.
16. N.T. Nguyen, Micromachined flow sensors - A review, Flow Meas. Instrum. 8 (1997) 7–16. https://doi.org/10.1016/S0955-5986(97)00019-8.
17. F. Ejeian, S. Azadi, A. Razmjou, Y. Orooji, A. Kottapalli, M. Ebrahimi Warkiani, M. Asadnia, Design and applications of MEMS flow sensors: A review, Sensors Actuators, A Phys. 295 (2019) 483–502. https://doi.org/10.1016/j.sna.2019.06.020.
18. W.J. Alvesteffer, D.C. Jacobs, D.H. Baker, Miniaturized thin film thermal vacuum sensor, J. Vac. Sci. Technol. A. 13 (1995) 2980–2985. https://doi.org/10.1116/1.579624.
19. S.B.K. Moorthy, Thin film structures in energy applications. (2015). Springer. https://doi.org/10.1007/978-3-319-14774-1.
20. S. He, Y. Zou, K. Chen, S.P. Jiang, A critical review of key materials and issues in solid oxide cells, Interdiscip. Mater. 2 (2023) 111–136. https://doi.org/10.1002/idm2.12068.
21. P. Vinchhi, M. Khandla, K. Chaudhary, R. Pati, Recent advances on electrolyte materials for SOFC: A review, Inorg. Chem. Commun. 152 (2023) 110724. https://doi.org/10.1016/j.inoche.2023.110724.
22. G. Chasta, M.S. Dhaka, TiO_2 incorporated YSZ thin films for SOFCs: Thermal evolution to the micro-structural, topographical and electrochemical properties, Surf. Coatings Technol. 458 (2023) 129318. https://doi.org/10.1016/j.surfcoat.2023.129318.
23. S.P.S. Badwal, Zirconia-based solid electrolytes: Microstructure, stability and ionic conductivity, Solid State Ionics. 52 (1992) 23–32. https://doi.org/10.1016/0167-2738(92)90088-7.
24. P.H. MILLER, The electrical, Phys. Rev. 60 (1941) 890.

25. L. Mathur, Y. Namgung, H. Kim, S.J. Song, Recent progress in electrolyte-supported solid oxide fuel cells: A review, J. Korean Ceram. Soc. 60 (2023) 614–636. https://doi.org/10.1007/s43207-023-00296-3.
26. G. Chasta, M.S. Dhaka, A comparative study of TiO_2 doped and undoped yttria stabilized zirconia thin films for solid oxide fuel cell application, J. Solid State Electrochem. 28 (2023) 1909–1917. https://doi.org/10.1007/s10008-023-05485-y.
27. E.S. Putna, J. Stubenrauch, J.M. Vohs, R.J. Gorte, Ceria-based anodes for the direct oxidation of methane in solid oxide fuel cells, Langmuir. 11 (1995) 4832–4837. https://doi.org/10.1021/la00012a040.
28. N. Jaiswal, K. Tanwar, R. Suman, D. Kumar, S. Uppadhya, O. Parkash, A brief review on ceria based solid electrolytes for solid oxide fuel cells, J. Alloys Compd. 781 (2019) 984–1005. https://doi.org/10.1016/j.jallcom.2018.12.015.
29. M.S. Mozumder, A.H.I. Mourad, H. Pervez, R. Surkatti, Recent developments in multifunctional coatings for solar panel applications: A review, Sol. Energy Mater. Sol. Cells. 189 (2019) 75–102. https://doi.org/10.1016/j.solmat.2018.09.015.
30. C. Li, J. Xia, Q. Wang, J. Chen, C. Li, W. Lei, X. Zhang, Photovoltaic property of a vertically aligned carbon nanotube hexagonal network assembled with CdS quantum dots, ACS Appl. Mater. Interfaces. 5 (2013) 7400–7404. https://doi.org/10.1021/am401725x.
31. S. Olson, K. Hummler, B. Sapp, Challenges in thin wafer handling and processing, ASMC (Advanced Semicond. Manuf. Conf. Proc. (2013) 62–65. https://doi.org/10.1109/ASMC.2013.6552776.
32. T. Kushida, S. Tanaka, C. Morita, T. Tanji, Y. Ohshita, Mapping of minority carrier lifetime distributions in multicrystalline silicon using transient electron-beam-induced current, J. Electron Microsc. (Tokyo). 61 (2012) 293–298. https://doi.org/10.1093/jmicro/dfs050.
33. M. Kaelin, D. Rudmann, A.N. Tiwari, Low cost processing of CIGS thin film solar cells, Sol. Energy. 77 (2004) 749–756. https://doi.org/10.1016/j.solener.2004.08.015.
34. H. Bi, W. Zhao, S. Sun, H. Cui, T. Lin, F. Huang, X. Xie, M. Jiang, Graphene films decorated with metal sulfide nanoparticles for use as counter electrodes of dye-sensitized solar cells, Carbon N. Y. 61 (2013) 116–123. https://doi.org/10.1016/j.carbon.2013.04.075.
35. N.N. Kariuki, W.J. Khudhayer, T. Karabacak, D.J. Myers, GLAD Pt-Ni alloy nanorods for oxygen reduction reaction, ACS Catal. 3 (2013) 3123–3132. https://doi.org/10.1021/cs400759u.
36. S.Y. Hsu, C.H. Tsai, C.Y. Lu, Y.T. Tsai, T.W. Huang, Y.H. Jhang, Y.F. Chen, C.C. Wu, Y.S. Chen, Nanoporous platinum counter electrodes by glancing angle deposition for dye-sensitized solar cells, Org. Electron. 13 (2012) 856–863. https://doi.org/10.1016/j.orgel.2012.01.035.
37. T.C. Wu, W.M. Huang, T.H. Meen, J.K. Tsai, Performance improvement of dye-sensitized solar cells with pressed TiO_2 nanoparticles layer, Coatings. 13 (2023) 907. https://doi.org/10.3390/coatings13050907.
38. A.A. Hussain, H. Abdulelah, A.H. Amteghy, R.A. Dheyab, B. Hamdan Almulla, Effect of multilayers CdS nanocrystalline thin films on the performance of dye-sensitized solar cells, J. Nanotechnol. 2023 (2023) 0–6. https://doi.org/10.1155/2023/7998917.
39. P. S. Kumar, Extraction and analysis of TCO coated glass from waste amorphous silicon thin film solar module, Sol. Energy Mater. Sol. Cells. 253 (2023) 112227. https://doi.org/10.1016/j.solmat.2023.112227.
40. V.A. Fonoberov, A.A. Balandin, ZnO quantum dots: Physical properties and optoelectronic applications, J. Nanoelectron. Optoelectron. 1 (2006) 19–38. https://doi.org/10.1166/jno.2006.002.
41. X. Sun, C. Zhou, M. Xie, H. Sun, T. Hu, F. Lu, S.M. Scott, S.M. George, J. Lian, Synthesis of ZnO quantum dot/graphene nanocomposites by atomic layer deposition with high lithium storage capacity, J. Mater. Chem. A. 2 (2014) 7319–7326. https://doi.org/10.1039/c4ta00589a.
42. J. Qin, C. He, N. Zhao, Z. Wang, C. Shi, E.Z. Liu, J. Li, Graphene networks anchored with Sn@Graphene as lithium ion battery anode, ACS Nano. 8 (2014) 1728–1738. https://doi.org/10.1021/nn406105n.
43. C. Tang, K. Hackenberg, Q. Fu, P.M. Ajayan, H. Ardebili, High ion conducting polymer nanocomposite electrolytes using hybrid nanofillers, Nano Lett. 12 (2012) 1152–1156. https://doi.org/10.1021/nl202692y.
44. X. Liu, Y. Su, R. Chen, Atomic-scale engineering of advanced catalytic and energy materials via atomic layer deposition for eco-friendly vehicles, Int. J. Extrem. Manuf. 5 (2023). https://doi.org/10.1088/2631-7990/acc6a7.
45. S.M. George, Atomic layer deposition: An overview, Chem. Rev. 110 (2010) 111–131. https://doi.org/10.1021/cr900056b.
46. J. Zhang, Y. Li, K. Cao, R. Chen, Advances in atomic layer deposition, Nanomanufacturing Metrol. 5 (2022) 191–208. https://doi.org/10.1007/s41871-022-00136-8.
47. S. Adhikari, S. Selvaraj, D.H. Kim, Progress in powder coating technology using atomic layer deposition, Adv. Mater. Interfaces. 5 (2018) 1–20. https://doi.org/10.1002/admi.201800581.
48. T. Suntola, J. Antson, Method for producing compound thin films, US Pat. 4,058,430. (1977).

49. H.J. Guo, Y. Sun, Y. Zhao, G.X. Liu, Y.X. Song, J. Wan, K.C. Jiang, Y.G. Guo, X. Sun, R. Wen, Surface Degradation of Single-crystalline Ni-rich Cathode and Regulation Mechanism by Atomic Layer Deposition in Solid-State Lithium Batteries. Angew Chem Int Ed Engl. 61(48) (2022) e202211626. doi: 10.1002/anie.202211626
50. A. Manthiram, X. Yu, S. Wang, Lithium battery chemistries enabled by solid-state electrolytes, Nat. Rev. Mater. 2 (2017) 1–16. https://doi.org/10.1038/natrevmats.2016.103.
51. W. Liu, X. Li, D. Xiong, Y. Hao, J. Li, H. Kou, B. Yan, D. Li, S. Lu, A. Koo, K. Adair, X. Sun, Significantly improving cycling performance of cathodes in lithium ion batteries: The effect of Al_2O_3 and $LiAlO_2$ coatings on LiNi0.6Co0.2$Mn_{0.2}O_2$, Nano Energy. 44 (2018) 111–120. https://doi.org/10.1016/j.nanoen.2017.11.010.
52. X. Li, J. Liu, M.N. Banis, A. Lushington, R. Li, M. Cai, X. Sun, Atomic layer deposition of solid-state electrolyte coated cathode materials with superior high-voltage cycling behavior for lithium ion battery application, Energy Environ. Sci. 7 (2014) 768–778. https://doi.org/10.1039/c3ee42704h.
53. M. Amirmaleki, C. Cao, B. Wang, Y. Zhao, T. Cui, J. Tam, X. Sun, Y. Sun, T. Filleter, Nanomechanical elasticity and fracture studies of lithium phosphate (LPO) and lithium tantalate (LTO) solid-state electrolytes, Nanoscale. 11 (2019) 18730–18738. https://doi.org/10.1039/c9nr02176k.
54. A. Masias, J. Marcicki, W.A. Paxton, Opportunities and challenges of lithium ion batteries in automotive applications, ACS Energy Lett. 6 (2021) 621–630. https://doi.org/10.1021/acsenergylett.0c02584.
55. D.E. Demirocak, S.S. Srinivasan, E.K. Stefanakos, A review on nanocomposite materials for rechargeable Li-ion batteries, Appl. Sci. 7 (2017) 1–26. https://doi.org/10.3390/app7070731.
56. U. Nisar, N. Muralidharan, R. Essehli, R. Amin, I. Belharouak, Valuation of surface coatings in high-energy density lithium-ion battery cathode materials. Energy Storage Mater. 38 (2021) 309–328. https://doi.org/10.1016/j.ensm.2021.03.015.
57. E. Korkut, M. Atlar, An experimental investigation of the effect of foul release coating application on performance, noise and cavitation characteristics of marine propellers, Ocean Eng. 41 (2012) 1–12. https://doi.org/10.1016/j.oceaneng.2011.12.012.
58. K. Kim, S.H. Bae, C.T. Toh, H. Kim, J.H. Cho, D. Whang, T.W. Lee, B. Özyilmaz, J.H. Ahn, Ultrathin organic solar cells with graphene doped by ferroelectric polarization, ACS Appl. Mater. Interfaces. 6 (2014) 3299–3304. https://doi.org/10.1021/am405270y.
59. H. Li, X. Liu, B. Yang, P. Wang, Influence of substrate bias and post-deposition Cl treatment on CdTe film grown by RF magnetron sputtering for solar cells, RSC Adv. 4 (2014) 5046–5054. https://doi.org/10.1039/c3ra44831b.
60. C. Yang, H. Bi, D. Wan, F. Huang, X. Xie, M. Jiang, Direct PECVD growth of vertically erected graphene walls on dielectric substrates as excellent multifunctional electrodes, J. Mater. Chem. A. 1 (2013) 770–775. https://doi.org/10.1039/c2ta00234e.
61. T. Matsui, M. Kondo, Advanced materials processing for high-efficiency thin-film silicon solar cells, Sol. Energy Mater. Sol. Cells. 119 (2013) 156–162. https://doi.org/10.1016/j.solmat.2013.05.056.
62. T. Kobayashi, T. Kumazawa, Z. Jehl Li Kao, T. Nakada, Cu(In,Ga)Se2 thin film solar cells with a combined ALD-Zn(O,S) buffer and MOCVD-ZnO:B window layers, Sol. Energy Mater. Sol. Cells. 119 (2013) 129–133. https://doi.org/10.1016/j.solmat.2013.05.052.
63. S.Y. Myong, L.S. Jeon, Improved light trapping in thin-film silicon solar cells via alternated n-type silicon oxide reflectors, Sol. Energy Mater. Sol. Cells. 119 (2013) 77–83. https://doi.org/10.1016/j.solmat.2013.05.033.
64. J.W. Lee, B.U. Ye, D. Kim, J.K. Kim, J. Heo, H.Y. Jeong, M.H. Kim, W.J. Choi, J.M. Baik, ZnO Nanowire-Based Antireflective Coatings with Double- Nanotextured Surfaces, ACS Appl. Mater. Interfaces. 6(3) (2014) 1375–1379. https://doi.org/10.1021/am4051734.
65. X. Liu, S. Jia, M. Yang, Y. Tang, Y. Wen, S. Chu, J. Wang, B. Shan, R. Chen, Activation of subnanometric Pt on Cu-modified CeO_2 via redox-coupled atomic layer deposition for CO oxidation, Nat. Commun. 11 (2020). https://doi.org/10.1038/s41467-020-18076-6.
66. Y. Zuo, Z. Wang, H. Zhao, L. Zhao, L. Zhang, B. Yi, W. Bao, Y. Zhang, L. Su, Y. Yu, J. Xie, Synthesis of a spatially confined, highly durable, and fully exposed Pd cluster catalyst via sequential site-selective atomic layer deposition, ACS Appl. Mater. Interfaces. 14 (2022) 14466–14473. https://doi.org/10.1021/acsami.2c00009.
67. X. Mao, A.C. Foucher, T. Montini, E.A. Stach, P. Fornasiero, R.J. Gorte, Epitaxial and strong support interactions between Pt and LaFeO3Films stabilize Pt dispersion, J. Am. Chem. Soc. 142 (2020) 10373–10382. https://doi.org/10.1021/jacs.0c00138.
68. S. Lee, C. Lin, S. Kim, X. Mao, T. Kim, S.J. Kim, R.J. Gorte, W. Jung, Manganese oxide overlayers promote CO oxidation on Pt, ACS Catal. 11 (2021) 13935–13946. https://doi.org/10.1021/acscatal.1c04214.

69. X. Wang, B. Jin, Y. Jin, T. Wu, L. Ma, X. Liang, Supported single Fe atoms prepared via atomic layer deposition for catalytic reactions, ACS Appl. Nano Mater. 3 (2020) 2867–2874. https://doi.org/10.1021/acsanm.0c00146.
70. M. Akazawa, T. Nakano, Layer deposition, Appl. Phys. Lett. 122110 (2012) 226–229.
71. H. Tian, Y. Ping, Y. Zhang, Z. Zhang, L. Sun, P. Liu, J. Zhu, X. Yang, Atomic layer deposition of silica to improve the high-temperature hydrothermal stability of Cu-SSZ-13 for NH3 SCR of NOx, J. Hazard. Mater. 416 (2021) 126194. https://doi.org/10.1016/j.jhazmat.2021.126194.
72. X. Qi, L. Han, J. Deng, T. Lan, F. Wang, L. Shi, D. Zhang, SO_2-tolerant catalytic reduction of NO_x via tailoring electron transfer between surface iron sulfate and subsurface ceria, Environ. Sci. Technol. 56 (2022) 5840–5848. https://doi.org/10.1021/acs.est.2c00944.
73. Y. Jin, H. Yu, X. He, X. Liang, Stabilizing the interface of all-solid-state electrolytes against cathode electrodes by atomic layer deposition, ACS Appl. Energy Mater. 5 (2022) 760–769. https://doi.org/10.1021/acsaem.1c03237.
74. L. Zhao, G. Chen, Y. Weng, T. Yan, L. Shi, Z. An, D. Zhang, Precise Al_2O_3 coating on LiNi0.5Co0.2$Mn_{0.3}O_2$ by atomic layer deposition restrains the shuttle effect of transition metals in Li-Ion capacitors, Chem. Eng. J. 401 (2020) 126138. https://doi.org/10.1016/j.cej.2020.126138.
75. X. Wang, J. Cai, Y. Liu, X. Han, Y. Ren, Atomic-scale constituting stable interface, Nanotechnology. 32 (2021) 115401.
76. J. Li, J. Xiang, G. Yi, Y. Tang, H. Shao, X. Liu, B. Shan, R. Chen, Reduction of surface residual lithium compounds for single-crystal LiNi0.6Mn0.2$Co_{0.2}O_2$ via Al_2O_3 atomic layer deposition and post-annealing, Coatings. 12 (2022). https://doi.org/10.3390/coatings12010084.
77. Y. Shi, Y. Xing, K. Kim, T. Yu, A.L. Lipson, A. Dameron, J.G. Connell, Communication—Reduction of DC resistance of Ni-rich lithium transition metal oxide cathode by atomic layer deposition, J. Electrochem. Soc. 168 (2021) 040501. https://doi.org/10.1149/1945-7111/abf17d.
78. Y. Tesfamhret, R. Younesi, E.J. Berg, Influence of Al_2O_3 coatings on HF induced transition metal dissolution from lithium-ion cathodes, J. Electrochem. Soc. 169 (2022) 010530. https://doi.org/10.1149/1945-7111/ac4ab1.
79. Y. Liu, W. Liu, M. Zhu, Y. Li, W. Li, F. Zheng, L. Shen, M. Dang, J. Zhang, Coating ultra-thin TiN layer onto $LiNi_{0.8}Co_{0.1}Mn_{0.1}O_2$ cathode material by atomic layer deposition for high-performance lithium-ion batteries, J. Alloys Compd. 888 (2021) 161594. https://doi.org/10.1016/j.jallcom.2021.161594.
80. S.H. Akella, S. Taragin, Y. Wang, H. Aviv, A.C. Kozen, M. Zysler, L. Wang, D. Sharon, S.B. Lee, M. Noked, Improvement of the electrochemical performance of $LiNi_{0.8}Co_{0.1}Mn_{0.1}O_2$ via atomic layer deposition of lithium-rich zirconium phosphate coatings, ACS Appl. Mater. Interfaces. 13 (2021) 61733–61741. https://doi.org/10.1021/acsami.1c16373.
81. B.Y. Lee, M. Krajewski, M.K. Huang, P. Hasin, J.Y. Lin, Spinel LiNi0.5Mn1.5O4 with ultra-thin Al2O3 coating for Li-ion batteries: Investigation of improved cycling performance at elevated temperature, J. Solid State Electrochem. 25 (2021) 2665–2674. https://doi.org/10.1007/s10008-021-05047-0.
82 E.R. Østli, M. Ebadi, Y. Tesfamhret, M. Mahmoodinia, M.J. Lacey, D. Brandell, A.M. Svensson, S.M. Selbach, and N.P. Wagner. On the Durability of Protective Titania Coatings on High-Voltage Spinel Cathodes. ChemSusChem, *15*(12) (2022) e202200324.
83. Y. Gao, X. He, L. Ma, T. Wu, J. Park, X. Liang, Understanding cation doping achieved by atomic layer deposition for high-performance Li-Ion batteries, Electrochim. Acta. 340 (2020) 135951. https://doi.org/10.1016/j.electacta.2020.135951.
84. O. Tiurin, N. Solomatin, M. Auinat, Y. Ein-Eli, Atomic layer deposition (ALD) of Lithium fluoride (LiF) protective film on Li-ion battery $LiMn_{1.5}Ni_{0.5}O_4$ cathode powder material, J. Power Sources. 448 (2020) 227373. https://doi.org/10.1016/j.jpowsour.2019.227373.
85. J. Gan, J. Zhang, B. Zhang, W. Chen, D. Niu, Y. Qin, X. Duan, X. Zhou, Active sites engineering of Pt/CNT oxygen reduction catalysts by atomic layer deposition, J. Energy Chem. 45 (2020) 59–66. https://doi.org/10.1016/j.jechem.2019.09.024.
86. Z. Song, Y.N. Zhu, H. Liu, M.N. Banis, L. Zhang, J. Li, K. Doyle-Davis, R. Li, T.K. Sham, L. Yang, A. Young, G.A. Botton, L.M. Liu, X. Sun, Engineering the low coordinated Pt single atom to achieve the superior electrocatalytic performance toward oxygen reduction, Small. 16 (2020) 1–12. https://doi.org/10.1002/smll.202003096.
87. X. Tang, S. Zhang, J. Yu, C. Lü, Y. Chi, J. Sun, Y. Song, D. Yuan, Z. Ma, L. Zhang, Preparation of platinum catalysts on porous titanium nitride supports by atomic layer deposition and their catalytic performance for oxygen reduction reaction, Wuli Huaxue Xuebao/Acta Phys. - Chim. Sin. 36 (2020) 1–7. https://doi.org/10.3866/PKU.WHXB201906070.

88. C. He, X. Wang, S. Sankarasubramanian, A. Yadav, K. Bhattacharyya, X. Liang, V. Ramani, Highly durable and active Pt/Sb-doped SnO_2 Oxygen reduction reaction electrocatalysts produced by atomic layer deposition, ACS Appl. Energy Mater. 3 (2020) 5774–5783. https://doi.org/10.1021/acsaem.0c00717.
89. J. Chen, Z. Li, Y. Chen, J. Zhang, Y. Luo, G. Wang, R. Wang, An enhanced activity of Pt/CeO_2/CNT triple junction interface catalyst prepared by atomic layer deposition for oxygen reduction reaction, Chem. Phys. Lett. 755 (2020) 137793. https://doi.org/10.1016/j.cplett.2020.137793.
90. W. ur rehman, Y. Xu, X. Du, X. Sun, I. Ullah, Y. Zhang, Y. Jin, B. Zhang, X. Li, Alumina-coated and manganese monoxide embedded 3D carbon derived from avocado as high-performance anode for lithium-ion batteries, Appl. Surf. Sci. 445 (2018) 359–367. https://doi.org/10.1016/j.apsusc.2018.03.112.
91. X. Tai, X. Li, A. Kakimov, S. Li, W. Liu, J. Li, J. Xu, D. Li, X. Sun, Optimized ALD-derived MgO coating layers enhancing silicon anode performance for lithium ion batteries, J. Mater. Res. 34 (2019) 2425–2434. https://doi.org/10.1557/jmr.2019.150.
92. Y. Jin, H. Yu, Y. Gao, X. He, T.A. White, X. Liang, $Li_4Ti_5O_{12}$ coated with ultrathin aluminum-doped zinc oxide films as an anode material for lithium-ion batteries, J. Power Sources. 436 (2019) 226859. https://doi.org/10.1016/j.jpowsour.2019.226859.

11 Multifunctional Thin Film for Health Care Industries

Sundara Subramanian Karuppasamy, N. Jeyaprakash, and Che-Hua Yang

11.1 INTRODUCTION

Over three decades, the medical industry has attained many tremendous advancements with the evolution of new technologies. These technologies have placed the conventional way of diagnosing and treating people on a new level. One of the key advancements in the medical industry is biomedical implants. Biomedical implants are artificial devices (materials) which are intended to perform like the natural organs in the human body. These materials are used to replace damaged organs or to act as supporting material until the damaged/fractured parts have healed. After healing, these implants can be removed from the body. Implants are a boon to humankind to overcome medical emergencies [1–3].

Since these implants are temporarily used in cases of bone fracture/bone misalignment or permanently attached to the human body in cases of organ replacement, special care is taken with the materials that are used to manufacture them. Only those materials with superior biomedical characteristics are adapted for manufacturing these implants in order to avoid conflicts. These characteristics are a mixture of chemical, mechanical, and bio-based features that include light weight, corrosion resistance, antifungal/bacterial resistance, biocompatibility, and so on [4–6]. Stainless steel is one of the most used implant materials due to its high degree of formability, strength, and corrosion resistance due to its high levels of chromium content. This alloy can be readily made into any form according to the clinical demands and economic needs. It is mostly employed in fabricating orthopedic implants required for bone repair and replacements [7, 8]. Titanium-based alloys have better biocompatibility than the stainless steel implants, but they are not economical. In particular, these alloys are used as fastening materials to provide joints between the bones and implant plates such as bone screws to avoid the risk of galvanic corrosion [9, 10]. On the other hand, cobalt-based alloys have superior antibacterial, antiwear, and anticorrosion features, and these alloys are implemented as the implant material in the case of heavy bone fractures/replacements such as knee bone, hip bone, and shoulder joints [11, 12]. Other than these three metal alloys, ceramics and polymers have been used to produce bio-based implants for short-term usage [13–15].

Even though the abovementioned metal alloys and nonmetal materials have better biocompatibility, in real-time implementation after prolonged usage, there is a reduction in their mechanical characteristics due to the chemical reactivity of implants under the skin. The pH of the blood serves as the root cause for the degradation of the properties they exhibit[16]. The implants made with stainless steel experience failures due to fatigue and high stress shielding with the bones [17, 18]. The aluminum and vanadium ions present in the Ti-based alloys are reported to be cytotoxic, and after a period of time, these ions will lead to several neurological disorders. Also, the bonding strength between the implant and tissue/bone tends to decrease, thereby resulting in misalignment or poor bone integration [19–21]. Moreover, the trace elements such as cobalt, chromium, and molybdenum present in the cobalt-based bioimplants induce severe localized corrosion attacks that could damage the other body organs, such as lungs, kidneys, blood cells, and so on [22, 23]. Thus, the temporary or permanent implants are prone to many attacks inside the human body that will degrade their service life.

To overcome the corrosion and other chemical effects of the implants and to enhance their working life, several surface treatment techniques are employed to enhance their mechanical behaviors.

DOI: 10.1201/9781032635347-11

Coatings are surface treatment processes where a thin film of materials is deposited on the implant's surface. This thin film is reported to act in multifunctional ways, such as protecting the implants against microbial invasion, corrosion, and other chemical reactions followed by improving the service life of the implants [24–27]. Moreover, a wide range of processes and a broad range of materials can be adapted to produce the multifunctional films on the implants. In this chapter, a diverse range of coating materials along with the coating processes implemented to coat the multifunctional films on the medical implants are discussed with relevant literature.

11.2 COATING METHODS

Based on the working environment, cost, and other factors, the multifunctional film can be produced on medical implants via a diverse range of coating methods. This section will briefly explain the most important methods in order to create multifunctional films on the implants.

11.2.1 Physical Vapor Deposition (PVD)

Physical vapor deposition is one of the most widely used techniques to create a thin multifunctional film on the substrates (implants). As the name implies, a fine coating layer can be obtained by depositing the coating material in a physical manner. This process takes place in a low-pressure or vacuum chamber depending on the needs, and a heat source is required to vaporize the coating material. During heating, the coating material is vaporized into atoms or a cluster of atoms, which is said to be condensed, in the vacuum chamber. After the condensation process, the condensed matter is deposited on the implant material as a film in the range of 1 to 10 μm [28–30]. A wide number of heat sources such as cathodic arc, electron beam, pulsed laser, thermal laser, pulsed electron beam, and so on, could be used to generate the atoms from the source (coating materials). Since it involves vaporization and condensation of the atoms to get deposited on the substrate, a broad range of organic and inorganic materials can be used for the coating process. Moreover, the film formed via the PVD process has better impact strength and abrasion resistance than the films formed by the traditional coating methods. Depending on the heating source, the physical deposition process is categorized into many types, and some of the major methods of PVD involved in creating the films on the implants are discussed next.

11.2.1.1 Thermal Evaporation Deposition (TED)

Thermal evaporation deposition belongs to the category of physical vapor deposition, and several reports say that this technique could be adapted for producing both multifunctional thin and thick films on a macroscale. In this approach, an evaporator (usually metals) in the form of filaments/shots/pellets is used as the heating source that evaporates the atoms present in the coating material. The evaporated atoms or molecules are said to be condensed once they reach the implant surface, thereby forming a thin or thick film on the implants, as in Figure 11.1. The whole process takes place in a vacuum chamber, and mostly pure metals and dielectric materials are used as the coating materials. Some process parameters have a significant effect during the coating process such as the vacuum pressure, evaporator source, and coating material's purity [31–33]. Highly efficient coatings can be obtained while using pure metals as the coating source at nominal vacuum pressure. On the other hand, greater deposition rates can be achieved at higher vacuum pressure, but the obtained film will have uneven thickness that will reduce the mechanical characteristics offered by the film. Moreover, the evaporator source containing the wire filaments is unable to produce thick films on the implant surface due to the low rates of evaporation and condensation of metal ions.

11.2.1.2 Electron Beam-Assisted Evaporation Deposition (EED)

The electron beam-assisted evaporation process is a variant of the previously mentioned thermal evaporation process where an electron beam is used to evaporate the metal ions. This process can

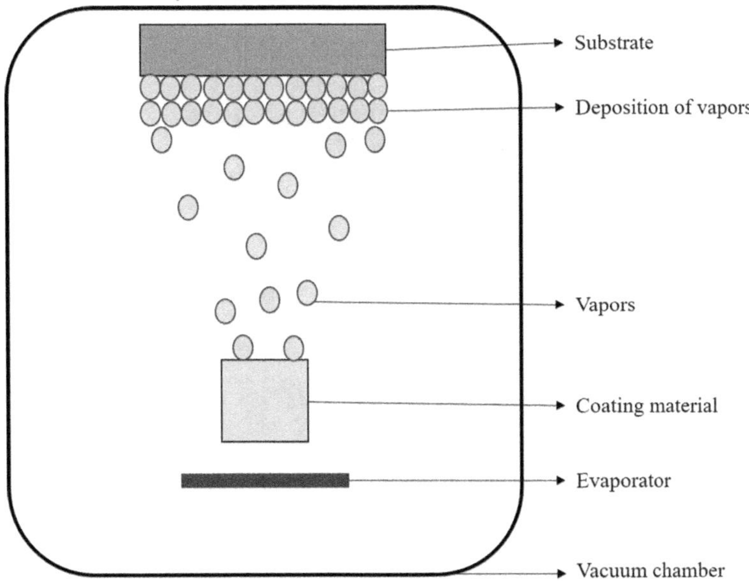

FIGURE 11.1 Schematic illustration of thermal evaporation PVD process.

produce films consisting of heavy and refractory metals. Here, the filament coil is heated to a very high temperature where the e^- are accelerated rapidly, thereby creating an electron beam. This beam is targeted at the source (coating material) where the kinetic energy of the beam is converted to the heat energy required to vaporize the source particles. These vaporized particles are deposited on the substrate and form a thin film, as shown in Figure 11.2 [34, 35]. The major advantages of using the electron beam over the TED are listed below:

i. This process could produce fine films with uniform thickness.
ii. Adequate control over the material deposition rates (0.1–100 μm/min).
iii. Fine coatings can be obtained even in a low-temperature environment.
iv. Efficient material usage.
v. Multilayer films can be produced without any defects by changing the source.
vi. Better method for the deposition of metals, metal oxides, dielectric oxides, and refractory metals.

Due to these advantages, the EED process holds good to create multifunctional films in other industries such as optical, aerospace, semiconductors, solar, and cutting-tool industries.

11.2.1.3 Sputtering

One of the most-used PVD processes, which adopts a diverse range of coating materials, is the sputtering process. This process is reported as the glow-discharge process, where the metal ions or atoms from the coating material are ejected by means of momentum transfer that takes place between the sputtering source and the coating source. By bombarding the highly reactive ions which are emitted from the sputtering source onto the coating material (source), the momentum exchange or transfer occurs. This exchange/transfer of momentum is able to eject the atoms or cluster of the atoms from the coating source and gets deposited on the implant's surface as films [36, 37]. Some of the key characteristics of the sputtering process are as follows:

i. A wide range of metals, metal oxides, metal alloys, refractory metals, and polymers can be used to create the films.

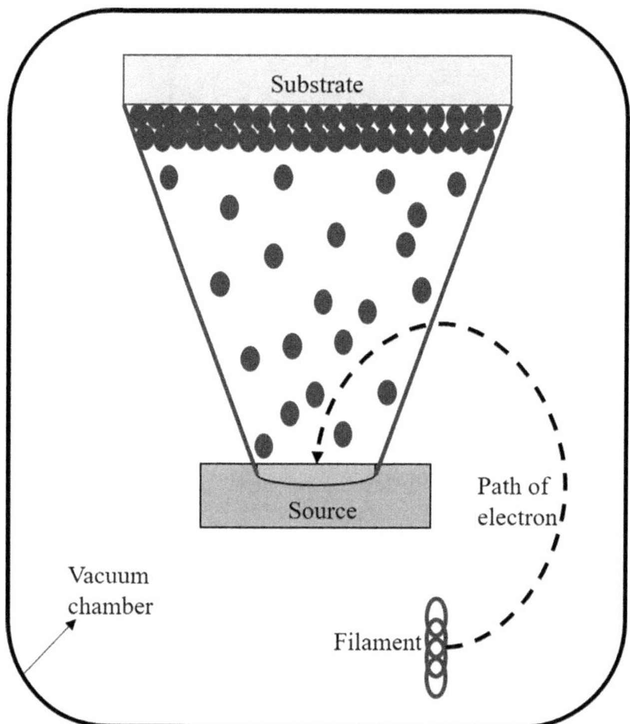

FIGURE 11.2 Schematic illustration of electron beam-assisted PVD process.

ii. Even thickness of films can be produced over a large surface area.
iii. Ease of control over the deposition rate and thickness of the films.
iv. Better service life.

Since the sputtering source determines the amount of momentum exchange which in turn decides the deposition rate, the sputtering process can be categorized into many subprocesses based on the sputtering source.

i. **Diode sputtering:** This type of sputtering process holds good for metal and metal alloys. Here, the metal ions are deposited by means of a diode (cathode and anode) setup. The coating material acts as the cathode, whereas the implant serves as the anode. Highly reactive ions are generated from the sputtering source (inert gas–Ar^+–incident ions) which are bombarded against the coating material. During bombardment, the momentum exchange occurs between the highly energetic Ar^+ ions and the coating material, and when the energy of the incident ion is higher than the binding energy of the atom in the coating material, the atom is released from the coating material and deposited on the surface of the implants, thereby creating a film on the substrate as in Figure 11.3 [38].
ii. **Triode sputtering:** This type of sputtering is similar to diode sputtering. To increase the deposition rate, other than two electrodes (anode and cathode), a thermionic filament is added to the diode setup. By means of this filament, more atoms are ejected from the coating source, which will improve the deposition rate of the produced film [39]. Moreover, triode sputtering holds good for small surface areas. Implementing triode sputtering over a large surface area results in uneven thickness of the deposited film.
iii. **Magnetron sputtering:** As the name implies, the sputtering process takes place in the presence of a magnetic field. Here, a magnetic source is combined with the cathode to

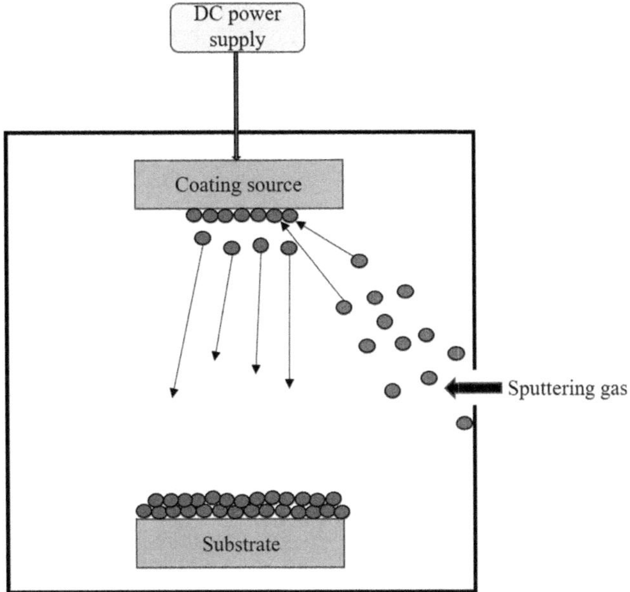

FIGURE 11.3 Diode sputtering process.

enhance the probability of bombardment. In most cases, permanent magnets are attached at a relatively small distance behind the cathode. During the sputtering process, the magnetic field is generated in the perpendicular, which in turn increases the probability of collision between the atoms [40, 41].

iv. **Reactive sputtering:** Here, the probability of collision is improved by adding a source to the sputter source (Ar^+). Mostly, oxygen serves as a reactive gas to promote deposition rates.
v. **Dual ion beam sputtering:** In this process, two ion beams are used to improve the bombardment rate. One ion beam serves as the sputtering source and the second one is for improving the collision rate of atoms [42].

11.2.2 Chemical Vapor Deposition (CVD)

In this process, the multifunctional films are formed on the substrate's surface by means of various chemical reactions. The coating material is supplied at the gaseous phase, which interacts with the heated substrate via chemical reactions followed by creating the film on the surface [43, 44]. A typical CVD setup consists of a reactant supply (materials to be coated in the vapor or gaseous state), deposition, and exhaust system. The steps involved in the CVD process (Figure 11.4) are as follows:

i. **Diffusion of precursors:** The coating materials present in the vapor phase are diffused from the mainstream of reactant supply. The reactant supply may be in horizontal or vertical directions depending on the requirements.
ii. **Adsorption:** The diffused precursors are adsorbed at the heated surface of the substrate to undergo chemical reactions such as pyrolysis, redox reactions, nitridation, hydrolysis, carburization, and so on.
iii. **Desorption:** The atom or cluster of atoms formed during the chemical reactions are deposited on the substrate's surface, leaving the by-products.
iv. **Diffusion of by-products:** With the aid of an exhaust system, the by-products are diffused along with the mainstream of reactant supply.

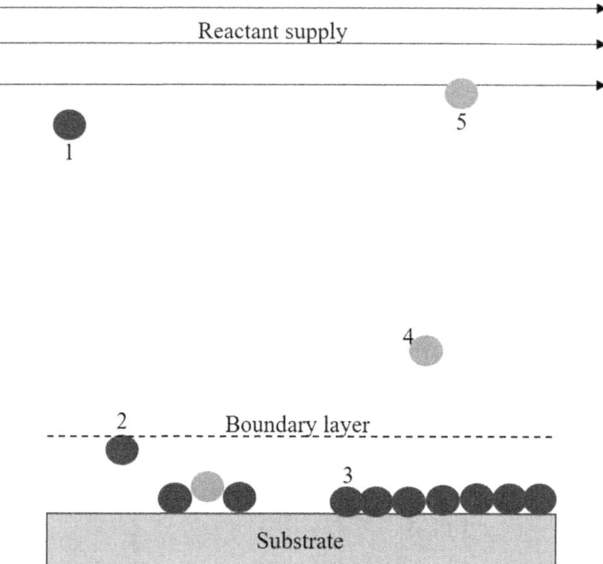

FIGURE 11.4 CVD process, 1 - diffusion of precursors, 2 - adsorption, 3 - surface diffusion (forming film), 4 - by-product, and 5 - diffusion of by-product.

11.2.3 Thermal Spray Deposition

Both the physical and chemical vapor deposition processes are employed to create a thin film on the substrate's surface. But if there is a need for thick coatings in the millimeter range, the abovementioned deposition techniques fail to create such thick films. Thermal spraying involves the creation of thick films over a large surface area by means of heating the coating material. Here, a thermal gun that serves as the source of heat is used to melt the coating feedstock in such a way that this melted feedstock is sprayed on the substrate's surface in the size of microparticles. After spraying, the microparticles solidify to form an even coating over the surface area [45, 46]. The feedstock can be either powder or wire, depending on the requirements. Based on the heat source, thermal spray deposition is classified into four major categories:

i. Plasma arc deposition
ii. Electric arc deposition
iii. Flame spraying
iv. Kinetic-based spraying

The abovementioned categories have subclassifications as shown in Figure 11.5.

Thermal spray			
Plasma arc	Electric arc	Flame	Kinetic
1. Air 2. Vacuum	1. Air 2. Vacuum	1. D-gun 2. High velocity	-

FIGURE 11.5 Thermal spray classification.

11.2.4 Sol-Gel Deposition

The sol-gel deposition process is one of the most efficient ways to produce commercial coatings made of alkoxide or metal oxide. This process is considered the wet chemical method, and the thin films are obtained based on chemical reactions such as hydrolysis, condensation, and agglomeration [47, 48]. The coatings produced via the sol-gel process exhibit many features (multifunctional behaviors), as in Figure 11.6. In this process, there are four steps involved in order to produce the coating on the substrate. They are (i) sol preparation, (ii) spraying (the prepared sol is sprayed on the substrate's surface), (iii) drying, and (iv) firing (the dried sol is fired for obtaining the adherent coating). The first step can be done in either of the following two ways:

i. Hydrolysis and then polymerization.
ii. Precipitation followed by peptization.

11.2.5 Laser Cladding

Laser cladding is reported as one of the most prominent ways to produce fine and uniform metal and metal alloy coatings on the substrate surface [49]. This technique has adequate advantages when compared with the PVD, CVD, thermal spray, and sol-gel deposition processes. In this process, the feedstock can be either powder or wire, depending on the type of laser cladding used to create the film. Based on the alignment of feedstock, the laser cladding process can be categorized into four types, namely (i) preplaced powder feedstock cladding, (ii) off-axis powder feedstock cladding, (iii) off-axis wire feedstock cladding, and (iv) coaxial powder feedstock cladding. Out of these four types, the coaxial powder feedstock system is usually preferred to produce efficient coatings on the substrate.

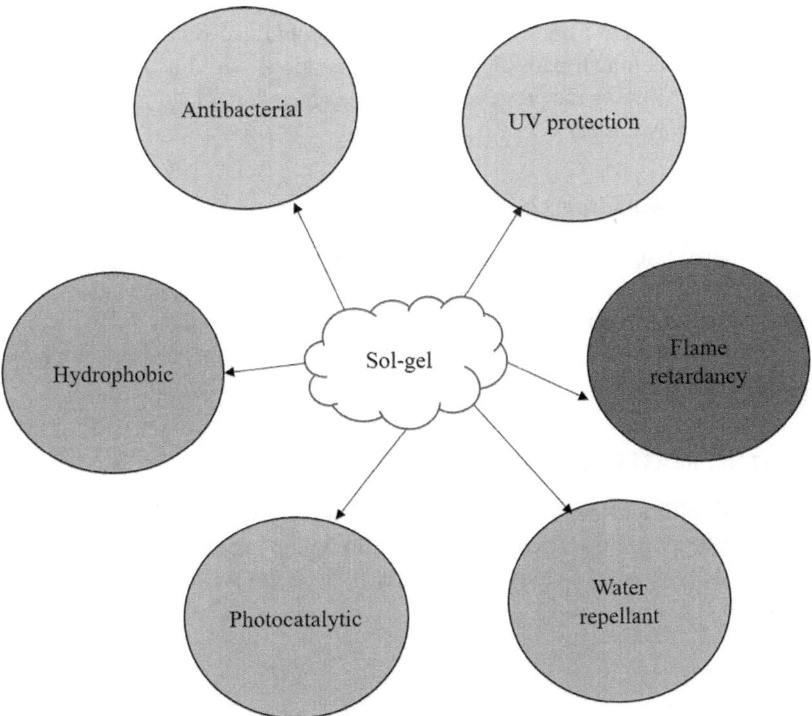

FIGURE 11.6 Multifunctional behaviors of sol-gel coatings.

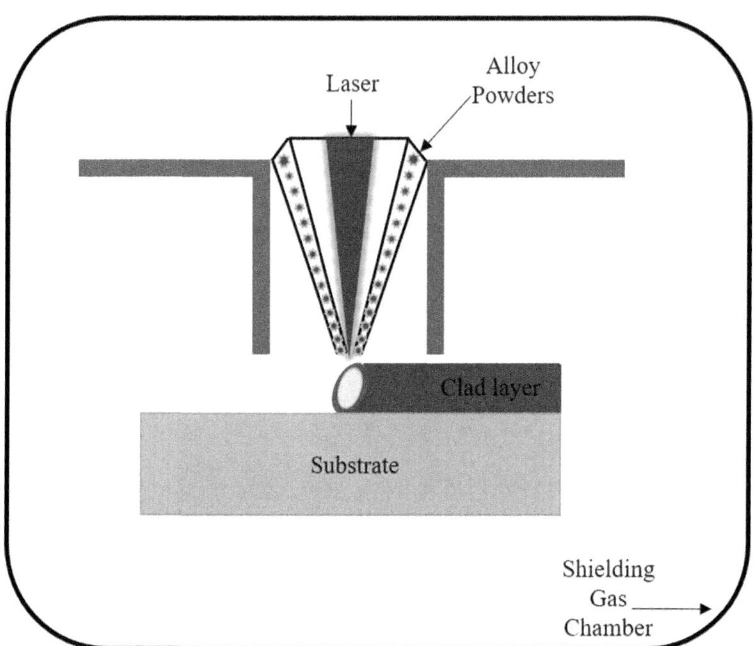

FIGURE 11.7 Schematic illustration of the laser cladding process.

In the laser cladding process (Figure 11.7), a high-energy continuous wave laser (Yb:YAG) is employed to irradiate the substrate's surface. During irradiation, the surface is heated and converted to a molten melt pool. The coating powder particles are deposited on the molten melt pool via a coaxial feeder system. A fine film of coated particles is obtained on the substrate after adequate solidification of the molten melt pool [50]. Pores-free coatings with fine grains and less distortion can be produced by optimizing the process parameters. Thus, laser cladding holds good for producing metal and metal alloy coatings.

11.2.6 Additive Manufacturing

Additive manufacturing technology has reversed the conventional way of making structures. In a layer-by-layer fashion, the components are manufactured with high precision and less wastage. The additive-manufactured parts show better mechanical characteristics than the parts made via traditional manufacturing methods [51]. In recent years, this technology has been implemented in the field of coatings to produce multifunctional films with superior performance [52].

11.3 COATING MATERIAL FOR IMPLANTS

For improving service life and to provide resistance against corrosion and other attacks on the implants, various coating materials are preferred to form the film on the implant's surface. A diverse range of coating materials that are adapted to produce a fine film on the surface of the medical implants is discussed next.

11.3.1 Nitrides

Coatings made by depositing various metal nitrides exhibit inertness against microbes, better melting point, and an acceptable level of adhesion characteristics. Some of the most commonly used metal nitrides used in treating medical implants are titanium nitride (TiN), tantalum nitride (TaN), zirconium nitride (ZrN), and niobium nitride (NbN). Kao et al. [53] analyzed the corrosion resistance characteristics of the titanium nitride coated on orthodontic metal brackets and found that the TiN coating exhibited better anticorrosion behaviors due to its chemical inertness against the tissues. Gobbi et al. [54] found that the TiN film serves as a diffusion barrier and provides better resistance against wear. Caiazzo et al. [55] compared the TiN films deposited by both physical and chemical vapor deposition processes. Their results concluded that the chemical vapor deposition (CVD) process holds good for depositing the TiN particles due to greater deposition rate and enhanced bonding strength between the coating and the implant's surface. DD Kumar et al. [56, 57] examined the hardness and wear resistance characteristics of both TiN and ZrN coatings obtained on the SS316L substrate. Their study revealed that the ZrN coating has better antiwear and higher hardness than the TiN coating. On the other hand, considering the thin films, the NbN and TaN showed better biocompatibility [58, 59].

11.3.2 Oxides

For enhancing the corrosion resistance of implants, it is better to coat the implant's surface with metal oxides. These metal oxides are reported to form a compact layer of passive film on the surface, thereby providing excellent corrosion resistance in the environments containing body fluids [60]. Several metal oxides, such as titanium oxide (TiO_2), tantalum oxide (Ta_2O_5), and ferric oxide (Fe_2O_3), are most commonly used to create oxide film in medical implants.

Aneta et al. [61] examined the corrosion resistance behavior of titanium oxide film formed on Mg alloy ($MgCa_4ZnGd$) by means of magnetron sputtering (PVD) and the sol-gel process. Their study concluded that both the PVD TiO_2 film and sol-gel TiO_2 film have a similar microstructure, but the grain size obtained in the sol-gel TiO_2 film was found to be larger than the PVD TiO_2 film. In the immersion test, the PVD TiO_2 film is said to be electrochemically inert, thereby providing greater resistance against corrosion when compared with the sol-gel TiO_2 film. Moreover, the sol-gel TiO_2 film has a higher surface roughness (Ra), which denotes that this film is more susceptible to corrosion attacks than the film formed via the magnetron sputtering process. Lin et al. [62] analyzed the biocompatibility and corrosion behavior of TiO_2 coating formed on the pure Ti substrate in an environment containing stimulated body fluid. Their results revealed that the TiO_2 film has better anticorrosion behaviors (nearly three times) when compared with the pure Ti substrate. Further, this film showed better biocompatibility than the substrate after 10 days of immersion in the stimulated body fluids (SBF). Paulo et al. [63] investigated the mechanical features such as hardness and elastic modulus of the TiO_2 film formed on the pure Ti implants. By means of indentation analysis, they observed that the deposited film has higher hardness and lower elastic modulus than the pure Ti implants. Jiang Xu et al. [64] compared the biocompatible nature of β-Ta_2O_5 deposited on the Ti-6Al-4V substrate and concluded that in the presence of SBF, β-Ta_2O_5 film has better biocompatibility due to the formation of a bonelike apatite layer. Thus the metal oxide coating could improve the service life of biomedical implants.

11.3.3 Oxynitrides

As the name implies, the oxynitride coatings are a combination of metal oxide-nitride elements. This combination paved the way for the coating materials to exhibit the combined properties of both nitride and oxide coatings such as excellent corrosion resistance, greater biocompatibility, better hardness, and higher resistance against wear with adequate antibacterial properties. Sami et al. [65] analyzed the bacterial inactivation kinetics by depositing titanium oxynitride and titanium oxynitride with Ag addition via the sputtering PVD process. Their study concluded that both TiON and TiON-Ag films have better antibacterial kinetics, and in particular, the antibacterial feature was enhanced by the addition of Ag. Kaliaraj et al. [66] compared the mechanical (hardness and wear) and biological (bacterial invasion) characteristics by depositing titanium oxynitride (TiON) and zirconium oxynitride (ZrON) on the stainless steel (SS 316L) substrate. Both films have a better hydrophilic nature when compared with the substrate. On evaluating the mechanical characteristics, the TiON film has better hardness and wear-resistant behaviors than the ZrON film. In addition, both films exhibited excellent corrosion resistance when immersed in the artificial blood plasma (ABP) solution, and there was a drastic reduction in the bacterial invasion on these films when compared to the SS316L substrate. Chang et al. [67] investigated antibacterial performance by depositing zirconic nitride (ZrN), zirconic oxide with silver addition (ZrO_2-Ag) and zirconic oxynitride with silver addition (ZrNO-Ag) on pure Ti medical implants. They concluded that the film consisting of ZrNo-Ag had greater antibacterial characteristics when compared with the ZrN, ZrO_2-Ag, and Ti substrates. Other than Ag, the Cu addition (ZrNO-Cu) also results in better antibacterial performance [68]. Thus, from the literature, the oxynitride coatings are capable of enhancing the mechanical, electrochemical, and biological characteristics of medical implants.

11.3.4 Carbon-Based Coatings

Other than metallic oxides, nitrides, and oxynitrides, several researchers focused on implementing carbon-based materials since these materials exhibit exceptional mechanical and biocompatible features. The most commonly used carbon materials are listed below:

i. Amorphous carbon nanostructures (diamond-like carbon (DLC) structures)
ii. Nanocrystalline diamond
iii. Graphene and its derivatives

Arkadiusz et al. [69] examined the film made by depositing the DLC on a pure Ti substrate and analyzed the mechanical properties offered by the film. The results showed that better hardness and other tribological properties were obtained on the DLC-coated film than the Ti substrate. Maciej et al. [70] compared the biocompatibility of the films made up of titanium nitride, zirconium nitride, and DLC nanocoatings on the Ti-6Al-4V implants. All three films have better biocompatibility, and considering the friction coefficient of all three films, the film consisting of DLC coating has the lowest value of friction coefficient, which might have a potential influence in the medical field. R. Hauert et al. [71] further confirmed that the DLC coatings have greater antibacterial and corrosion resistance characteristics. O. Medina et al. [72] compared the antibacterial (bactericide and bacterial anti-adhesive) features of both nanocrystalline diamond film and microcrystalline diamond film along with their tribological properties. Their study revealed that the surface roughness (Ra) of nanocrystalline diamond coating is comparatively lower than the microcrystalline diamond film. Moreover, this low surface roughness value accounts for the superhydrophobic nature followed by enhancing the antibacterial characteristics of the nanocrystalline diamond film. Mohamed S. Selim et al. [73] synthesized the hybrid film consisting of polydimethylsiloxane (PDMS)/graphene oxide with zinc oxide nanorods (PDMS/GO-ZnO), which are coated on the substrate via an air-assisted thermal spray process. The authors concluded that this hybrid film exhibits multifunctional features such as self-cleaning, anticorrosion, biocompatibility, and hydrophobicity, and it could be implemented in producing coatings on medical implants. Further, Lai Suo et al. [74] examined the hybrid film coated on the pure Ti implants. This hybrid film is a combination of graphene oxide, chitosan, and hydroxyapatite. They reported that enhanced biocompatibility is reported in the GO/CS/HA film and could be implemented in dental implants.

11.3.5 Bioactive Ceramics

The abovementioned oxides, nitrides, oxynitrides, and carbon-based coatings exhibit superior characteristics such as biocompatibility, antibacterial activity, and wear and corrosion resistance, thereby improving the service life of biomedical implants. These coatings are focused on enhancing the working life of the implants, but they do not aid in improving the healing time of the damaged bones or tissues. As soon as the tissues grow or bone heals, people can return to their normal lives. Hence, researchers work on certain materials that could make the healing time as fast as possible [75]. Bioactive ceramics are a special class of ceramic materials that are reported to interact with the cells and tissues and enhance their growth. These materials are also termed bioactive glasses. Bioactive ceramics are a combination of calcium oxide (CaO), silicon dioxide (SiO_2), phosphorous pentaoxide (P_2O_5), and sodium oxide (Na_2O). While interacting with the body cells, the Ca, Si, P, and Na ions are released from the ceramics and are reported to enhance the healing time and hasten tissue/bone growth.

Szymon et al. [76] examined the bioactive characteristics of the bioactive glasses deposited on the pure Ti substrate via the laser cladding process. Their results revealed that the coating consists of fine grains containing α' acicular martensite structure. After immersion in the SBF solution, they compared the bioactive responses of both Ti substrate and bioactive glass coating and concluded that the bioactive ceramics coating showed better bioactive responses than the Ti substrate. Vaibhav et al. [77] compared the in vivo and corrosion characteristics of both hydroxyapatite and bioglass films coated on Ti substrates through an additive manufacturing process. They concluded that the existence of ceramic phases such as silicate and phosphate had improved the hardness and wear resistance of the bioglass film. The in vivo study and the immersion test further confirmed that the bioglass film had better anticorrosion and bioactivity than the hydroxyapatite film. Further, this bioactive ceramic coating resulted in enhanced bone growth [78]. Though these coatings exhibit superior bioactivity, their amorphous structure results in the formation of defects such as pores or voids in the film [79]. This can be overcome by adding oxides or graphene derivatives along with bioactive ceramics. P. Bargavi et al. [80] investigated the bioactivity and antibacterial characteristics of ZiO_2 reinforced with the bioactive ceramic coating on titanium implants. Their study revealed that the biofilm consisting of ZrO_2 remained stable up to 14 days with better in vitro performance after the immersion test carried out in SBF solution. Balakumar et al. [81] prepared a hybrid film consisting of graphene oxide reinforced with the bioactive ceramics via the sol-gel method and concluded that the synthesized hybrid film showed better bioactive characteristics [82].

11.3.6 Smart Polymeric Coatings

There is a certain class of polymers that shows multifunctional characteristics once exposed to body fluids. These polymers are termed smart polymers, and their multifunctionality is represented in Figure 11.8. The smart polymers can be categorized based on their stimuli-responsive function as follows:

i. **Temperature-responsive polymers** [83, 84] Examples: PNIPPAm, linoleic acid, PLGA-PEG-PLGA, etc.
ii. **pH-responsive polymers** [85, 86] Examples: Polyvinylamine, PEG-PPG-PEG, etc.
iii. **Light-responsive polymers** [87] Examples: PEG, etc.
iv. **Electric field–responsive polymers** [88, 89] Examples: vinyl alcohol (C_2H_4O), vinyl-acrylic acid ($C_5H_6O_2$), allyl amide (C_3H_6N), methacrylic acid ($C_4H_6O_2$), etc.
v. **Enzyme-responsive polymers** [90].

These polymers show excellent performance during drug delivery and tend to be bioactive in nature.

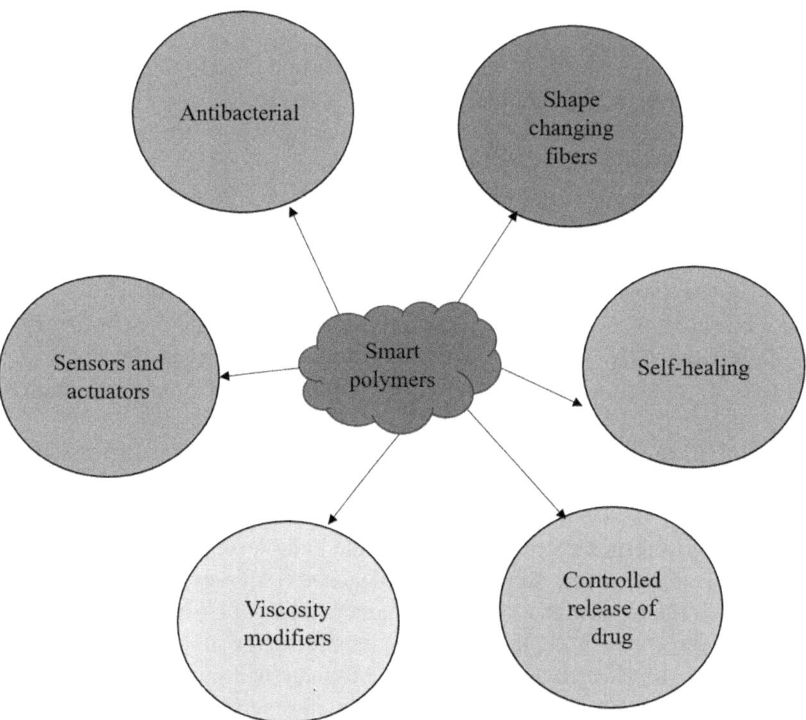

FIGURE 11.8 Multifunctional behaviors of smart polymers.

11.4 CONCLUSION

Medical implants are a boon to humankind since they replace or act as a support to organs in case of failures. In most cases, these implants are made up of metal alloys with superior mechanical characteristics such as higher hardness, excellent corrosion and wear resistance, and better biocompatibility. But in real-world applications, these implants are prone to various attacks such as corrosion and wear in the presence of body fluids, which reduce the service life of the implants. The working life of the implants could be improved by coating them with suitable materials. This chapter gives an overview of the multifunctional films that will be suitable for medical implants. It details the most prominent coating methods for producing films on the implants. Further, this chapter deals with a diverse range of coating materials that could be adapted for medical implants.

REFERENCES

1. Rebelo R, Fernandes M, Fangueiro R. Biopolymers in medical implants: a brief review. Procedia Engineering. 2017 Jan 1;200:236–43.
2. Yan Y, editor. Bio-tribocorrosion in biomaterials and medical implants. Elsevier; 2013 Sep 30.
3. Sharma Y, Mehta A, Vasudev H, Jeyaprakash N, Prashar G, Prakash C. Analysis of friction stir welds using numerical modelling approach: a comprehensive review. International Journal on Interactive Design and Manufacturing (IJIDeM). 2023 May 6:1–4.
4. Ibrahim MZ, Sarhan AA, Yusuf F, Hamdi M. Biomedical materials and techniques to improve the tribological, mechanical and biomedical properties of orthopedic implants–A review article. Journal of Alloys and Compounds. 2017 Aug 15;714:636–67.
5. Manivasagam G, Dhinasekaran D, Rajamanickam A. Biomedical implants: corrosion and its prevention-a review. Recent Patents on Corrosion Science. 2010 May 24;2(1):40–54.
6. Hermawan H, Ramdan D, Djuansjah JR. Metals for biomedical applications. Biomedical engineering - From theory to applications. InTech; 2011 Aug 29;1:411–30.

7. Bekmurzayeva A, Duncanson WJ, Azevedo HS, Kanayeva D. Surface modification of stainless steel for biomedical applications: Revisiting a century-old material. Materials Science and Engineering: C. 2018 Dec 1;93:1073–89.
8. Jeyaprakash N, Yang CH, Sivasankaran S. Laser cladding process of Cobalt and Nickel based hard-micron-layers on 316L-stainless-steel-substrate. Materials and Manufacturing Processes. 2020 Jan 25;35(2):142–51.
9. Elias CN, Lima JH, Valiev R, Meyers MA. Biomedical applications of titanium and its alloys. JOM. 2008 Mar;60:46–9.
10. Jeyaprakash N, Yang CH, Tseng SP. Characterization and tribological evaluation of NiCrMoNb and NiCrBSiC laser cladding on near-α titanium alloy. The International Journal of Advanced Manufacturing Technology. 2020 Jan;106:2347–61.
11. Aherwar A, Singh AK, Patnaik A. Cobalt Based Alloy: A Better Choice Biomaterial for Hip Implants. Trends in Biomaterials & Artificial Organs. 2016 Jan 1;30(1):50–55.
12. Zhang E, Zhao X, Hu J, Wang R, Fu S, Qin G. Antibacterial metals and alloys for potential biomedical implants. Bioactive Materials. 2021 Aug 1;6(8):2569–612.
13. Catledge SA, Fries MD, Vohra YK, Lacefield WR, Lemons JE, Woodard S, Venugopalanc R. Nanostructured ceramics for biomedical implants. Journal of Nanoscience and Nanotechnology. 2002 Jul 1;2(3–4):293–312.
14. Colilla M, Manzano M, Vallet-Regí M. Recent advances in ceramic implants as drug delivery systems for biomedical applications. International Journal of Nanomedicine. 2008 Dec 1;3(4):403–14.
15. Riveiro A, Maçon AL, del Val J, Comesaña R, Pou J. Laser surface texturing of polymers for biomedical applications. Frontiers in Physics. 2018 Feb 27;6:16.
16. Virtanen S. Corrosion of biomedical implant materials. The journal Corrosion Reviews. 2008; 26 (2–3):147–171. https://doi.org/10.1515/corrrev.2008.147
17. Hansen DC. Metal corrosion in the human body: the ultimate bio-corrosion scenario. The Electrochemical Society Interface. 2008 Jun 1;17(2):31.
18. Radhi NS, Al-Khafaji ZA. Investigation biomedical corrosion of implant alloys in physiological environment. International Journal of Mechanical and Production Engineering Research and Development. 2018;8(4):247–56.
19. Jeyaprakash N, Yang CH, Karuppasamy SS, Rajendran DK. Correlation of microstructural with corrosion behaviour of Ti-6Al-4V specimens developed through selective laser melting technique. Proceedings of the Institution of Mechanical Engineers, Part E: Journal of Process Mechanical Engineering. 2022 Oct;236(5):2240–51.
20. Izman S, Abdul-Kadir MR, Anwar M, Nazim EM, Rosliza R, Shah A, Hassan MA. Surface modification techniques for biomedical grade of titanium alloys: oxidation, carburization and ion implantation processes. Titanium Alloys - Towards Achieving Enhanced Properties for Diversified Applications. InTech; 2012 Mar 16;42:201–28.
21. Bedi RS, Beving DE, Zanello LP, Yan Y. Biocompatibility of corrosion-resistant zeolite coatings for titanium alloy biomedical implants. Acta Biomaterialia. 2009 Oct 1;5(8):3265–71.
22. Michel R, Nolte M, Reich M, Löer3 F. Systemic effects of implanted prostheses made of cobalt-chromium alloys. Archives of Orthopaedic and Trauma Surgery. 1991 Feb;110(2):61–74.
23. Chen Q, Thouas GA. Metallic implant biomaterials. Materials Science and Engineering: R: Reports. 2015 Jan 1;87:1–57.
24. Jeyaprakash N, Karuppasamy SS, Yang CH. 1 Application of Wear-Resistant Laser Claddings. In Handbook of Laser-Based Sustainable Surface Modification and Manufacturing Techniques. CRC Press; 2023 Jul 5;1–26.
25. Goldschmidt A, Streitberger HJ. BASF Handbook on Basics of Coating Technology. William Andrew; 2003.
26. Driver M, editor. Coatings for biomedical applications. Elsevier; 2012 Feb 22.
27. Yoshida M, Langer R, Lendlein A, Lahann J. From advanced biomedical coatings to multifunctionalized biomaterials. Journal of Macromolecular Science, Part C: Polymer Reviews. 2006 Dec 1;46(4):347–75.
28. Schmitz T. Functional coatings by physical vapor deposition (PVD) for biomedical applications (Doctoral dissertation, Universität Würzburg). 2016.
29. Geyao L, Yang D, Wanglin C, Chengyong W. Development and application of physical vapor deposited coatings for medical devices: A review. Procedia CIRP. 2020 Jan 1;89:250–62.
30. Gabor R, Cvrček L, Doubková M, Nehasil V, Hlinka J, Unucka P, Bušil M, Podepřelová A, Seidlerová J, Bačáková L. Hybrid coatings for orthopaedic implants formed by physical vapour deposition and microarc oxidation. Materials & Design. 2022 Jul 1;219:110811.

31. Moore B, Asadi E, Lewis G. Deposition methods for microstructured and nanostructured coatings on metallic bone implants: a review. Advances in Materials Science and Engineering. 2016; 2017: 1–9.
32. Sharma MC, Tripathi B, Kumar S, Srivastava S, Vijay YK. Low cost CuInSe2 thin films production by stacked elemental layers process for large area fabrication of solar cell application. Materials Chemistry and Physics. 2012 Jan 5;131(3):600–4.
33. Lukaszkowicz K. Review of nanocomposite thin films and coatings deposited by PVD and CVD technology. In Nanomaterials. InTech Rijeka, Croatia; 2011 Dec 22;145–162.
34. Singh J, Quli F, Wolfe DE, Schriempf JT, Singh J. An overview: Electron beam-physical vapor deposition technology - Present and future applications. Applied Research Laboratory, Pennsylvania State University; 1999.
35. Esmaeili MM, Mahmoodi M, Imani R. Tantalum carbide coating on Ti-6Al-4V by electron beam physical vapor deposition method: Study of corrosion and biocompatibility behavior. International Journal of Applied Ceramic Technology. 2017 May;14(3):374–82.
36. Schäfer E. Effect of physical vapor deposition on cutting efficiency of nickel-titanium files. Journal of Endodontics. 2002 Dec 1;28(12):800–2.
37. Yarlagadda PK, Tesfamichael T, Schuetz M, Valiveti L, Li T. Nanoscale texture on glass and Titanium substrates by physical vapor deposition process. Procedia Engineering. 2014 Jan 1;97:1506–11.
38. Matthews A, Rohde SL. Coatings and Surface Engineering: Physical Vapor Deposition. Mechanical Engineers' Handbook, Manufacturing and Management. Wiley publisher; 2015 Feb 2;3:235.
39. Anner GE, Anner GE. Physical Vapor Deposition; Sputtering. Planar Processing Primer. 1990: 493–534.
40. Hein M, Dias NF, Kokalj D, Stangier D, Hoyer KP, Tillmann W, Schaper M. On the influence of physical vapor deposited thin coatings on the low-cycle fatigue behavior of additively processed Ti-6Al-7Nb alloy. International Journal of Fatigue. 2023 Jan 1;166:107235.
41. Stuart BW, Stan GE. Physical Vapour Deposited Biomedical Coatings. Coatings. 2021 May 21;11(6):619.
42. Hirvonen JK. Ion beam assisted thin film deposition. Materials Science Reports. 1991 Jul 1;6(6):215–74.
43. Piszczek P, Radtke A. Silver nanoparticles fabricated using chemical vapor deposition and atomic layer deposition techniques: Properties, applications and perspectives: Review. In Seehra, MS, Bristow, AD, (Eds), Noble and Precious Metals. Intech publisher; 2018 Jul 4;187–213.
44. Liu Z, Jiang X, Li Z, Zheng Y, Nie JJ, Cui Z, Liang Y, Zhu S, Chen D, Wu S. Recent progress of photoexcited antibacterial materials via chemical vapor deposition. Chemical Engineering Journal. 2022 Feb 23:135401.
45. Robotti P, Zappini G. Thermal plasma spray deposition of titanium and hydroxyapatite on PEEK implants. In PEEK Biomaterials Handbook. William Andrew Publishing; 2019 Jan 1;147–177.
46. Yeh CH, Jeyaprakash N, Yang CH. Temperature dependent elastic modulus of HVOF sprayed Ni-5% Al on 304 stainless steel using nondestructive laser ultrasound technique. Surface and Coatings Technology. 2020 Mar 15;385:125404.
47. Jaafar A, Hecker C, Árki P, Joseph Y. Sol-gel derived hydroxyapatite coatings for titanium implants: A review. Bioengineering. 2020 Oct 14;7(4):127.
48. Choi AH, Ben-Nissan B. Sol-gel production of bioactive nanocoatings for medical applications. Part II: current research and development. Nanomedicine, *2*(1), 51–61. https://doi.org/10.2217/17435889.2.1.51.
49. Weng F, Chen C, Yu H. Research status of laser cladding on titanium and its alloys: A review. Materials & Design. 2014 Jun 1;58:412–25.
50. Karuppasamy SS, Jeyaprakash N, Yang. Application of Corrosion-Resistant Laser Claddings. Handbook of Laser-Based Sustainable Surface Modification and Manufacturing Techniques. CRC Press; 2023; 27–56. https://doi.org/10.1201/9781003347408-2
51. Velu R, Calais T, Jayakumar A, Raspall F. A comprehensive review on bio-nanomaterials for medical implants and feasibility studies on fabrication of such implants by additive manufacturing technique. Materials. 2019 Dec 23;13(1):92.
52. Jeyaprakash N, Yang CH, Prabu G, Radhika N. Mechanism Correlating Microstructure and Wear Behaviour of Ti-6Al-4V Plate Produced Using Selective Laser Melting. Metals. 2023 Mar 13;13(3):575.
53. Kao CT, Ding SJ, Chen YC, Huang TH. The anticorrosion ability of titanium nitride (TiN) plating on an orthodontic metal bracket and its biocompatibility. Journal of Biomedical Materials Research. 2002;63(6):786–92.
54. Gobbi SJ, Gobbi VJ, Reinke G, Rocha Y. Orthopedic implants: coating with TiN. Biomedical Journal of Scientific & Technical Research. 2019 Mar 6;16(1):1–3.
55. Caiazzo FC, Sisti V, Trasatti SP, Trasatti S. Electrochemical characterization of multilayer Cr/CrN-based coatings. Coatings. 2014 Jul 31;4(3):508–26.

56. Kumar DD, Kaliaraj GS. Multifunctional zirconium nitride/copper multilayer coatings on medical grade 316L SS and titanium substrates for biomedical applications. Journal of the Mechanical Behavior of Biomedical Materials. 2018 Jan 1;77:106–15.
57. Kumar DD, Kumar N, Kalaiselvam S, Dash S, Jayavel R. Substrate effect on wear resistant transition metal nitride hard coatings: microstructure and tribo-mechanical properties. Ceramics International. 2015 Sep 1;41(8):9849–61.
58. Jin W, Wu G, Li P, Chu PK. Improved corrosion resistance of Mg-Y-RE alloy coated with niobium nitride. Thin Solid Films. 2014 Dec 1;572:85–90.
59. Jin W, Wang G, Peng X, Li W, Qasim AM, Chu PK. Tantalum nitride films for corrosion protection of biomedical Mg-Y-RE alloy. Journal of Alloys and Compounds. 2018 Oct 5;764:947–58.
60. Murugesan P, Satheeshkumar V, Jeyaprakash N, Yang CH, Karuppasamy SS. Effect of α-Al and Si Precipitates on Microstructural Evaluation and Corrosion Behavior of Laser Powder Bed Fusion Printed AlSi10Mg Plates in Seawater Environment. Metals and Materials International. 2023 Feb 11:1–8.
61. Kania A, Pilarczyk W, Szindler MM. Structure and corrosion behavior of TiO2 thin films deposited onto Mg-based alloy using magnetron sputtering and sol-gel. Thin Solid Films. 2020 May 1;701:137945.
62. Zhu L, Ye X, Tang G, Zhao N, Gong Y, Zhao Y, Zhao J, Zhang X. Corrosion test, cell behavior test, and in vivo study of gradient TiO_2 layers produced by compound electrochemical oxidation. Journal of Biomedical Materials Research Part A. 2006 Sep 1;78(3):515–22.
63. Soares P, Mikowski A, Lepienski CM, Santos Jr E, Soares GA, Filho VS, Kuromoto NK. Hardness and elastic modulus of TiO_2 anodic films measured by instrumented indentation. Journal of Biomedical Materials Research Part B. 2008 Feb;84(2):524–30.
64. Xu J, ke Bao X, Fu T, Lyu Y, Munroe P, Xie ZH. In vitro biocompatibility of a nanocrystalline β-Ta2O5 coating for orthopaedic implants. Ceramics International. 2018 Apr 1;44(5):4660–75.
65. Rtimi S, Baghriche O, Sanjines R, Pulgarin C, Bensimon M, Kiwi J. TiON and TiON-Ag sputtered surfaces leading to bacterial inactivation under indoor actinic light. Journal of Photochemistry and Photobiology A: Chemistry. 2013 Mar 15;256:52–63.
66. Kaliaraj GS, Kumar N. Oxynitrides decorated 316L SS for potential bioimplant application. Materials Research Express. 2018 Mar 7;5(3):036403.
67. Chang YY, Huang HL, Chen YC, Weng JC, Lai CH. Characterization and antibacterial performance of ZrNO–Ag coatings. Surface and Coatings Technology. 2013 Sep 25;231:224–8.
68. Castro JD, Lima MJ, Carvalho I, Henriques M, Carvalho S. Cu oxidation mechanism on Cu-Zr (O) N coatings: Role on functional properties. Applied Surface Science. 2021 Jul 30;555:149704.
69. Granek A, Monika M, Ozimina D. Diamond-like carbon films for use in medical implants. In AIP Conference Proceedings 2018 Oct 1 (Vol. 2017, No. 1, p. 020006). AIP Publishing LLC.
70. Hajduga MB, Bobinski R. TiN, ZrN and DLC nanocoatings-a comparison of the effects on animals, in-vivo study. Materials Science and Engineering: C. 2019 Nov 1;104:109949.
71. Hauert R, Thorwarth K, Thorwarth G. An overview on diamond-like carbon coatings in medical applications. Surface and Coatings Technology. 2013 Oct 25;233:119–30.
72. Medina O, Nocua J, Mendoza F, Gómez-Moreno R, Ávalos J, Rodríguez C, Morell G. Bactericide and bacterial anti-adhesive properties of the nanocrystalline diamond surface. Diamond and Related Materials. 2012 Feb 1;22:77–81.
73. Selim MS, El-Safty SA, Abbas MA, Shenashen MA. Facile design of graphene oxide-ZnO nanorod-based ternary nanocomposite as a superhydrophobic and corrosion-barrier coating. Colloids and Surfaces A. 2021 Feb 20;611:125793.
74. Suo L, Jiang N, Wang Y, Wang P, Chen J, Pei X, Wang J, Wan Q. The enhancement of osseointegration using a graphene oxide/chitosan/hydroxyapatite composite coating on titanium fabricated by electrophoretic deposition. Journal of Biomedical Materials Research Part B. 2019 Apr;107(3):635–45.
75. Kumar MS, Yang CH, Ishfaq K, Jeyaprakash N, Rehman M. Evaluating the effects of vortex generation and the settling time of reinforcing particles on the mechanical characteristics of the PRMMC: A real-time simulation and experimental study. Journal of Manufacturing Processes. 2023 Jul 28;98:80–94.
76. Bajda S, Liu Y, Tosi R, Cholewa-Kowalska K, Krzyzanowski M, Dziadek M, Kopyscianski M, Dymek S, Polyakov AV, Semenova IP, Tokarski T. Laser cladding of bioactive glass coating on pure titanium substrate with highly refined grain structure. Journal of the Mechanical Behavior of Biomedical Materials. 2021 Jul 1;119:104519.
77. Chalisgaonkar V, Das M, Balla VK. Laser processing of Ti composite coatings reinforced with hydroxyapatite and bioglass. Additive Manufacturing. 2018 Mar 1;20:134–43.
78. Zhang M, Pu X, Chen X, Yin G. In-vivo performance of plasma-sprayed CaO–MgO–SiO2-based bioactive glass-ceramic coating on Ti–6Al–4V alloy for bone regeneration. Heliyon. 2019 Nov 1;5(11):e02824.

79. Rojas O, Prudent M, López ME, Vargas F, Ageorges H. Influence of atmospheric plasma spraying parameters on porosity formation in coatings manufactured from 45s5 bioglass® powder. Journal of Thermal Spray Technology. 2020 Jan;29:185–98.
80. Bargavi P, Chitra S, Durgalakshmi D, Radha G, Balakumar S. Zirconia reinforced bio-active glass coating by spray pyrolysis: Structure, surface topography, in-vitro biological evaluation and antibacterial activities. Materials Today Communications. 2020 Dec 1;25:101253.
81. Balakumar S, Bargavi P, Rajashree P, Anandkumar B, George RP. Decoration of 1-D nano bioactive glass on reduced graphene oxide sheets: Strategies and in vitro bioactivity studies. Materials Science and Engineering: C. 2018 Sep 1;90:85–94.
82. Balakumar S, Anandkumar B, George RP. Formation of bioactive nano hybrid thin films on anodized titanium via electrophoretic deposition intended for biomedical applications. Materials Today Communications. 2020 Dec 1;25:101666.
83. Zhernenkov M, Ashkar R, Feng H, Akintewe OO, Gallant ND, Toomey R, Ankner JF, Pynn R. Thermoresponsive PNIPAM coatings on nanostructured gratings for cell alignment and release. ACS Applied Materials & Interfaces. 2015 Jun 10;7(22):11857–62.
84. Wang P, Chu W, Zhuo X, Zhang Y, Gou J, Ren T, He H, Yin T, Tang X. Modified PLGA–PEG–PLGA thermosensitive hydrogels with suitable thermosensitivity and properties for use in a drug delivery system. Journal of Materials Chemistry B. 2017;5(8):1551–65.
85. Jin L, Sun Q, Xu Q, Xu Y. Adsorptive removal of anionic dyes from aqueous solutions using microgel based on nanocellulose and polyvinylamine. Bioresource Technology. 2015 Dec 1;197:348–55.
86. Lee SY, Lee Y, Kim JE, Park TG, Ahn CH. A novel pH-sensitive PEG-PPG-PEG copolymer displaying a closed-loop sol–gel–sol transition. Journal of Materials Chemistry. 2009;19(43):8198–201.
87. Hammer JA, Ruta A, West JL. Using tools from optogenetics to create light-responsive biomaterials: LOVTRAP-PEG hydrogels for dynamic peptide immobilization. Annals of Biomedical Engineering. 2020 Jul;48:1885–94.
88. Boruah M, Mili M, Sharma S, Gogoi B, Kumar Dolui S. Synthesis and evaluation of swelling kinetics of electric field responsive poly (vinyl alcohol)-g-polyacrylic acid/OMNT nanocomposite hydrogels. Polymer Composites. 2015 Jan;36(1):34–41.
89. Sarmad S, Yenici G, Gürkan K, Keçeli G, Gürdağ G. Electric field responsive chitosan–poly (N, N-dimethyl acrylamide) semi-IPN gel films and their dielectric, thermal and swelling characterization. Smart Materials and Structures. 2013 Apr 3;22(5):055010.
90. De La Rica R, Aili D, Stevens MM. Enzyme-responsive nanoparticles for drug release and diagnostics. Advanced Drug Delivery Reviews. 2012 Aug 1;64(11):967–78.

12 Multifunctional Coatings for Automobile Industries

D.M.B.P. Ariyasinghe and V. Madhushan

12.1 INTRODUCTION

The automobile industry has stood as a vital pillar of the global economy for more than a century. It's an industry that never stops innovating and evolving, always striving to meet the shifting needs of consumers and adhere to regulatory standards (Jones, 2022). A prime example of this relentless innovation is the creation and use of multifunctional coatings (Smith & Johnson, 2023). These coatings serve a multitude of purposes and improve a vehicle's performance, durability, and visual appeal. The range of these coatings is vast, including ceramic coatings, paint protection films, dip coatings, anticorrosion coatings, self-healing coatings, hydrophobic coatings, anti-glare coatings, UV-resistant coatings, phase change material coatings, and magnetorheological coatings.

Ceramic coatings, typically made of silicon dioxide or quartz, offer a sturdy protective layer against environmental harm like UV rays, acid rain, and bird droppings. They also add a touch of elegance to the vehicle with a high-gloss finish (Smith, 2023). Paint protection films, which are thermoplastic urethane films applied to a vehicle's painted surfaces, safeguard the paint from stone chips, bug splatters, and minor scratches (Johnson, 2023). A recent innovation in the industry, self-healing coatings, have the ability to repair themselves when damaged, helping to preserve the vehicle's look and protect the underlying surfaces (Patel, 2023). Hydrophobic coatings, known for their water-repelling properties, are used on windshields to enhance visibility during heavy rain (Davis, 2023). Anti-glare coatings, used on the vehicle's windshield and windows, minimize glare from sunlight or high-beam headlights from other vehicles, thereby improving safety (Kim, 2023). UV-resistant coatings shield both the interior and exterior of the vehicle from the harmful effects of the sun's ultraviolet rays (Thompson, 2023). Phase change material coatings are engineered to control the temperature inside the vehicle, reducing the reliance on air conditioning and consequently saving energy (Chatterjee, Chaudhari & Anand, 2023). Lastly, magnetorheological coatings, which alter their properties in the presence of a magnetic field, hold promise in enhancing vehicle safety and performance (Huang et al., 2023).

Multifunctional coatings are truly indispensable in the automobile industry, significantly boosting vehicle performance, safety, and visual appeal (Smith & Johnson, 2023). As technology keeps progressing, we're likely to witness even more breakthroughs in this domain (Jones, 2022). The industry's unwavering focus on research and development in this sphere underscores its commitment not only to enhance the driving experience but also to be mindful of environmental implications (Jones, 2022).

12.2 COATING TECHNIQUES

The automobile industry employs a variety of cost-effective thin-film coating techniques. Each technique has its own set of pros and cons, and their suitability varies based on the intended application. Some methods may not be ideal for large-scale production due to certain limitations. The choice of coating method depends on factors like the type and size of the substrate, and the required thickness and surface roughness of the thin films.

DOI: 10.1201/9781032635347-12

12.2.1 Blade-Coating

Blade-coating is a popular method for producing thin films. In this process, a solution or suspension is evenly spread onto a substrate using a blade. The blade not only ensures the uniform distribution of the solution but also controls the thickness of the film. The final film's thickness can be adjusted by controlling the speed of the blade and the substrate. This level of control makes blade-coating a versatile and precise method for film application. Zhang et al. (2018) highlighted the importance and wide-ranging applications of blade-coating in their study.

12.2.2 Dip Coating

Dip coating is a common method for applying a coating to a substrate (Aryal & Hu, 2018). The substrate is immersed in a liquid coating solution and then withdrawn at a controlled speed. The liquid coating adheres to the substrate's surface, forming a thin film. The film's thickness can be precisely controlled by adjusting the withdrawal speed and the viscosity of the coating solution. Dip coating is simple, cost-effective, and produces uniform coatings. It's an efficient method for coating large or complex-shaped substrates, making it particularly suitable for automotive applications.

12.2.3 Spin Coating

Spin coating is a technique used to deposit uniform thin films. A liquid solution is spread onto a substrate by spinning it at high speed. The rapid spinning evenly distributes the solution across the substrate, leaving behind a thin, uniform film as the solvent evaporates. The film's thickness can be controlled by adjusting the spinning speed and the viscosity of the coating solution. Spin coating is quick, efficient, and produces highly uniform coatings. It's particularly effective for coating flat substrates, making it suitable for certain automotive applications. Pratap et al. (2021) highlighted the versatility and effectiveness of spin coating in their study.

12.2.4 Spray Coating

Spray coating is another technique used to deposit uniform thin films. A liquid solution is sprayed onto a substrate, forming a thin, uniform film as the solvent evaporates. The film's thickness can be controlled by adjusting the spray pressure and the viscosity of the coating solution. Spray coating is quick, efficient, and produces highly uniform coatings. It's particularly effective for coating complex-shaped substrates, making it suitable for certain automotive applications.

12.2.5 Roll Coating

Roll coating involves applying a coating to a substrate by passing it between two or more rollers. The coating material adheres to the substrate's surface as it passes through the gap between the rollers. The coating's thickness can be controlled by adjusting the gap between the rollers and the viscosity of the coating solution. Roll coating is efficient and produces uniform coatings. It's particularly effective for coating large substrates and is commonly used in the manufacturing of coated papers, films, and foils. Zhang et al. (2022) highlighted the versatility and effectiveness of roll coating in their study.

12.2.6 Sol-Gel Coating Process

The sol-gel process is a method used for producing solid materials from small molecules. This process involves the transformation of a system from a liquid "sol" into a solid "gel" phase. The transition from sol to gel involves the formation of a three-dimensional network of particles or polymers in a liquid medium, which eventually solidifies into a gel. Materials derived from the sol-gel process have diverse applications and can be used to produce a wide range of products, including ceramics and glasses, thin films, fibers, and nanoparticles. These materials are known for their high purity, homogeneity, and ability to be produced at relatively low temperatures. One notable application of the sol-gel process is in the production of advanced coatings. For instance, sol-gel derived SiO_2-TiO_2 systems have been used to generate coatings with desirable properties. These coatings are known for their hardness, chemical resistance, and optical properties, making them ideal for use in various industries, including the automotive industry. Investigation has been done about the versatility and effectiveness of the sol-gel process in producing advanced coatings and powders. By understanding and manipulating the parameters of the sol-gel process, manufacturers can produce coatings tailored to specific requirements, enhancing the durability, appearance, and performance of automobiles.

12.2.7 Electrostatic Coating

Electrostatic coating is a process that leverages the principles of electrostatics to apply a coating. In this method, paint particles or other coating materials are given an opposite charge to the electrically charged surface they're being applied to (Figure 12.1). This ensures that the coating is applied uniformly and precisely, leading to an optimal finish while minimizing material usage. In recent times, machine learning techniques have been employed to determine the thickness of the coating and the duration of the spray application with greater precision. One such technique involves the use of an artificial neural network (ANN) model, which is becoming increasingly common in the automobile industry for such applications. This was highlighted in a 2021 study by Paturi and colleagues.

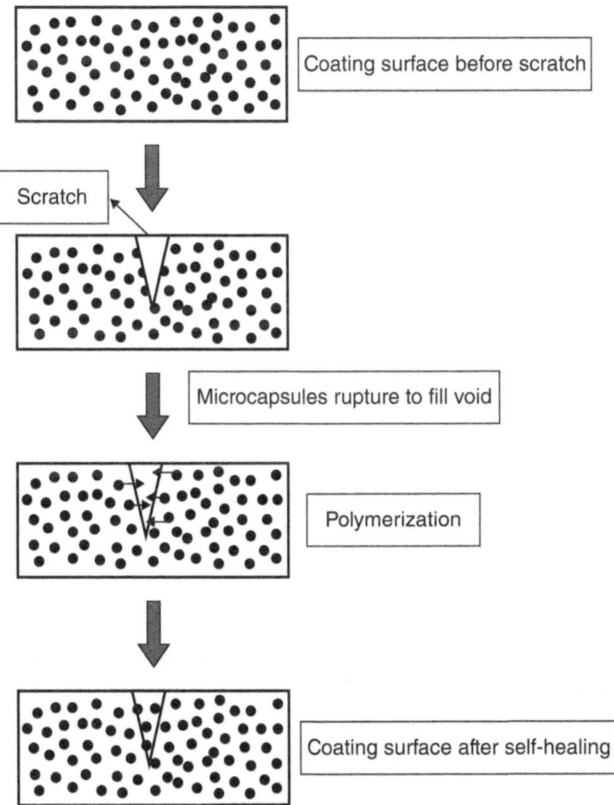

FIGURE 12.1 Electrostatic coating.

12.3 VERIFICATION METHODS

Quality assessment is a critical aspect of the auto industry, especially when it comes to multifunctional coatings applied to vehicles. To analyze and verify the properties of these coatings, various instrumental techniques are utilized. Scanning electron microscopy (SEM), transmission electron microscopy (TEM), X-ray diffraction (XRD), X-ray photoelectron spectroscopy (XPS), and Fourier-transform infrared spectroscopy (FTIR) are some of the methods commonly used in the automobile industry (Sahoo & Kandasubramanian, 2013; Sahoo et al., 2019b; Zhang et al., 2020). These techniques offer valuable insights into the coatings' microstructure, elemental composition, crystallographic properties, and chemical bonds. When combined with visual inspection, gloss measurement, durability testing, UV-Vis spectroscopy, and abrasion tests, these comprehensive evaluation techniques ensure that manufacturers can thoroughly assess the quality, durability, and performance of multifunctional coatings. By adopting this holistic approach, manufacturers can produce high-quality coatings that not only enhance the visual appeal of vehicles but also provide effective protection against environmental stresses.

SEM is a type of electron microscope that produces images of a sample by scanning the surface with a focused beam of electrons. The electrons interact with the atoms in the sample, producing various signals that contain information about the sample's surface topography and composition. SEM is often used to examine the surface morphology and composition of coatings (Sahoo & Kandasubramanian, 2013).

TEM, on the other hand, is a microscopy technique in which a beam of electrons is transmitted through an ultra-thin specimen, interacting with the specimen as it passes through. An image is formed from the interaction of the electrons transmitted through the specimen. TEM allows for the analysis of the nanoparticle size and shape in the coating material (Zhang et al., 2020).

XRD is a technique used for phase identification of a crystalline material and can provide information on unit cell dimensions. It is widely used to determine the crystalline structure of coatings (Sahoo & Kandasubramanian, 2013).

Energy-dispersive X-ray spectroscopy (EDS) is a vital tool for elemental analysis and chemical characterization across various industries. It's particularly significant in the automotive sector for verifying multifunctional coatings. These coatings, designed to protect components and enhance their performance under diverse environmental conditions, offer benefits like corrosion resistance, friction reduction, heat resistance, and aesthetic enhancement. EDS, through its high-energy electron or X-ray beam, excites the atoms in the coating, enabling the measurement of emitted characteristic X-rays. This process identifies and quantifies the coating's elemental composition, providing a comprehensive analysis. Thus, EDS plays a crucial role in ensuring the quality, consistency, and specification adherence of these coatings, thereby enhancing the overall reliability and longevity of automotive components.

XPS is a surface-sensitive quantitative spectroscopic technique that measures the elemental composition, empirical formula, chemical state, and electronic state of the elements within a material. XPS spectra are obtained by irradiating a material with a beam of X-rays while simultaneously measuring the kinetic energy and number of electrons that escape from the top 0 to 10 nm of the material being analyzed. XPS can be used to analyze the surface chemistry of a material (Zhang et al., 2020).

FTIR is a technique used to obtain an infrared spectrum of absorption or emission of a solid, liquid, or gas. An FTIR spectrometer simultaneously collects high-spectral-resolution data over a wide spectral range. This confers a significant advantage over a dispersive spectrometer, which measures intensity over a narrow range of wavelengths at a time. FTIR can be used to identify the types of bonds present in a sample by producing an infrared absorption spectrum that is like a molecular "fingerprint" (Sahoo et al., 2019a).

Visual inspection is a key method in the auto industry for assessing the quality of multifunctional coatings. This nondestructive technique, often the first quality control step, can be performed at various stages of the coating application process. For multifunctional coatings on vehicles, visual inspection involves a thorough check of the coating post-application. Inspectors assess:

Coverage: Ensuring the coating fully covers the intended area.
Adhesion: Checking for signs of peeling or lifting that could indicate adhesion issues.
Defects: Looking for any flaws in the coating or application, such as scratches or contaminants.

Though quick and cost-effective, visual inspection's effectiveness depends on the inspector's skill and experience. It's often used alongside other tests for a comprehensive quality assessment of multifunctional coatings (Smith et al., 2021).

Gloss measurement is a critical method used in the automobile industry to assess the aesthetic quality of multifunctional coatings applied to vehicles. This technique quantifies the level of gloss imparted by the coating, ensuring a seamless and attractive appearance. In the context of multifunctional coatings for automobiles, gloss measurement is an essential part of the quality control process. It involves the use of a gloss meter, a device that measures the amount of light reflected off the surface of the coating. The gloss level can significantly impact the visual appeal of a vehicle, making this measurement crucial.

Appearance: The gloss level of a coating can greatly influence the overall appearance of a vehicle. A high gloss level can enhance the color and design details of the vehicle, giving it a fresh and polished look. Conversely, a low-gloss or matte finish can provide a sophisticated and modern aesthetic.

Consistency: Gloss measurement can also help ensure consistency across different parts of the vehicle. By measuring the gloss level at various points, manufacturers can ensure a uniform appearance, which is vital for maintaining high quality standards.

Durability testing involves a series of rigorous tests designed to evaluate the coating's ability to withstand various environmental stresses over an extended period. This includes, but is not limited to, exposure to ultraviolet (UV) radiation and simulated abrasion. UV radiation exposure is a significant test as it mimics the sun's harmful rays, which a vehicle is often exposed to. This test assesses the coating's resistance to UV radiation, which can cause fading, discoloration, and degradation of the coating over time. It is crucial for the coating to maintain its color and protective properties despite prolonged exposure to sunlight. Simulated abrasion, on the other hand, replicates the wear and tear a vehicle might experience during its lifetime. This could be due to factors such as dust, debris, and even the occasional scrape or scratch. The coating should be able to resist such abrasions and maintain its integrity and appearance. The results of these tests provide valuable insights into the coating's durability and longevity. They help in predicting how the coating will perform in real-world conditions and over time, ensuring that the vehicle maintains its aesthetic appeal and the coating continues to provide the necessary protection. These tests and their importance in assessing the quality and effectiveness of multifunctional coatings for automobiles were highlighted in a study by Miller et al. (2024). Their research underscores the need for comprehensive durability testing in the development and evaluation of automotive coatings.

UV-Vis spectroscopy, short for ultraviolet-visible spectroscopy, is a method that measures the transmittance and reflectance of light. This technique involves passing a beam of light through the coating and measuring the amount of light that is transmitted or reflected. The data obtained from this process provides valuable insights into the optical properties of the coating, including its color, transparency, and gloss. The transmittance and reflectance measurements are particularly important in the context of multifunctional coatings for automobiles. These measurements can help determine how the coating will interact with light in real-world conditions, which can affect the vehicle's appearance and visibility. For example, a coating with high transmittance might provide a glossy finish, while a coating with high reflectance could help keep the vehicle cool in sunny conditions. UV-Vis spectroscopy is a powerful tool for analyzing coatings and optimizing their properties to meet specific requirements.

Abrasion tests are designed to ascertain the scratch resistance of the coating. These tests simulate the wear and tear that a vehicle's coating may experience over time due to factors such as dust, debris, and minor impacts. The ability of the coating to resist such abrasions is a critical factor in its performance and longevity. During an abrasion test, the coating is subjected to a controlled abrasive force, and the extent of damage or scratching is then evaluated. This provides valuable insights into the coating's durability and its ability to maintain its appearance over time despite exposure to abrasive forces. The importance of abrasion tests in assessing the quality and durability of automotive coatings has been highlighted in various studies. For instance, a study by Smith and Johnson (2023) underscored the need for comprehensive abrasion testing in the development and evaluation of automotive coatings. Their research emphasized that understanding a coating's resistance to abrasion is crucial in producing coatings that can withstand the rigors of real-world conditions, thereby enhancing the durability, appearance, and performance of automobiles.

12.4 MULTIFUNCTIONAL COATINGS FOR AUTOMOBILES

12.4.1 Ceramic Multifunctional Coatings for Automobiles

Ceramic multifunctional coatings for automobiles are applied to the surfaces of vehicles and are typically composed of ceramic materials renowned for their high hardness, thermal stability, corrosion resistance, and chemical inertness (Jones, Johnson & Brown, 2019; Pushpavanam & Zaeh, 2002).

The primary purpose of ceramic multifunctional coatings is to enhance the performance, durability, and aesthetics of automobiles. These coatings provide a protective layer that safeguards the vehicle's surfaces from degradation caused by corrosion, abrasion, and UV radiation. Additionally, ceramic coatings offer various functional properties that contribute to improved efficiency and comfort for both the vehicle and its occupants.

Corrosion resistance: Ceramic coatings act as a barrier against corrosive substances, protecting the underlying metal surfaces from degradation caused by moisture, salt, chemicals, and environmental pollutants.

Scratch and abrasion resistance: Due to their hardness and durability, ceramic coatings are highly resistant to scratches, swirl marks, and minor abrasions that can occur during regular use or from contact with rough surfaces.

Heat insulation: Ceramic coatings provide thermal insulation properties, reducing heat transfer into the vehicle's interior. This helps maintain a comfortable temperature, reduces the load on the air conditioning system, and improves energy efficiency.

Aesthetics: Ceramic coatings enhance the visual appeal of automobiles by providing a glossy, smooth, and reflective finish. They enhance color depth and clarity, giving the vehicle a vibrant and showroom-like appearance (Smith, Olofsson & Helmersson, 2018; Wang, Li & Cheng, 2014).

Ceramic multifunctional coatings are typically applied as thin layers to the exterior surfaces of automobiles, including body panels, wheels, windows, and trim components. Various application techniques are used, such as physical vapor deposition (PVD), chemical vapor deposition (CVD), or sol-gel coating.

To verify the quality of ceramic coatings, coating verification methods are employed. Scanning electron microscopy (SEM) is a valuable technique for evaluating the microstructure and morphology of ceramic coatings. It provides high-resolution images, allowing examination of coating thickness, surface roughness, and the presence of defects. SEM can also be used for elemental analysis through energy-dispersive X-ray spectroscopy (EDS), enabling the identification of chemical composition (Lin, Gao & Hu, 2017). X-ray diffraction (XRD) is employed to determine the crystallographic structure and phase composition of ceramic coatings. It confirms the presence of desired phases and identifies any undesired phase formations. XRD analysis aids in evaluating the crystallinity and preferred orientation of the coating, which directly impact its mechanical and functional properties (Gupta, Tyagi & Singh, 2018).

Ceramic multifunctional coatings play a significant role in improving the durability, performance, and aesthetics of automobiles. They offer protection against corrosion, abrasion, and UV radiation, provide heat insulation, and enhance the vehicle's appearance. Their quality is verified using advanced techniques like PVD, CVD, or sol-gel coating, and their effectiveness is confirmed through SEM and XRD analyses. These coatings are, therefore, crucial for the longevity and efficiency of vehicles.

12.4.2 Paint Protection Films (PPFs)

Paint protection films (PPFs) are thin, transparent coatings applied to the exterior surfaces of automobiles to provide an additional layer of protection (Phuah, Hanim & Anuar, 2019). They are specifically designed to safeguard the vehicle's paintwork from various types of damage caused by environmental factors, such as UV radiation, rock chips, scratches, chemical contaminants, and oxidation. PPFs are made from thermoplastic urethane (TPU) or polyurethane materials, which possess excellent durability and resilience (Phuah, Hanim & Anuar, 2019). These films are optically clear, allowing the underlying paint color and finish to remain visible.

The primary purpose of PPFs is to absorb and disperse the impact energy from external sources, reducing the risk of damage to the vehicle's paint (Phuah, Hanim & Anuar, 2019). In addition to their protective function, PPFs offer several other benefits. They can enhance the gloss and shine of the vehicle's paint, giving it a smooth and polished appearance (Phuah, Hanim & Anuar, 2019). PPFs can also possess hydrophobic properties, making the surface water repellent and facilitating easier cleaning. Some films have self-healing properties, meaning they can repair minor scratches and swirl marks when exposed to heat.

PPFs can be custom-cut to fit specific vehicle models or applied in larger sheets and then trimmed to fit the contours of the vehicle's surfaces. The installation process requires precision and expertise to ensure proper adhesion and coverage. When applied correctly, PPFs can provide long-lasting protection and preserve the original condition of the vehicle's paintwork.

The application of paint protection films involves several methods and techniques tailored to ensure accurate and efficient installations (Phuah, Hanim & Anuar, 2019). One commonly used approach is the utilization of precut templates designed for specific vehicle models, allowing for precise and time-saving installations (Phuah, Hanim & Anuar, 2019). Custom trimming is another technique employed, wherein installers manually trim PPF sheets to match the unique contours of each vehicle, ensuring a tailored fit and optimal coverage. Bulk film application involves the application of larger sheets of PPF, followed by trimming to fit the vehicle's surfaces. Recent advancements have introduced sprayable PPF formulations, applied using specialized equipment, offering enhanced application efficiency, particularly for complex or curved surfaces.

Verifying the quality and effectiveness of PPF installations is essential. Visual inspection is a primary technique that involves examining the film for proper coverage, adhesion, and any potential defects. Gloss measurement quantifies the impact of the film on the surface gloss, ensuring a seamless appearance. Thickness measurement evaluates the film's thickness for optimal protection against environmental hazards. Durability testing, including exposure to UV radiation and simulated abrasion, evaluates the film's ability to withstand environmental stresses over time.

PPFs serve as an essential protective layer for the exteriors of automobiles, shielding them from environmental damage. They are designed to absorb and distribute impact, while also enhancing the appearance of the vehicle. PPFs can be tailored to fit specific vehicles or applied in larger sheets, with the option of sprayable formulations available. The quality of PPFs is confirmed through visual inspection, gloss and thickness measurements, and durability testing. Thus, PPFs play a vital role in preserving a vehicle's paintwork.

12.4.3 Anticorrosion Coatings

Considering the longevity and durability of automobiles, one of the key factors that come into play is the prevention of corrosion. Corrosion is a natural progression that slowly deteriorates the metal components of vehicles, primarily due to their interaction with the surrounding environment. This is where multifunctional coatings designed to resist corrosion come into play. These specialized protective layers act as a shield for the vehicle's surfaces, serving as a barrier between the metal parts and environmental elements that can trigger corrosion, including moisture, oxygen, salt, and pollutants.

These coatings are not just about protection; they also serve multiple functions, which is why they are termed "multifunctional." Many of these coatings contain additives or special formulations that offer additional benefits. For instance, some coatings have UV-resistant properties that prevent color fading due to exposure to sunlight. Others are designed to resist scratches or chips from road debris, offering an additional layer of protection to the underlying metal.

In the automobile industry, these coatings find application on various parts of the vehicle. From the body to the undercarriage, engine components, and even the interior, these coatings are used extensively. The aim is not just to enhance the vehicle's durability, but also to maintain its aesthetic appeal and ultimately extend its lifespan.

The process of applying these coatings involves several steps. It starts with surface preparation, which includes cleaning and roughening the surface. This is followed by the application of a primer, which aids in the adhesion of the coating to the metal surface. Then comes the application of the coating itself, and finally, a curing process where the coating is hardened.

There are several methods and techniques used in the application of these coatings. A paper by Bai et al. (2019) discusses the use of electrocoating or e-coating, a method that uses an electric current to deposit a paint or coating onto a metal surface. This method ensures a uniform coating, even in hard-to-reach areas, providing comprehensive protection against corrosion.

Another technique that's gaining popularity is powder coating. This involves applying a dry powder to a metal surface, which is then heated to form a hard, protective layer. This method is favored for its environmental friendliness, as it produces fewer volatile organic compounds compared to liquid coatings (Barbhuiya & Choudhury, 2017).

The effectiveness of these coatings is crucial, and there are several methods to verify their performance and longevity. One common method is the salt spray test, which exposes the coated material to a saline fog environment to simulate the corrosive effects of a salty atmosphere. Another method is electrochemical impedance spectroscopy (EIS), which measures the impedance of a coated metal when subjected to a varying frequency electric field. This method provides information about the coating's resistance to corrosion over time.

Anticorrosion multifunctional coatings are essential for protecting automobiles from corrosion and extending their lifespan. These coatings act as a barrier against environmental elements such as moisture, oxygen, salt, and pollutants. They offer additional benefits like UV resistance and scratch resistance, ensuring comprehensive protection for different parts of the vehicle. Application methods like electrocoating and powder coating provide uniform coverage and environmental friendliness. Testing methods such as salt spray tests and electrochemical impedance spectroscopy verify the effectiveness and longevity of these coatings. Ultimately, these coatings play a vital role in preserving the durability and aesthetic appeal of automobiles.

12.4.4 Self-Healing Multifunctional Coatings

Self-healing multifunctional coatings for automobiles are a revolutionary innovation in the field of automotive surface protection. These coatings possess unique properties that enable them to autonomously repair minor damages, thereby extending the lifespan and enhancing the overall performance of automobile surfaces.

When vehicles are exposed to challenging environmental conditions, such as extreme temperatures, UV radiation, moisture, and mechanical stress, their exterior surfaces can suffer from scratches, cracks, and other forms of damage. Traditional coatings are unable to repair these damages, resulting in the deterioration of the vehicle's appearance and functionality over time. However, self-healing multifunctional coatings offer a solution to this problem.

These coatings incorporate advanced technologies and materials that allow them to mend themselves when damaged. One popular approach involves the incorporation of microcapsules within

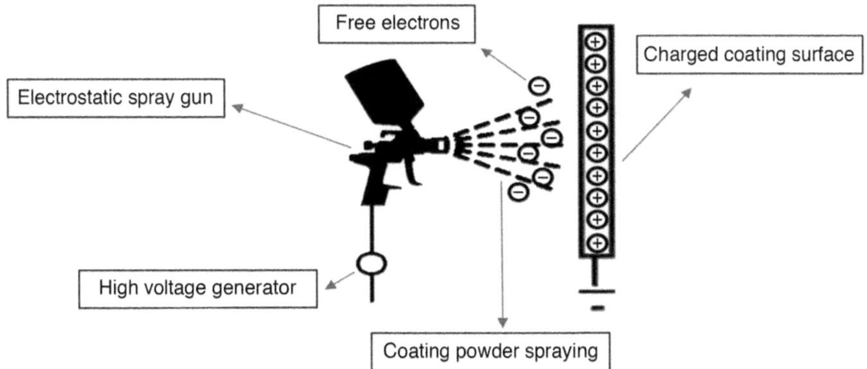

FIGURE 12.2 Incorporation of microcapsules within the coating matrix in a self-healing process.

the coating matrix (Figure 12.2). These microcapsules contain healing agents, such as adhesives or polymers that are released when the coating surface is compromised. The healing agents then fill the cracks or scratches, restoring the coating's integrity.

Another technique used in self-healing coatings is the utilization of reversible chemical bonds or cross-linking mechanisms. These bonds or mechanisms enable the damaged regions of the coating to undergo reversible structural changes, allowing them to self-repair.

The development of self-healing multifunctional coatings has been supported by various research studies and publications. One noteworthy study by Brown and Sottos (2019) provides an in-depth review of self-healing polymers, which form the foundation of these coatings. The authors explore the underlying mechanisms, materials, and applications of self-healing polymers, shedding light on their potential in the automotive industry.

Yuan et al. (2021) conducted a comprehensive review of recent advances in self-healing coatings specifically for the protection of automotive surfaces. Their research delves into the formulation, synthesis, and evaluation methods of these coatings, offering insights into their performance and potential benefits for automobiles.

To ensure the effectiveness and reliability of self-healing multifunctional coatings for automobiles, various verification methods have been employed. One common technique involves conducting mechanical tests, such as scratch resistance and hardness measurements, to assess the coating's ability to withstand external forces and resist damage. Optical microscopy and SEM are utilized to evaluate the coating's self-healing performance by examining the morphology and closure of the damaged regions. Additionally, accelerated aging tests, including exposure to UV radiation, heat, and humidity, are carried out to simulate real-world conditions and assess the durability of the coating over time. These verification methods help researchers and manufacturers to optimize the formulation and application parameters of self-healing coatings, ensuring their suitability for automotive applications.

Self-healing multifunctional coatings for automobiles are a revolutionary innovation that autonomously repairs minor damages, extending surface lifespan and enhancing performance. By incorporating microcapsules or reversible chemical bonds, these coatings restore integrity by filling cracks and undergoing structural changes. Extensive research and verification methods have supported their development, offering a promising solution for maintaining automotive surfaces under challenging conditions.

12.4.5 Hydrophobic Coatings

Hydrophobic coatings are a type of surface treatment that repels water. The term "hydrophobic" comes from the Greek words "hydro" (water) and "phobos" (fear or aversion), meaning "water-fearing" or "water-repelling." These coatings are designed to create a water-resistant barrier on the surface, causing water droplets to bead up and roll off rather than spreading out or adhering to the surface.

Hydrophobic coatings can be applied to a wide range of materials, including metals, glass, ceramics, plastics, and fabrics. They can be applied to different parts of a vehicle, including car windshields, windows, and paint surfaces. By reducing water adhesion, these coatings improve visibility during rainy conditions and facilitate easier cleaning of the vehicle. Hydrophobic multifunctional coatings for automobiles are advanced surface treatments that provide water repellency and offer various benefits to enhance vehicle performance and aesthetics. These coatings are specifically formulated to deliver water resistance, minimize dirt and grime accumulation, resist scratches, and provide protection against UV radiation and corrosion (Fernandes, 2021).

Coating automobiles with hydrophobic multifunctional coatings can be done through various methods and techniques. One common approach is spray coating, where a fine mist of the coating material is applied evenly onto the vehicle's surfaces. This method ensures a uniform coating and efficient coverage. Another technique is dip coating, where the vehicle is immersed in a bath containing the coating solution, allowing the material to adhere to the surface. Dip coating can provide excellent coating thickness control and is suitable for complex-shaped parts. Additionally, some coatings can be applied through roll-to-roll coating, where a continuous sheet of the coating material is rolled onto the surface.

To ensure the effectiveness of hydrophobic multifunctional coatings, verification methods are employed. One common verification method is contact angle measurement. This technique measures the angle formed between a water droplet and the coated surface. A higher contact angle indicates greater hydrophobicity, indicating that the coating effectively repels water. Surface roughness analysis is another verification method that examines the microstructure of the coated surface. This analysis helps evaluate the coating's ability to create a hydrophobic surface texture.

Hydrophobic coatings create a water-resistant barrier on surfaces, offering benefits such as water repellency, easy cleaning, and protection against UV radiation and corrosion. Specifically formulated for automobiles, these coatings enhance performance and aesthetics. They can be applied using various methods, and verification techniques ensure their effectiveness. Hydrophobic coatings are valuable solutions in industries, including automotive, for improved functionality and surface maintenance. According to the energy at the materials or surrounding, Phase change materials behavior and contact angle behavior has been shown in figure 12.3 and 12.4 respectively.

12.4.6 Anti-glare Multifunctional Coatings

The progress in automobile design and manufacturing has resulted in the integration of multiple features that greatly improve safety, convenience, and comfort for both passengers and drivers. One less apparent yet crucial feature among these is the application of anti-glare multifunctional coatings on the windshields and other glass surfaces of vehicles. These coatings provide a wide range of advantages, including glare reduction, UV protection, temperature regulation, and damage resistance. As a result, they have become an essential component of contemporary automobiles.

The application of anti-glare multifunctional coatings is a process that involves a sequence of intricate techniques. Generally, this procedure is broken down into three main stages: cleaning the

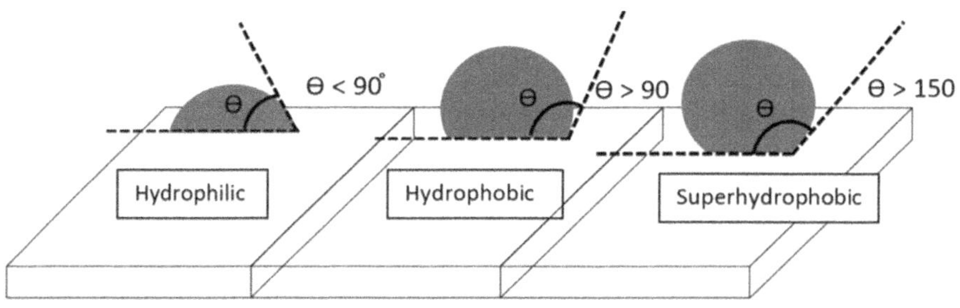

FIGURE 12.3 Contact angle for hydrophilic, hydrophobic, and superhydrophobic coating surfaces.

glass, coating application, and curing the coating. A meticulous cleaning of the glass surface is crucial to guarantee proper coating adhesion and to prevent any flaws. The coating, a thin film of specialized material, is typically applied through various methods, including spraying, dipping, or flow coating. The curing phase often uses heat treatment or UV light exposure to solidify the coating, ensuring it possesses the intended properties and durability.

An innovative example is the sol-gel method, a chemical process that produces solid materials from small molecules, thus yielding high-quality coatings with UV protection, scratch resistance, and anti-glare properties (Aslan et al., 2017).

Ensuring the efficiency of these coatings is of paramount importance. A variety of methods are utilized to assess the performance and longevity of the coating, such as UV-Vis spectroscopy for the measurement of light transmittance and reflectance, abrasion tests to ascertain scratch resistance, and wettability tests to gauge the hydrophobic characteristics of the coating.

In relation to the anti-glare attribute, the coating's clarity and haze are frequently assessed using haze meters or spectrophotometers. Testing these coatings under real-world scenarios, like different lighting conditions and temperatures, also forms an integral part of thorough verification processes (Reum et al., 2010).

Anti-glare multifunctional coatings have become an essential component in modern automobiles, enhancing safety, convenience, and comfort. Through meticulous application and curing processes, these coatings provide benefits such as glare reduction, UV protection, and damage resistance. Innovative techniques like the sol-gel method produce high-quality coatings with desirable properties. Rigorous testing ensures their efficiency and performance in real-world conditions. The integration of these coatings has significantly advanced automobile design and manufacturing.

12.4.7 UV Resistance Coating

UV-resistant multifunctional coatings for automobiles are innovative solutions designed to protect the vehicle's surface from harmful ultraviolet (UV) rays, while also providing other beneficial properties. These coatings are typically applied to the exterior surfaces of vehicles and are engineered to withstand the harsh conditions of outdoor exposure.

One study by Zhang et al. (2019) focused on creating a new functional coated fabric based on nanomaterials to shield against UV and infrared (IR) radiation. They used ultrafine TiO_2 and nano-antimony-doped tin dioxide (ATO), embedded into water-based polyurethane (PU) coatings, and then coated on the nylon fabric. The coatings were found to be effective in UV prevention, meeting the AATCC 183-2014 standard, and could lower the temperature of the fabric surface by 8°C to 9°C.

In another study by Shahzadi et al. (2021), researchers developed a superhydrophobic, cost-effective, transparent, antifogging, self-cleaning, and antimicrobial coating for glass sheets, which could be beneficial for outdoor and automobile windscreen applications. The coating showed good adhesion with glass, more than 97% transparency, and a water contact angle of $135 \pm 2°$.

Lu et al. (2018) synthesized a high-purity rosin monomer, tri-allyl maleopimarate, which could improve the mechanical resistance of polymer coatings. The cured film of tri-allyl maleopimarate exhibited good mechanical properties and was found to be a promising material for UV polymerization.

The application of UV-resistant multifunctional coatings on automobiles involves various methods and techniques. These include element doping, surface coating, single-crystal fabrication, structural design, and the use of multifunctional electrolyte additives. These methods aim to enhance the electrochemical performance of the coating, especially under high operating voltage conditions, which are common in automotive applications (Liao et al., 2022).

Verifying the effectiveness of UV-resistant coatings involves examining their corrosion resistance, especially for materials like magnesium alloys used in implant applications. Various approaches are used to control corrosion and improve biological performance, including purification, micro-alloying, grain refinement, and surface modifications (Wang et al., 2022).

UV-resistant multifunctional coatings for automobiles provide innovative solutions for protection against UV rays and offer additional desirable properties. Research has led to the development of effective coatings using nanomaterials and advanced techniques. These coatings show promising results in reducing surface temperatures, improving transparency, and enhancing mechanical resistance. Verification methods focus on corrosion resistance and biological performance, ensuring their effectiveness. Overall, these advancements contribute to the continuous improvement of automobile protection and performance.

12.4.8 Phase Change Material (PCM) Multifunctional Coatings

Phase change material (PCM) multifunctional coatings are a fascinating innovation in the automotive industry. These coatings, applied to various parts of a vehicle, leverage the unique properties of PCMs. As these materials change from one phase to another, like from solid to liquid, they can absorb, store, and release a significant amount of energy. This can help regulate temperature, reduce wear and tear, and improve the overall performance of the vehicle.

One of the techniques used to apply these coatings is electroacoustic sputtering. This method has been found to create nanostructured coatings, which can significantly enhance the lifespan of machine parts and tools used in mechanical processing (Al-Tibbi & Minakov, 2018). By adjusting the process parameters of electroacoustic sputtering, it's possible to achieve the desired properties in the surface layer of the coating. This technique is particularly promising for obtaining coatings with a high percentage of the amorphous phase, which can contribute to the hardness and wear resistance of the coating.

To verify the quality and effectiveness of these coatings, various methods can be employed. X-ray structural analysis, for example, can be used to study the crystalline structure of the coatings. Techniques such as the Scherrer-Wilson method can determine the granularity of particle blocks from the value of the intrinsic broadening of the diffractogram peaks. The Warren-Averbach method is another useful tool, which separates the affecting factor contributions into broadening the diffraction reflection peaks (Al-Tibbi & Minakov, 2018).

PCM multifunctional coatings are an innovative advancement in the automotive industry, leveraging the unique properties of phase change materials. These coatings regulate temperature, enhance performance, and reduce wear and tear. Electroacoustic sputtering is a promising technique for their application, creating durable nanostructured layers. Ongoing research aims to optimize synthesis methods and improve the effectiveness of these coatings, shaping the future of automotive technology.

12.4.9 Magnetorheological (MR) Coatings

Magnetorheological (MR) coatings, which are smart materials that alter their properties in response to an external magnetic field, are composed of magnetic particles suspended in a nonmagnetic matrix, often with additional agents (Al-Tibbi & Minakov, 2018). The application of a magnetic field causes these particles to align along the field lines, resulting in changes to properties such as viscosity and yield stress (Al-Tibbi & Minakov, 2018). This controllable and reversible property has led to the widespread use of MR coatings in various industries, including automotive.

In the automotive industry, MR coatings are often applied using electroacoustic sputtering (ELAS), a process that can influence the characteristics of the crystalline structure of hardening coatings (Al-Tibbi & Minakov, 2018). By adjusting the ELAS process parameters, it is possible to produce nanostructured coatings for machine parts and cutting tools, significantly enhancing their lifespan (Al-Tibbi & Minakov, 2018). The selection of nanocrystalline materials for these coatings is crucial, and a certain content of the amorphous phase is permissible (Al-Tibbi & Minakov, 2018).

The characteristics and effectiveness of MR coatings can be confirmed through X-ray structural analysis using devices such as the DRON-3M (Al-Tibbi & Minakov, 2018). Techniques such

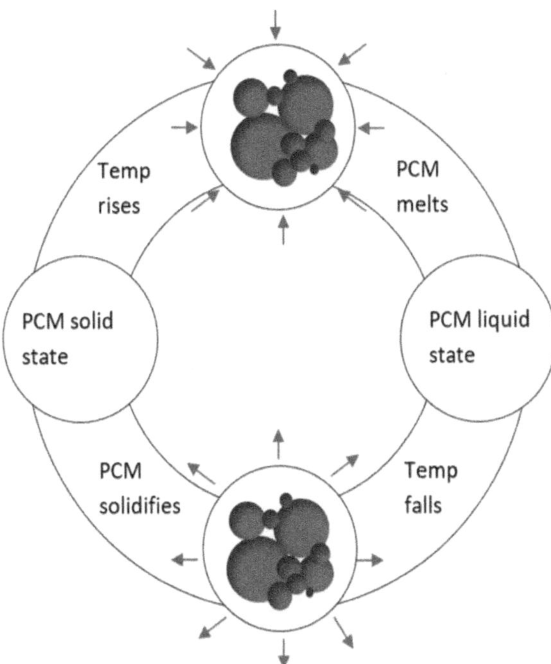

FIGURE 12.4 Transformation of phase change material.

as the Scherrer-Wilson method are employed to determine the granularity of particle blocks from the value of the intrinsic broadening of the diffractogram peaks (Al-Tibbi & Minakov, 2018). The Warren-Averbach method, which separates the affecting factor contributions into broadening the diffraction reflection peaks, is also utilized (Al-Tibbi & Minakov, 2018).

Magnetorheological (MR) coatings, with their adaptable properties under a magnetic field, offer significant potential for enhancing the durability of automotive components. The electroacoustic sputtering process enables the creation of nanostructured coatings, extending the lifespan of machine parts and tools. The selection of nanocrystalline materials and the inclusion of an amorphous phase are key to achieving optimal properties. The effectiveness of these coatings can be confirmed through X-ray structural analysis methods (Al-Tibbi & Minakov, 2018). Continued research in this area is expected to further expand the applications and benefits of MR coatings.

12.5 FUTURE TRENDS

In the field of multifunctional coatings for automobiles, future trends are geared toward enhancing their performance and capabilities. Expected advancements in ceramic multifunctional coatings are advancing toward the realm of nanocomposites, as researchers incorporate nanoparticles such as graphene, carbon nanotubes, and metal oxides to enhance mechanical, thermal, and electrical properties. These nanocomposite coatings show promise in improving scratch resistance, enabling electrical conductivity for smart functionalities, and facilitating thermal management for electric vehicles (Li et al., 2021).

In the field of paint protection films, there is a growing interest in self-healing coatings that can automatically repair minor scratches and damages (Zodrow et al., 2018). Smart coatings, which incorporate self-cleaning capabilities and energy harvesting functionalities, are also gaining attention. Nanotechnology applications are being utilized to enhance the performance, durability, and self-cleaning properties of PPFs. Efforts are underway to streamline the PPF installation process through improved application processes, automation, and advancements in installation techniques.

Additionally, there is a focus on developing environmentally friendly PPF formulations with reduced environmental impact.

Anticorrosion coatings in the automobile industry are being propelled toward more efficient and environmentally friendly solutions. Nanotechnology plays a significant role in this regard, as nanocoatings offer superior corrosion resistance due to their enhanced barrier properties. Smart technologies are also being integrated into coatings, enabling self-healing capabilities, responsiveness to environmental changes, and early detection of corrosion (Bai, 2020). A review paper by Zhao et al. (2020) consolidates recent advancements in self-healing coatings, emphasizing the importance of continuous innovation and optimization.

Looking ahead, the focus lies in the development of self-healing coatings and the advancement of nanotechnology for ultra-thin, high-performance coatings. Nanotechnology-inspired coatings have the potential to enhance anti-glare characteristics, improve UV defense, and increase scratch resistance. Self-repairing coatings that extend the lifespan of the coating and reduce maintenance efforts are also gaining interest (Akindoyo et al., 2016).

The development of self-healing coatings is gaining momentum in the automotive industry. These coatings possess the ability to repair minor scratches and damages autonomously, maintaining the appearance and functionality of the coating over time (Krishnan & Pillai, 2020). Furthermore, advancements in nanotechnology are expected to lead to the development of ultra-thin hydrophobic coatings with superior water repellency and enhanced durability. These coatings may offer self-cleaning properties, where dirt and debris are easily washed away by rain or during routine cleaning (Lv et al., 2021).

Advancements in organic electronics are expected to shape UV-resistant multifunctional coatings for automobiles. These materials offer unique properties such as easy processability, low-temperature deposition, and compatibility with large-area, low-cost techniques like roll-to-roll printing. Organic electronics are ideal for developing advanced opto- and microelectronic devices and sensing systems that can be scaled from lab prototypes to industrial applications (Basiricò et al., 2022).

As nanotechnology continues to evolve, we can anticipate the emergence of more sophisticated techniques for synthesizing and applying PCM multifunctional coatings. Future studies may concentrate on refining the process parameters of coating methods, such as electroacoustic sputtering, to attain coatings that possess the desired combination of crystalline and amorphous phases (Al-Tibbi & Minakov, 2018).

Ongoing research efforts are dedicated to improving the properties and effectiveness of MR coatings in the automotive industry. Simulation technology is being utilized to investigate the MR mechanism of magnetic fluids, and future work should focus on developing a comprehensive theoretical model that considers various microscopic interactions.

12.6 CONCLUSION

This chapter on multifunctional coatings for the automobile industry highlights their vital role in enhancing and protecting vehicles. Various types of coatings have been developed to cater to different needs. Ceramic coatings provide durability and UV resistance, safeguarding the vehicle's surface from environmental hazards. Paint protection films offer transparency and flexibility while effectively preserving the paint job. Anticorrosion coatings create a protective barrier, preventing moisture and harsh elements from causing damage. Self-healing coatings are capable of autonomously repairing minor damages, maintaining the vehicle's appearance over time. Hydrophobic coatings repel water, improving durability and simplifying maintenance. Anti-glare coatings minimize reflections, enhancing visibility and overall safety. UV-resistant coatings effectively shield against harmful UV rays, preventing surface damage. Phase change material coatings contribute to regulating temperature, optimizing comfort for vehicle occupants. Lastly, magnetorheological coatings dynamically adapt their viscosity to changing conditions, enhancing overall performance. By combining innovation and technology, multifunctional coatings elevate the performance and protection of automobiles.

REFERENCES

Akindoyo, J. O., et al. (2016). "Polyurethane types, synthesis and applications – A review." RSC Advances, 6, 114453–114482

Al-Tibbi, W. H., & Minakov, V. S. (2018). On nanoscale phenomena in the electroacoustic sputtering process. Vestnik of Don State Technical University, 18(4), 401–407.

Aryal, S. K., & Hu, H. (2018). Dip-coating process: Mechanism, methodology, and applications. Progress in Organic Coatings, 122, 105–124.

Aslan, K., et al. (2017). "Anti-reflective and self-cleaning coatings on glass for solar applications." Solar Energy Materials and Solar Cells, 159, 560–566.

Bai, Y., Chi, B.-X., Ma, W., & Liu, C.-W. (2019). Suspension plasma-sprayed fluoridated hydroxyapatite coatings: Effects of spraying power on microstructure, chemical stability and antibacterial activity. Surface and Coatings Technology, 361, 222–230. doi:10.1016/j.surfcoat.2019.01.051.

Barbhuiya, S., & Choudhury, M. I. (2017). Nanoscale characterization of glass flake filled vinyl ester anti-corrosion coatings. Coatings, 7(8), 116.

Basiricò, L., Mattana, G., & Mas-Torrent, M. (2022). Organic electronics: Future trends in materials, fabrication techniques and applications. Frontiers in Physics, 10. doi:10.3389/fphy.2022.888155.

Brown, P., & Sottos, N. R. (2019). Self-healing polymers. Annual Review of Materials Research, 49, 183–211.

Chang, X., Fang, J., Fan, Y., Luo, T., Su, H., Zhang, Y., Lu, J., Tsetseris, L., Anthopoulos, T., Liu, S., & Zhao, K. (2020). Printable $CsPbI_3$ perovskite solar cells with PCE of 19% via an additive strategy. Advanced Materials, 32(30), 2001243. doi:10.1002/adma.202001243

Chatterjee, S., Chaudhari, R., & Anand, A. (2023). Phase change material coatings: A novel approach to temperature regulation in vehicles. Journal of Coating Technology and Research, 20(8), 86–95.

Davis, E. (2023). Hydrophobic coatings for windshields: Improving visibility and safety. Journal of Coating Technology and Research, 20(5), 56–65.

Fernandes, D. (2021). Hydrophobic coatings for automotive applications. Coatings, 11(3), 319.

Gupta, M., Tyagi, A., & Singh, V. (2018). X-ray diffraction: Technique and applications. Journal of Materials Education, 40(3–4), 167–184.

Huang, Y., Lee, C., Rath, M., Ferrari, V., Yu, H., Woehl, T., Ni, J., Takeuchi, I., & R'ios, C. (2023). Tunable structural transmissive color in fano-resonant optical coatings employing phase-change materials. Material Advances.

Johnson, C. (2023). Paint protection films: A review of their use in the automobile industry. Journal of Coating Technology and Research, 20(3), 36–45.

Jones, A. (2022). The evolution of the automobile industry: A century of innovation. Journal of Automotive Technology, 35(2), 1–15.

Jones, A. B., Johnson, C. L., & Brown, M. A. (2019). Multifunctional ceramic coatings for automotive applications. Advanced Materials and Processes, 177(3), 33–38.

Kim, J. (2023). Anti-glare coatings: Enhancing safety in automotive applications. Journal of Coating Technology and Research, 20(6), 66–75.

Krishnan, A., & Pillai, P. K. (2020). Multifunctional self-cleaning coatings for automotive applications: A review. Journal of Coatings Technology and Research, 17(2), 411–434.

Li, C., Chen, Z., Zhang, W., Zhu, Y., & Zhang, X. (2021). Development and prospect of ceramic nanocomposite coatings. Surface Engineering, 37(4), 304–317.

Liao, C., Li, F., & Liu, J. (2022). Challenges and Modification Strategies of Ni-Rich Cathode Materials Operating at High-Voltage. Nanomaterials, 12, 1888. doi:10.3390/nano12111888.

Lin, J., Gao, L., & Hu, H. (2017). Characterization of thermal barrier coatings by scanning electron microscopy and energy-dispersive X-ray spectroscopy. Journal of Materials Science & Technology, 33(5), 501–510.

Lu, Y., et al. (2018). Synthesis of a multifunctional hard monomer from rosin: The relationship of allyl structure in maleopimarate and UV-curing property. Scientific Reports, 8, 2399. doi:10.1038/s41598-018-20695-5.

Lv, W., He, L., Guo, L., & Deng, Z. (2021). Recent advances in hydrophobic coatings for self-cleaning applications. ACS Applied Polymer Materials, 3(8), 3346–3368.

Miller, H. A., Wallace, J. Q., Li, H. Li, X.-Z., de Bettencourt-Dias, A., & Kievit, F. M. (2024). Sensitization of europium oxide nanoparticles enhances signal-to-noise over autofluorescence with time-gated luminescence detection, ACS Omega, 9(28).

Patel, D. (2023). Self-healing coatings: A new frontier in automotive coating technology. Journal of Coating Technology and Research, 20(4), 46–55.

Phuah, K. L., Hanim, M. A. A., & Anuar, M. R. K. (2019). Recent advances in automotive paint protection coatings. Progress in Organic Coatings, 134.

Pratap, S., Babbe, F., Barchi, N. S., Yuan, Z. K., Luong, T., Haber, Z. J., Song, T. B., Slack, J., Stan, C., Tamura, N., Sutter-Fella, C. M., & Müller-Buschbaum, P. (2021). Out-of-equilibrium processes in crystallization of organic-inorganic perovskites during spin coating. Nature Communications, 12(1), 1–11.

Pushpavanam, M., & Zaeh, M. F. (2002). Ceramic coatings for automotive applications: A state-of-the-art review. Surface Engineering, 18(2), 89–98.

Reum, N., Fink-Straube, C., Klein. T., Hartmann, R. W., Lehr, C.-M., & Schneider, M. (2010). Multilayer coating of gold nanoparticles with drug–polymer coadsorbates. Langmuir, 26(22).

Sahoo, P., & Kandasubramanian, B. (2013). Progress in organic coatings: Review on test methods and characterization of properties of coatings. Progress in Organic Coatings, 76(2-3), 364–377.

Sahoo, P., Rana, S., Cho, J. W., Li, L., & Chan, S. H. (2019a). Polymer nanocomposites based on functionalized carbon nanotubes. Progress in Polymer Science, 34(9), 984–1011.

Sahoo, S. K., Thakur, V. K., Sekhar, S. C., & Farzana, M. (2019b). Advances in paint protection films: Materials. Technologies, and Applications. Coatings, 9(5), 325.

Shahzadi, P., et al. (2021). Transparent, self-cleaning, scratch resistance and environment friendly coatings for glass substrate and their potential applications in outdoor and automobile industry. Scientific Reports, 11, 20743. doi:10.1038/s41598-021-00230-9.

Smith, B. (2023). The science of ceramic coatings: Applications in the automotive industry. Journal of Coating Technology and Research, 20(2), 26–35.

Smith, B., & Johnson, C. (2023). Multifunctional coatings in the automobile industry: A comprehensive review. Journal of Coating Technology and Research, 20(1), 1–25.

Smith, J. R., Lamprou, D. A., Larson, C., & Upson, S. J. (2021). Biomedical applications of polymer and ceramic coatings: a review of recent developments. Transactions of the IMF, 100(1), 25–35. doi:10.1080/00202967.2021.2004744.

Smith, R. M., Olofsson, U., & Helmersson, U. (2018). Hard and wear resistant coatings for automotive applications. Surface Engineering, 34(9), 673–680.

Thompson, L. (2023). UV-resistant coatings in the automobile industry: Protecting against the Sun's damaging rays. Journal of Coating Technology and Research, 20(7), 76–85.

Wang, L., et al. (2022). Review: Degradable magnesium corrosion control for implant applications. Materials, 15, 6197. doi:10.3390/ma15186197.

Wang, J., Li, W., & Cheng, Y. (2014). Ceramic coatings for the automotive industry. Progress in Organic Coatings, 77(9), 1410–1417.

Yuan, L., et al. (2021). A review of recent advances in self-healing coatings for the protection of automotive surfaces. Coatings, 11(6), 705.

Zhang, H., Yao, H., Hou, J., Zhu, J., Zhang, J., Li, W., Yu, R., Gao, B., & Zhang, S. (2018). Overcoming the interface incompatibility for the blade-coating of high-performance organic solar cells. Advanced Materials, 30(52), 1804811.

Zhang, S., & Wang, C. (2023). Precise Analysis of Nanoparticle Size Distribution in TEM Image. Methods and Protocols, 6, 63. doi:10.3390/mps6040063.

Zhang, Y., Tian, W., Liu, L., Cheng, W., Wang, W., Liew, K. M., Wang, B., & Hu, Y (2019). Eco-friendly flame retardant and electromagnetic interference shielding cotton fabrics with multi-layered coatings. Chemical Engineering Journal, 372, 1077–1090. doi:10.1016/j.cej.2019.05.012.

Zhang, Y., Liu, Y., Li, Y., & Liu, Y. (2022). Computational fluid dynamics simulation of the roll-to-roll coating process for the production of thin film composite membranes including validation. Advanced Engineering Materials, 24(1), 2100766.

13 Multifunctional Coatings for Aerospace Industries

Samson Rwahwire and Ivan Ssebagala

13.1 INTRODUCTION

The aerospace industry is one of the most geometrically growing and capital-intensive industries; huge investments are needed to become established. Furthermore, this industry provides air transport for both humans and cargo, and so in aircraft applications, safety is always a priority since accidents associated with airplanes can be fatal and cause huge losses. Above all, the loss of people's lives must be prevented at all costs. Airplanes ranging from private jets to military/defense aircraft are designed in most cases to last for about 30 to 40 years. However, due to their increasing costs coupled with decreasing budgets, sometimes aircraft continue to be used beyond their service life expectancy [1]. Some of the obstacles to the continued and proper use of aircraft are reduction in mechanical strength, reduction in fatigue strength, and corrosion. This, therefore, calls for regular inspection, maintenance, and repair of individual aircraft components, parts, and entire airplanes. It should be noted that most aircraft are primarily made from aluminum alloys, with about 70% of the metal details being aluminum owing to its high strength, good electrical conductivity, and lightweight properties. However, aluminum alloys being metals and the fact that aircraft operate mostly in adverse environmental conditions like high temperatures and damp areas, corrosion of the aluminum alloys tends to occur because of the additive components of iron, magnesium, silicon, and copper, hence deteriorating the serviceability of the aircraft.

According to IUPAC, corrosion refers to the physical-chemical reaction between the metal and the environment that eventually results in variations in the metal properties, and this can render the metal unable to perform its primary function or the system formed by that metal. Corrosion can be classified into uniform corrosion and localized corrosion, as shown in Figure 13.1.

Uniform corrosion generally results from the continuous shifting of cathode and anode regions of a metal surface that is in contact with an electrolyte. A perfect example is the rusting of steel in seawater. Localized corrosion refers to the intense attack at particular or localized regions on the metal surface, for example at reinforcement regions like bolts or welded joints.

Corrosion in airplanes is simply rust that occurs on the surface of the unprotected substrate of the airplane due to encountering oxygen during application (Figure 13.2). Corrosion in airplanes is mostly caused by the damp conditions of the atmosphere during operations coupled with the very

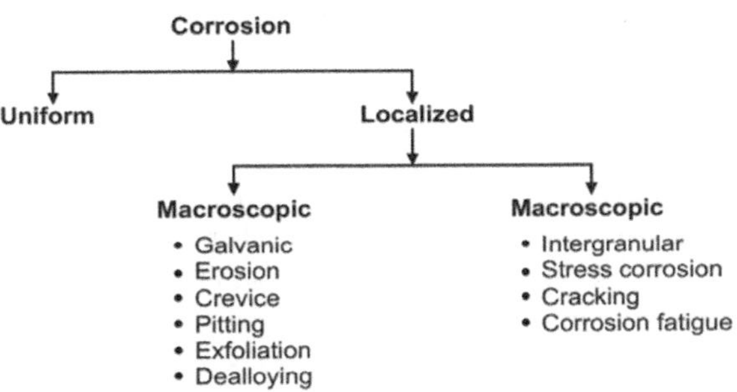

FIGURE 13.1 The basic classifications of corrosion [2].

DOI: 10.1201/9781032635347-13

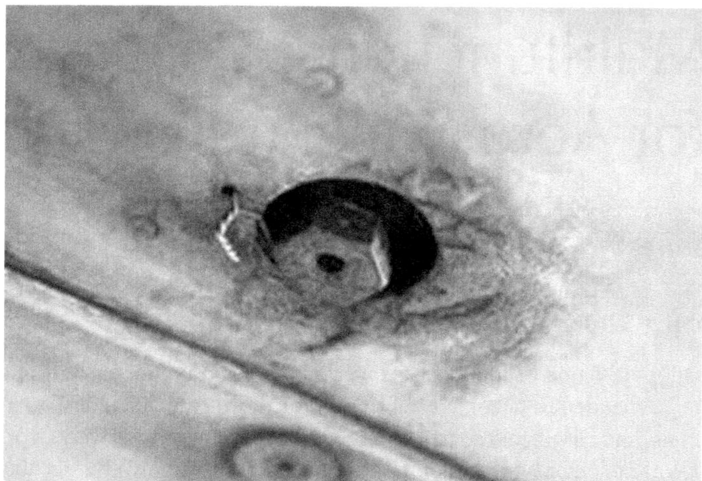

FIGURE 13.2 Typical corrosion of bolts on an airplane.

high temperatures associated with airplane engines and compressors that cause thermal corrosion. Corrosion of fastener heads (bolts, rivets, or pins) or any skin joints is common due to continued stress cycling caused by periodic movements.

The application of coatings on airplane surfaces is primarily for preventing corrosion to ensure enhanced lifetime service of the structure. Coatings are used as barriers between the environment and the coated substrate in addition to their use for purposes of decoration. Furthermore, coatings are used to impart some special desired properties like light absorption, electrical properties, hardness and abrasive resistance, oxidation resistance, and radar cross-section reduction.

As opposed to traditional conventional coatings that only served one or two functions of decoration plus substrate protection from the environment, for a material to be referred to as multifunctional, it must perform more than one function either simultaneously or in a sequence during application. Such materials are based on polymers, metals, or ceramics backed up by engineering physics, materials science, and mathematics. These multifunctional materials make it easy to design and develop stronger, lighter, and more resistant devices and components in relation to their application environments. Some of the additional functional performances of multifunctional coatings are superhydrophobicity, self-cleaning, self-healing of corrosion, self-healing repair of scratches, thermochromic responsiveness, antifouling, high heat and electricity conductivity, to mention but a few. Coatings are considered protective because their primary role is to prevent damages to the substrate from the surrounding environment, and it should be noted that coatings serve according to the prevailing environmental conditions. For instance, some coatings could be suitable for regions with low temperatures, and yet they cannot be used in areas that are moist. Hence attention should be given to the manufacturers' instructions. All multifunctional coatings are categories in Figure 13.3.

13.2 PRETREATMENT OF SUBSTRATE (METAL) SURFACE

The base substrate to be coated must be first pretreated since it always contains contaminants like oils, dust, grease, rust, old paint coating residues, and other foreign material. If it is coated without pretreatment, a discontinuous and porous coating would result, hence deteriorating the coating quality. Therefore, pretreatment is done to produce a very neat, smooth, and uniform coating by enhancing the bonding between the coating and the metal surfaces, and it's done by one or more of the following pretreatments methods.

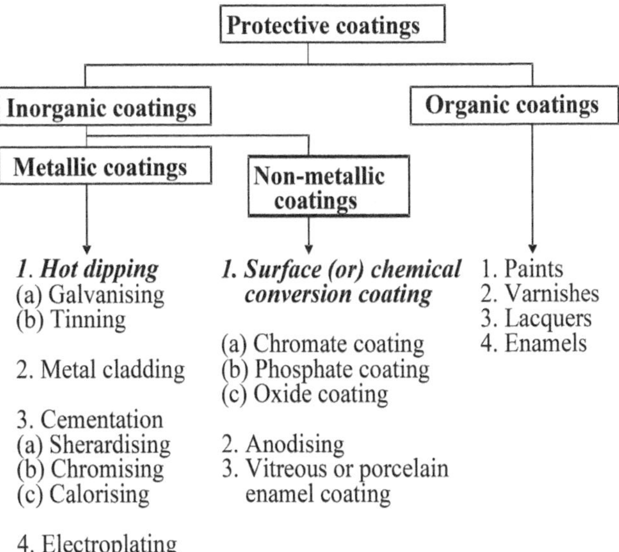

FIGURE 13.3 Different categories of coatings.

13.2.1 Mechanical Method

This technique is mainly used for the removal of oxides and some loose scale. Some of the applied approaches under the mechanical method include wire brushing, scraping, hammering, pneumatic blast, and sandblasting. The mechanical method is used more often when reasonably rough surfaces are desired.

13.2.2 Chemical Method

This technique is mostly applied for the removal of grease, oils, and waxes. The chemical method can take on any of the following approaches.

Solvent cleaning: Solvents like alcohols, toluene, chlorinated hydrocarbons, and xylene are generally used in this case, which is then followed by rinsing with hot water and steam containing wetting agents to dissolve any remaining contaminants.

Alkali cleaning: In alkali cleaning, chemicals like sodium hydroxide, trisodium phosphate, sodium silicate, and soda ash are used. This is useful in removing oil paints. This cleaning is followed by washing with 1% chromic acid solution.

Acid pickling: Acids like H_2SO_4, HCl, HF, H_3PO_4 and HNO_3 in dilute solutions are used for ferrous metals. For nonferrous metals, HNO_3 with other acids is mainly used. The metals are dipped inside the solution at a higher temperature.

13.2.3 Electrochemical Method

This method is used where the oxide scales cannot be removed by other methods. The metal whose surface must be cleaned is made either anode (cathode pickling) or cathode (anode pickling). The electrolyte is usually in acid solution or an alkali solution; on passing a direct current, the dissolution of the oxide scales at anode or cathode takes place.

13.3 COATING APPLICATION TECHNIQUES

There is a range of techniques used during the application of coatings on equipment, structures, or components, and they are selected based on the physical nature as well as the thermal and chemical properties of the coating material. In addition, the techniques cut across different industries like automotive, industrial equipment manufacturing, and aerospace, and in this chapter we shall focus our attention on the aerospace industry. Following are some of the general techniques used in the application of coatings in airplanes.

13.3.1 Thermal Spraying

This is a coating process where melted materials are deposited on the substrate surface. Heating is done either electrically (arc or plasma) or chemically using a combustion flame. A very wide range of coating thicknesses can be achieved from about 25 μm to many millimeters (mm), depending on the feedstock, and the process can be employed at a relatively higher rate of deposition as compared to other techniques like electroplating and chemical deposition. This technique is also very good for the formation of composite coatings containing a wide range of varying properties. Materials that can be applied with thermal spraying include metals, plastics, composites, ceramics, and alloys, where they are either fed in the form of wire or powder, then heated to molten or semimolten form, and finally accelerated to the substrate in the form of micrometer particle sizes [3, 4]. The quality of the coating improves with good pretreatment of the substrate's surface and with increasing deposition velocity of the coating particles. Thermal spraying is possible for even flammable surfaces since the surface doesn't heat up a lot during the coating process. Table 13.1 shows some of the coating materials for thermal spraying plus their respective functional uses beyond decoration and corrosion control. In airplanes, thermal spraying is applied on exhaust heat systems, engine compressors, etc.

Thermal spraying is further divided into various methodologies: Plasma spraying uses a plasma jet at a high temperature (>15,000 K) generated by an arc discharge. It is used mostly for coating refractory materials like molybdenum and other oxides. Wire-arc and flame spraying generally have low coating particle velocity of deposition of less than 160 m/s, and their coating materials should be in molten form. High-velocity oxy-fuel (HVOF) spraying produces a relatively thin but strong adhesive coating, highly corrosion resistant though not suitable for very high temperatures. Other methodologies are cold spraying, warm spraying, and spray and fuse.

TABLE 13.1
Examples of Multifunctional Coatings for Thermal Spraying

Function	Typical Coating Material
Abrasive wear resistance	WC-Co, Cr_3C_2-NiCr, Cr_2O_3, Al_2O_3/TiO_2, cobalt alloy, Cu and Cu alloy, Ti and Ti alloy
Antifretting	WC-Co, Cr_3C_2-NiCr, Cu-Ni-In alloy
Erosion resistance	WC-Co, Cr_3C_2-NiCr
Thermal barriers	Y_2O_3-ZrO_2
Clearance control/abradables	Ni/graphite, Si-Al/polyester, MCrAlY/polyester, YSZ/polyester, Ni (Cr,Al)/bentonite, NiCrAlY-graphite, NiCrFe-hBN
Corrosion resistance	Al and Al alloy, Zn and Zn alloy, stainless steel, Cr_2O_3, MCrAlY
Antioxidation	Superalloy, MCrAlY
Restoration of dimension and repair	Ni-5%Al, Ni-Fe-5%Al, Ni-Cr-5%Al, stainless steel
Near-net shape forming	Ta, Mo, Nb, NiAl, TiAl, Ti alloy, superalloy, metal-metal matrix composite (MMC), ceramic matrix composite (CMC), fiber-reinforced metal-matrix composites (FRM)

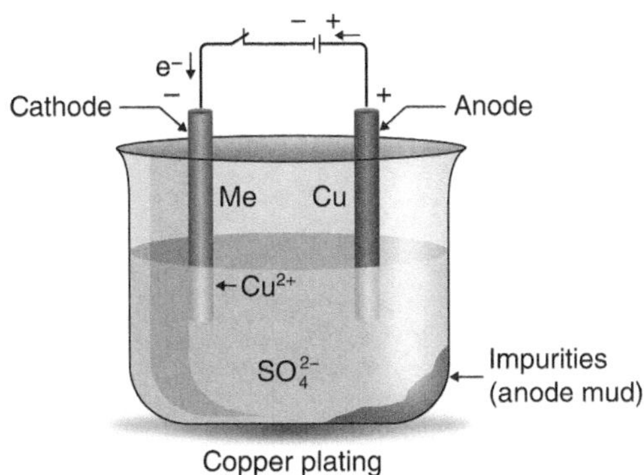

FIGURE 13.4 Typical electroplating of copper.

13.3.2 Electroplating

This technique involves the migration and deposition of metal ions from the positive electrode to the negative electrode through a solution (Figure 13.4). The flowing current in the solution accelerates the coating of the metal at the cathode, and this is called hydrolysis. Its primary principle is to build the thickness of a metal, though in so doing a coating can be deposited on a substrate for corrosion resistance, abrasion resistance, and lubricity. However, for aerospace applications, electroplating is done to increase surface hardness for improved abrasion resistance, electrical conductivity, and body thickness building for safety purposes.

Methodologies used in electroplating include barrel plating for high-volume plating, rack plating for fragile or large and complex parts, and reel-to-reel plating for plating strips of manufactured goods or reels of materials before being used for parts assembly. Some common metals used in the electroplating process are copper, tin, silver, gold, palladium, and nickel.

13.3.3 Chemical Deposition

This is sometimes referred to as chemical conversion coating, and it involves reacting the outer metal or substrate atoms with the anion of a given medium, hence forming an insoluble coating compound. The process basically changes metals into thin adherent coatings of insoluble compounds and serves the purposes of anticorrosion, lubricity, thermal insulation, electrical insulation, absorption of light, and improving aesthetic properties. Chemical conversion coatings are applied to a variety of material surfaces such as steel, brass, and aluminum. The most common forms of chemical coating are discussed below, and they can be used as precoating, post-coating, or finishing coating processes [5].

13.3.3.1 Chromate Coating

This involves the application of a solution of chromate salt or chromic acid to the surface of the metal, rinsing, and finally drying it off. Chromate coating is mainly used when in addition to corrosion protection, abrasion resistance and some chemical resistances are required (Figure 13.5).

13.3.3.2 Phosphate Coating

This involves the application of phosphates on the surface of steel so that a thin adherent layer of manganese, iron, or zinc phosphate is formed (Figure 13.6). This type of coating can promote lubricity and act as a base for further coatings.

FIGURE 13.5 Typical zinc chromate (green color) coated airplane on the production floor.

FIGURE 13.6 Typical phosphate-coated gears and rods.

13.3.3.3 Blackening (Black Oxide)

This involves the treatment of black oxide with oils or waxes to produce maximum resistance to corrosion. This method can be used on copper, zinc, silver solder, stainless steel, and all ferrous metals (Figure 13.7).

13.4 FUNCTIONAL COATINGS

These coatings are designed and manufactured with the aim of serving a particular special purpose in addition to the primary functions of corrosion protection plus decorative surface finish. Some of the special functions include electrical and thermal conductivity or insulation, reflective surfaces, light absorbers, chemical resistance, abrasive resistance, and surface hardness, to mention but a few,

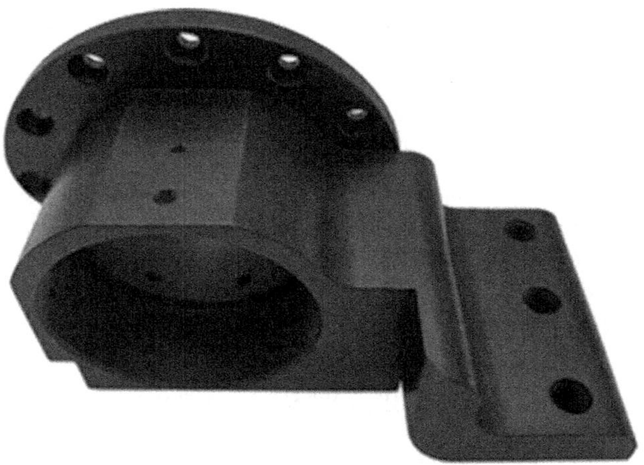

FIGURE 13.7 An aluminum component coated with black oxide.

and if a coating is designed and engineered to perform one or more functions of the above, then it is referred to as a multifunctional coating. Next we will discuss some of these multifunctional coatings, classifying them as self-healing coatings, smart coatings, nanostructured coatings, and anti-icing and de-icing coatings.

13.4.1 Nanostructured Coatings

These coatings can take various forms, for instance, multilayer nanocoatings, nano-graded coatings, nanocomposite coatings, and superlattice coatings. In addition, several application methods of the coatings are employed, for example, deposition by a solution of sol-gel, polymer physisorption, thermal spraying, deposition by electrochemical method, and physical vapor deposition. Some of the common properties intended to be imparted to the substrate surface are crystallinity and thermal, optical, and electrical properties [6].

Aleksandra et al. [7] developed a nanocomposite coating based on chitosan where they added a bioactive sol-gel rich in strontium (Sr) plus some potential antimicrobial element of tin(IV) oxide (SnO_2) nanoparticles and applied it on a titanium substrate using the deposition by cathodic electrophoretic technique. Results revealed a rougher coated surface than the original titanium surface, and there was an improvement in the surface wettability, the surface free energy increased by over 32%, and there was improvement in the corrosion resistance of the titanium substrate in Ringer's solution (Figure 13.8). The improvements were attributed to better adhesion strength, good morphology,

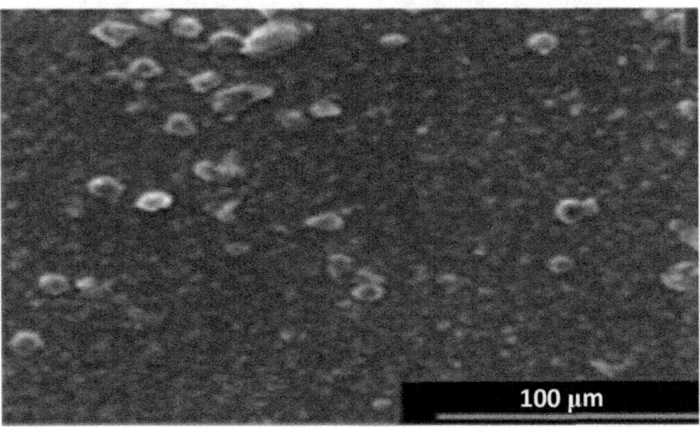

FIGURE 13.8 SEM image showing the deposition of tin(IV) oxide nanoparticles on titanium substrate.

high chemical corrosion resistance, and higher wettability of the coating, and it showed a direction for the application in antibacterial coating.

Rana et al. [8] presented a nanocrystalline and a microcrystalline coating (NiCrAIY) for superalloys prepared by high-velocity oxide fuel (HVOF) and sputtering, respectively, and they prevented corrosion by the formation of protective scales on the surface [9]. It was observed that the nanocrystalline coating exhibited superior resistance to high-temperature corrosion in air, which was attributed to the formation of nonporous, adherent, and uniform protective scales like the $NiCr_2O_4$ and the Cr_2O_3. By comparison, in the microcrystalline coating the scales produced were nonuniform and discontinuous, hence not able to seal the substrate completely. The parameters that were used to deposit the NiCrAIY coating are:

Pressure	Temperature	Voltage	Current	Time
10^{-4} Torr	250°C	300 V	200 mA	30 minutes

With a detection of change in temperature, the SEM image in Figure 13.9 shows the reaction of the coating.

Mamudu et al. [10] presented two nanocomposites for coating purposes: one developed from the nanocrystalline cellulose (NNC) obtained through the hydrolysis of microcrystalline cellulose (MCC) and the other by sulfating NCC to form S-NCC (sulfated-NCC), and they were all applied for mild steel surface protection against corrosion. The two composite coatings were formulated by loading the NCC and S-NCC into an epoxy matrix. Results showed that the sulfation of NCC led to an improvement in the anticorrosion functionality of the composite as compared to neat NCC composites [9].

Palmolahti et al. [11] developed a 30 nm atomic-layer deposit of TiO_2 film on Si wafer from tetrakis (dimethylamino) titanium plus water. The TiO_2 film proved to be very good at resisting chemical degradation as well as dissolution at the grain boundaries in an alkaline environment, hence rendering it a good protective coating to photoelectrodes.

Zhang et al. [12] developed an abrasion-resistant amorphous as well as nanocrystalline composite, a coating made from Al-Ni-Mm-Fe, and applied it on an AZ91 alloy of magnesium. The coating exhibited a dense structure having a low porosity of about 2% and a Vickers hardness level of about 330 $HV_{0.1}$, about five times that of AZ91 magnesium while about four times that of pure aluminum coating. Conclusively, the Al-Ni-Mm-Fe coating showed good resistance to abrasion.

The Al-Ni-Mm-Fe coating exhibited a low porosity of about 2% since it was very dense as compared to the pure Al coating, which had porosity of up to 5%, hence the modified coating had fewer defects as compared to the pure Al coating (Figure 13.10).

Hyie et al. [13] also developed a nanocrystalline coating based on pure cobalt, and it was deposited on a stainless steel substrate at pH values of 3, 5, and 7 so as to investigate three major factors of its

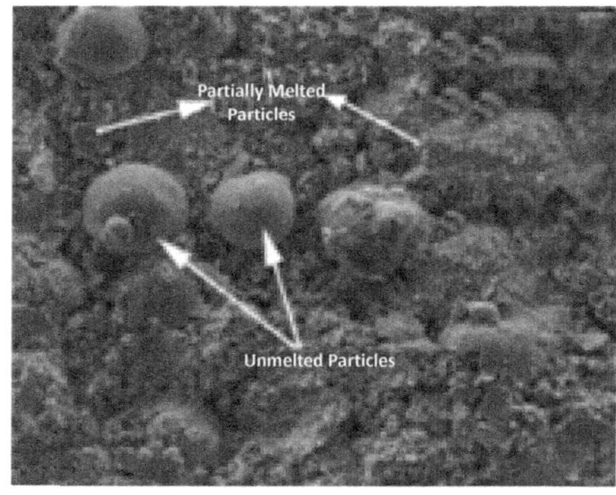

FIGURE 13.9 FESEM images of the surface of the HVOF sprayed mc NiCrAlY coatings.

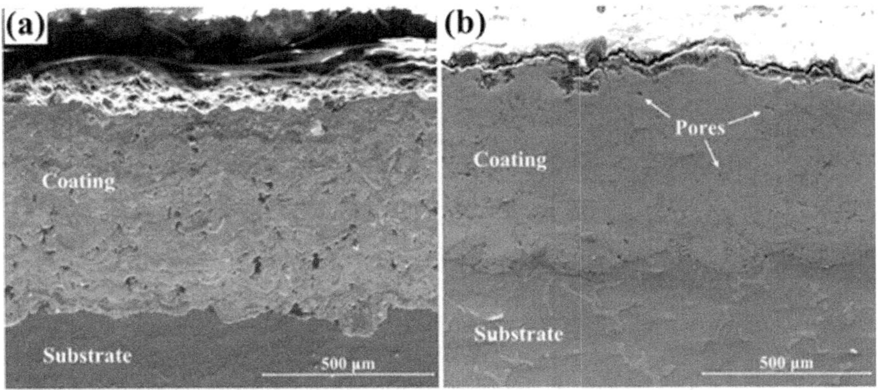

FIGURE 13.10 SEM images of the coatings (a) Al-Ni-Mm-Fe composite coating; (b) pure Al coating.

hardness, magnetic behavior, and corrosion properties with respect to pH [14]. Results revealed that the coating deposited at the pH of 3 exhibited the highest hardness of about 240 HV as well as having the lowest rate of corrosion. Additionally, it showed the lowest coercivity plus the highest magnetization saturation. Hence, consideration of pH is crucial in the application of the coating to substrates.

Jang et al. [15] also attempted to reduce the imbalance between the wear of a tetrahedral carbon (amorphous) or ta-C film and its counterpart since ta-C is very hard, and so it can adversely affect its counterpart in terms of wear. This balancing was done by developing a nanocomposite film made of nanocrystalline copper and tetrahedral carbon (NC-Cu/ta-C), which allowed for the reductions in wear of the nanocomposite film and its counterpart to about 88% and 99% respectively [16].

13.4.2 Self-Healing Coatings

The concept of self-healing is basically copied from natural organisms that can automatically recover after sensing a change in the material matrix to repair any broken part on their own. For instance, when a person bites their tongue, the body reacts by releasing chemicals that recover the wound without any external intervention. The use of self-healing coatings based on appropriate encapsulated self-healing compounds/agents (smart carriers) has continually increased due to its advantages. Some of these include continuous release and repairs of damaged parts like scratches and notches, plus lower maintenance requirements for the aerospace vessels. The self-healing compounds encapsulated in these coatings are released with the occurrence of some driving force such as changes in temperature, pH, moisture, dissolved oxygen, or light, to mention but a few. Therefore, on top of inhibiting corrosion, the self-healing coatings contain self-healing properties to enhance the repair of damages on the coated substrate, such as scratches. Some work done around this topic is discussed below.

Habib et al. [17] presented a smart coating based on polyolefin for inhibition of corrosion that contained modified hybrid-based particles and the hybrid (ZnO@β-CD) particles containing zinc oxide plus β-cyclodextrin and was then modified by 2-mercaptobenzothiazole (2-MBT) (Figure 13.11). The modified (ZnO@β-CD-2MBT) and unmodified (ZnO@β-CD) polyolefin coatings were then applied on a steel substrate by the dip-coating method to form a smart coating composite. The results showed that the modified polyolefin coating exhibited better inhibition of corrosion up to 98 GΩ cm^2 as compared to the unmodified coating with corrosion inhibition of 87 GΩ cm^2 in a solution of 0.56M NaCl after 20 days. The high corrosion inhibition of the modified polyolefin is attributed to the presence of zinc oxide and 2-MBT combined.

Wang et al. [18] synthesized a Fe_3O_4-MBT epoxy-based smart coating by incorporating Fe_3O_4 nanoparticles as a photothermal converter and 2-MBT as a corrosion inhibitor. When applied on a

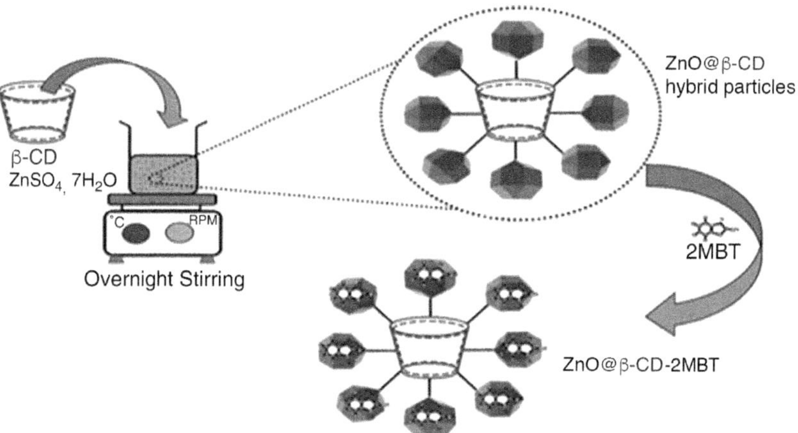

FIGURE 13.11 Schematic representation of development of unmodified and modified ZnO@β-CD hybrid particles.

scratch of about 80 μm, it reduced to approximately 5 μm after just 30 s triggered by near infrared light radiation (NIR) (Figure 13.12). This is believed to have been accelerated by the property of photothermal conversion in Fe_3O_4. Therefore, the Fe_3O_4-MBT epoxy-based smart coating provides a great deal of corrosion inhibition for a long time according to results.

Ma et al. [19] also synthesized an epoxy-based self-healing coating incorporated with magnetic particles as the function particles plus emulsifiers and microcapsules of tung oil. The electrochemical impedance spectroscopy results revealed that the magnetically driven self-healing coating reduced over 50% of the migration paths of tung oil released plus a quick rate of release of tung oil following mechanical damage to the coating. Hence the magnetically driven coating can be utilized as a self-healing coating due to its ability to repair damages such as scratches. The bonding strength of the coating is, however, affected by the percentage of the microcapsules, where an increase in the fraction of microcapsules reduces the bonding strength.

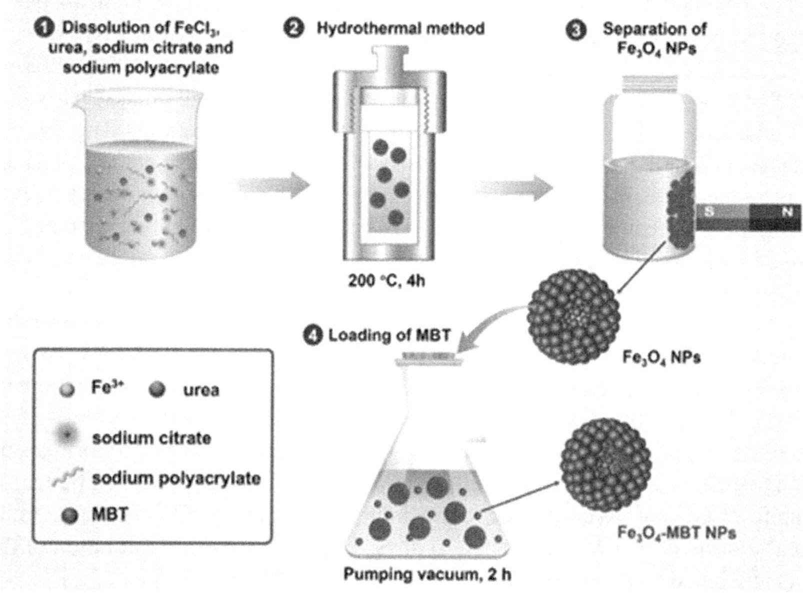

FIGURE 13.12 Synthesis flow of Fe_3O_4-MBT nanoparticles.

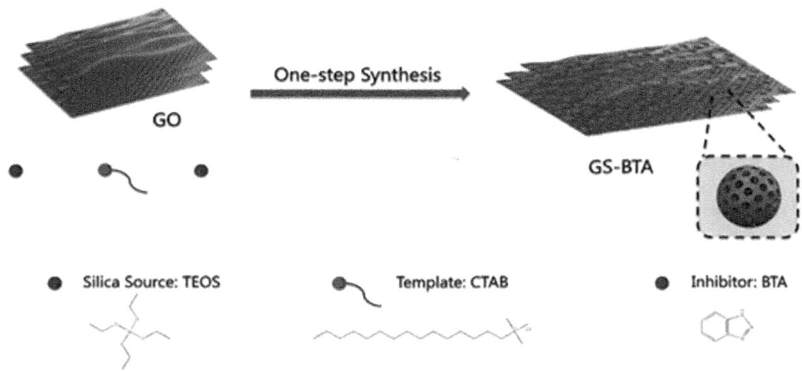

FIGURE 13.13 A one-step synthesis of GO-SiO_2 nanocontainers loaded with inhibitors.

Jin et al. [20] also presented a novel epoxy-based self-healing coating incorporated with graphene oxide/silicon oxide (GO/SiO_2) nanocontainers containing benzotriazole (BTA) using a one-step technique. The electrochemical impedance spectroscopy and wire beam electrode impedance scanning (WBE-EIS) results revealed that the coating exhibited good long-term resistance to corrosion as compared to pure coating of epoxy alone. The impedance results of GS-BTA epoxy coating changed little with respect to time, and later the value increased by a factor of three times on the 48th day as compared to the pure epoxy coating (Figure 13.13). Additionally, the self-healing property on an artificial scratch part was very significant owing to the release of the benzotriazole and its high adsorption at sites of anodic corrosion. Hence an epoxy incorporated with graphene oxide and benzotriazole is a potential self-healing coating for aerospace applications.

Yin et al. [21] used a mesoporous SiO_2 as a nanocontainer for storage of active compounds/agents for inhibition of corrosion in a self-healing manner. The mesoporous SiO_2 contained PVB coatings and was applied on a zinc substrate so it could prevent corrosion as well as self-heal the already delaminated layers. The principle was that the mesoporous SiO_2 would degrade under high alkaline pH at a delaminated part and then form a silicate interface that could block the under-film corrosion plus reestablish a protective and a resistant delamination interface (Figure 13.14). Results revealed that the mesoporous SiO_2 layer could reduce the rate of cathodic delamination formation.

Ma et al. [22] also developed a photothermal, superhydrophobic self-healing coating (EP@DTMS@PDA@CNSs40/CNTs80) based on carbon nanotubes (CNTs)/carbon nanospheres (CNSs), dodecyltrimethoxysilane (DTMS), epoxy (EP) and polydopamine (PDA). The results revealed that the coating was superhydrophobic since it had a water-contact angle of 160.7° plus a sliding angle of 4°. The coating also converted light energy to heat energy in a very short time following simulated sunlight irradiation (Figure 13.15). Lastly, the treatment of the coating with O_2 plasma changed it to superhydrophilic, but the long hydrophobic chains that were presented by DTMS moved to the coating's upper layer under the driving force of thermodynamics. Hence a superhydrophobic, photothermal self-healing coating was achieved.

Huang et al. [23] developed a self-healing coating with a triple action to prevent corrosion on carbon steel using a shape memory polymer epoxy coating plus polyaniline/benzotriazole nanocapsules (PANI/BTA). The shape memory coating was thermo-responsive and was capable of

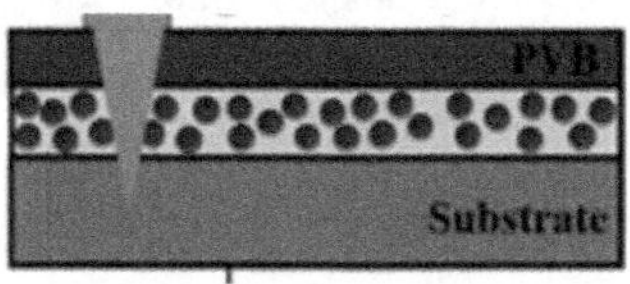

FIGURE 13.14 PVB coating for self-healing embedded into SiO_2 (red capsules).

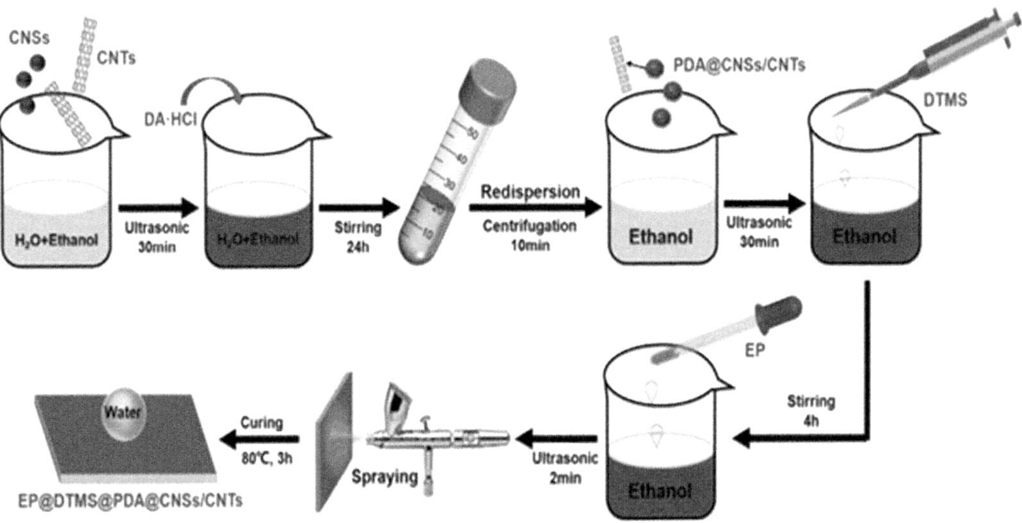

FIGURE 13.15 The preparation process of EP@DTMS@PDA@CNSs40/CNTs.

narrowing scratch size after sensing heat, meaning the number of inhibitors required for covering the substrate surface reduces. The PANI/BTA performed two major functions: first, PANI acted as saline that could control the release rate of the inhibitors to be adsorbed on the substrate's surface for corrosion inhibition, and second, it could prompt the formation of a layer of iron(ii) oxide as a passive layer for corrosion control. Results indicated that self-healed coating of 0.5 wt% PANI/BTA exhibited an excellent property of self-healing, and it was higher than 96% (Figure 13.16). The iron(ii) oxide passive layer that formed around scratched areas resisted the corrosion of steel.

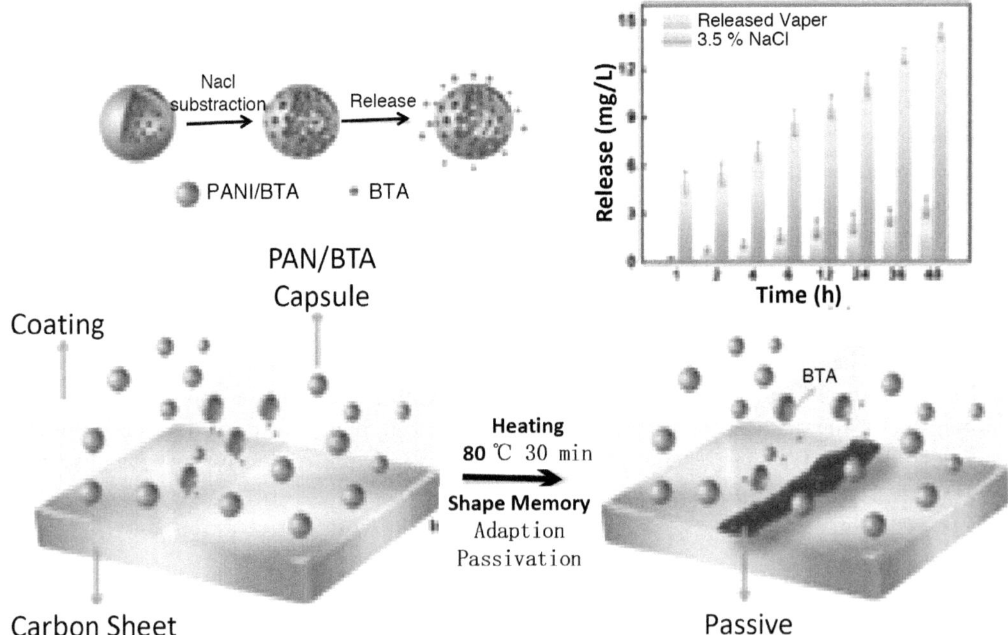

FIGURE 13.16 The functionality of a saline-responsive self-healing coating with triple actions.

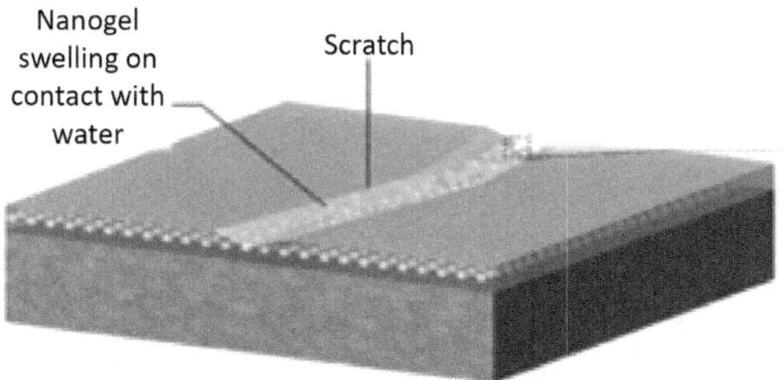

FIGURE 13.17 The action of a titanium oxide nanogel composite self-healing coating.

Ghomi et al. [24] developed a super self-healing coating embedded with titanium dioxide (TiO_2), 2-acrylamido-2-methyl propane sulfonic acid (AMPS), acrylamide (AAm), 3- (trimethoxysilyl) propyl methacrylate (MPS) in nanogel composite in epoxy which was sensitive to pH and salt concentration that triggered the release of the titanium oxide to act upon the corrosion ions the moment cracks or scratches occurred (Figure 13.17). Results indicated that the nanogel composite was good for corrosion resistance, and the incorporation of NTFM/AAm/AMPS always led to uniform dispersion, accelerated self-healing, and acting as anticorrosion agents [25].

Binbin Zhang et al. [26] presented a thermally activated self-healing coating based on epoxy while embedded with 2-aminophenyl disulfide (APD) as a hardener, as shown in Figure 13.18.

13.4.3 Anti-icing and De-icing (Superhydrophobicity) Coatings

In the de-icing mode, the system operates in cycles where a very small noncritical layer of ice is allowed to form on the substrate's surface before it can be removed, whereas in anti-icing mode, continuous heating is done on the surface of the substrate, hence not allowing any trace of ice to form. The principle behind the two mechanisms is that they use heat for raising the substrate's temperature such that the ice can be melted and flows off the surface. The anti-icing equipment is basically designed to hinder the formation of ice on the substrate's surface, while the de-icing equipment is for the removal of ice formed on the surface of any material [27]. These are commonly applied in the aerospace industry, and some of the common areas where de-icing and anti-icing equipment and/or coatings are applied include windshields, propeller blades, leading edge of aircraft wings,

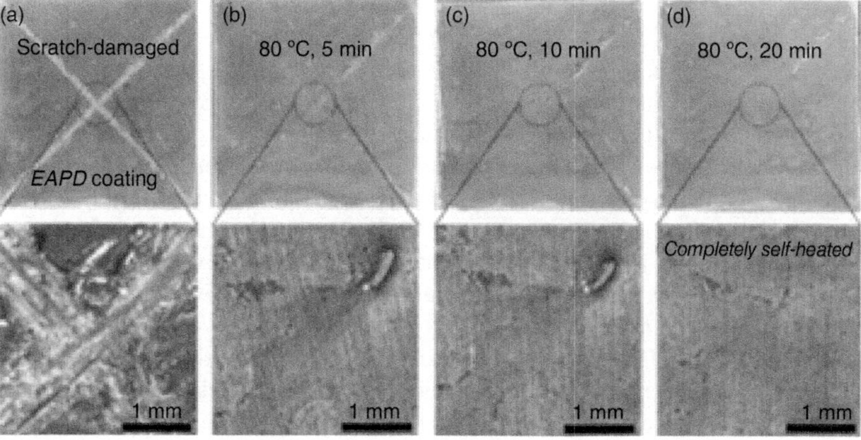

FIGURE 13.18 Snapshots and microscopy images of self-healing coating triggered by temperature increase.

FIGURE 13.19 Typical icing in aircraft.

tail surfaces, pitot, vents for fuel tanks, and devices for stall warning. Aircraft icing occurs mainly when the vehicle is flying close to the top of cold air but below the deep layer of warm air. The raindrops present in such an area are bigger than the cloud droplets, so there is a much faster rate of catch, and in very low temperatures, the droplets form ice (Figure 13.19).

Additionally, superhydrophobic surfaces can be developed by imitating nature such that the contact angle or area between a surface and liquids is reduced, hence limiting the interaction between the two [28]. Various artificial superhydrophobic surfaces and/or coatings have been created and applied in different industries as separators of oil and water and anti-icing, self-cleaning, and anticorrosion agents or equipment. Some work that has been done around superhydrophobicity, deicing, and anti-icing is presented below:

Cherubin et al. [29] discovered that adding amines to de-sugared beet molasses and winery lees enhances their anticorrosive qualities. Therefore, studies on weight loss clearly demonstrate that adding a small number of amines to molasses or lees significantly inhibits corrosion on carbon and galvanized steel. The results of ice-melting tests showed that the de-icing agent's characteristics are unaffected by the presence of molasses/lees and triethanolamine. The inhibitory effect of the molasses/triethanolamine mixture resulted in a reduction of 88% for galvanized steel and 65% for carbon steel when corrosion on probes made of galvanized and carbon steel was evaluated under environmental circumstances.

Grishaev et al. [30] considered the interaction of commercially available Newtonian and pseudoplastic anti-icing fluids with superhydrophilic and hydrophobic aluminum surfaces. At 10% ice coating of the surfaces, freezing rain models showed no appreciable surface influence on fluid endurance times. Surface tensions of the fluid, contact angle hysteresis of test plates, and fluid viscosity (the latter of which is unimportant for Newtonian fluids) can all contribute to the discrepancy with earlier investigations. On hydrophobic surfaces, the physical adsorption of fluid causes the water's contact angles to drift away, weakening its ability to repel water. So, for ice mitigation systems contacting aircraft anti-icing fluids, smooth hydrophobic surfaces are probably the best option.

Vertuccio et al. [31] focused on the design of bulk nanomaterials that can provide ice protection in a variety of applications, including civil, aeronautical, and automotive engineering. Electrically conductive nanoparticles were created as part of bulk nanomaterials. The heating performed through the Joule effect represented an efficient strategy to rapidly contrast extreme cold and humid conditions, reduce environmental pollution, and control rheological properties during the process. The Joule effect's efficacy has been assessed for a single resin that has low viscosity values, incorporates carbon nanotubes, and contains two grades of expanded graphite. The comparison of the various fillers shows that the unidimensional filler-containing nanocomposite achieves greater temperatures at lower applied voltage levels.

According to Khan et al. [32], the superhydrophobic silver metal surface on the copper substrate (Cu@Ag-SHS) was successfully created using a chemical etching procedure using silver nanoparticles that resemble fern leaves. The fabricated product is characterized by PXRD, EDX, SEM, and XPS techniques. Cu@Ag-SHS had special qualities, such as long-term stability, super buoyancy force (a drop of water may bounce off the surface of the balloon, exhibiting exceptional nonsticking properties), anticorrosion properties, self-cleaning, anti-abrasion, and durability. Additionally, the photodegradation of 4-NP into 4-AP was used to assess the photocatalytic abilities of Cu@Ag-SHS. Furthermore, Cu@AgSHS displayed morphology that was distinct and resembled fern leaves. As a result, it has been proven using different characterizing approaches that the created surfaces of superhydrophobic goods and their characteristics guarantee their suitability for use in the fabric and metallurgical industries.

Wu et al. [33] established that ice buildup on moving parts frequently results in catastrophic collisions or accidents. He emphasizes that the durability of Land Information System [LIS] constructions remains a big challenge, including mechanical vulnerability and rapid depletion of lubricants. Therefore, according to his experiments, an all-encompassing strategy was put forth to include microporous metallic scaffolds in the development of LIS to boost its adaptability and robustness and to raise the likelihood that LIS will be effective at mitigating ice. Microporous Ni scaffolds were chosen to integrate with polydimethylsiloxane modified by silicone oil addition. In cyclic tests for icing and de-icing, the new LIS architecture showed noticeably increased endurance. It also provided a remedy for quick oil depletion by limiting the matrix material's capacity to deform.

According to Sánchez-Romate et al. [34], it was proven that multifunctional coatings based on a GNP/epoxy system have been produced, and their heating and strain sensing the potential for anti-icing and de-icing applications were investigated. It was found that a higher GNP percentage causes the gauge factor to decline (from 5.75 at 8% to 2.49 at 12%) because the nanoparticles are less sensitive and have a smaller interparticle distance. However, the GF values under bending conditions are considerably more than typical metallic gauges (which are about 2). A higher number of conducting routes, which enables a more effective Joule heating effect, causes the resistive heating to be more effective when the GNP content is increased, as is expected.

Khaleque and Goel [35] established that the necessity for ongoing maintenance caused by ice accumulation on surfaces is a significant economic challenge in the energy and transportation sectors. The use of ice-phobic surfaces inspired by nature is a more environmentally friendly alternative than the conventional de-icing methods, which involve the use of electrical, mechanical, chemical, or combinatorial approaches. Ice-phobic coatings, which are permanent anti-icing coatings, can be used to prevent icing as well as antifreezing fluids, which can be sprayed on a regular basis.

Laroche et al. [36] presented ice-phobic coatings made from silicone nanofilament (SNF) networks grown on anodic metal oxide surfaces. It has been demonstrated that many innovative ice-phobic coatings have little adhesion to ice produced at zero or low velocity [37]. Few of them have been demonstrated to likewise have weak adhesion to ice formed by the collision of high-velocity supercooled water droplets. Even fewer of them have been demonstrated to have weak adherence to ice formed under a variety of environmental conditions. Those who have exhibited this conduct have been restricted due to their vulnerability to specific UV exposure bands. As a result, ice-phobic coatings display good durability under the tested settings and low ice adhesion strength under a variety of impact icing scenarios. Thus, the developed coatings are a viable option for enabling hybrid ice protection systems on airplanes, which would lower the energy required for anti- and de-icing.

Balordi et al. [38] discovered that growing ZnO nanorods in a brief hydrothermal treatment at 90°C can produce zinc surfaces that are both superhydrophobic and ice phobic. It was established that steel is frequently zinc plated in various industrial applications to provide a barrier against corrosion and environmental attack, although this process has wetting and icing problems. Surfaces with static contact angles higher than 165° and tilting angles as low as 1° result after additional coating with stearic acid or fluoroalkylsilane. All of the superhydrophobic samples demonstrated exceptional ice-phobic performances, attaining shear stress values lower than 10 kPa, and a clear association between wettability and ice adhesion was discovered.

13.4.4 Smart Coatings

These coatings are able to sense or detect changes in operational parameters like temperature, pressure, light radiation, and pH [39]. This acts as a driving force that triggers the functional actions of the coating, e.g., self-cleaning, self-de-icing, self-healing, etc. These coatings contain a material or agent that is responsible for the designed function, and it should be responsive to the variation in the predetermined operational parameter. Smart coatings are incorporated with metal or metal oxide particles at a nano scale. Nanoparticles possess fascinating properties mechanically, physically, optically, chemically, and electrically which are responsible for the functions found in smart coatings. Next we discuss some of the successful attempts toward smart coating development.

Qingshi Meng et al., developed a multifunctional, smart, and mechanically strong nanocomposite by way of combining functionalized graphene (F-GNPs) nanoplatelets together with polyurea using in situ polymerization (Figure 13.20). The nanocomposite at 0.2 wt% of F-GNPs showed improvements in the mechanical strength (tensile) up to 60.7% and 92.1% for elongation [40]. Additionally,

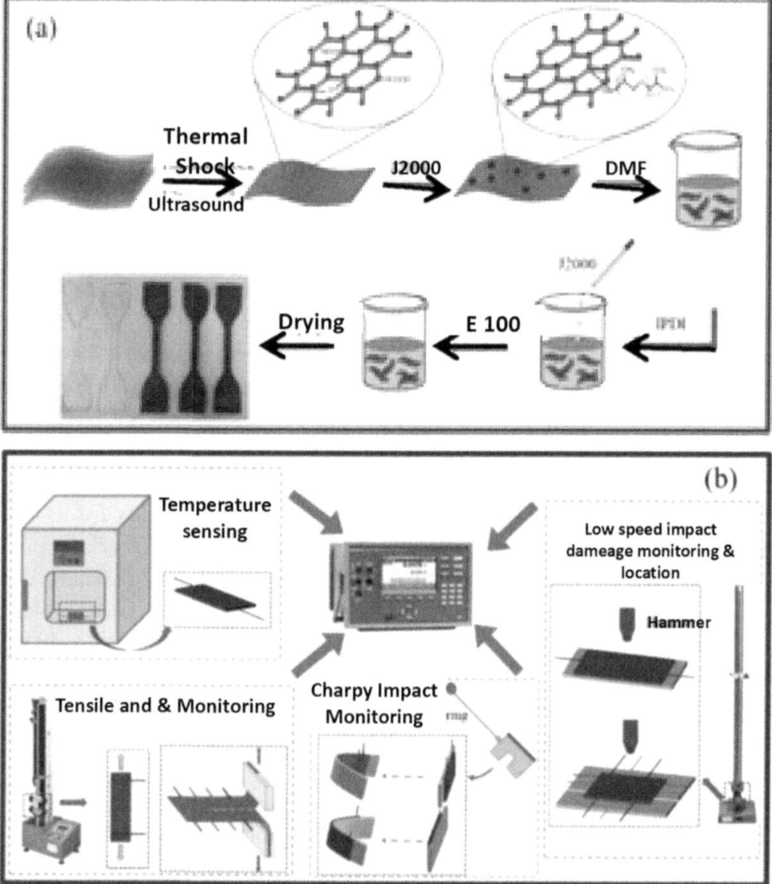

FIGURE 13.20 (a) The preparation of a polyurea/F-GNP nanocomposite and (b) the multichannel sensing system.

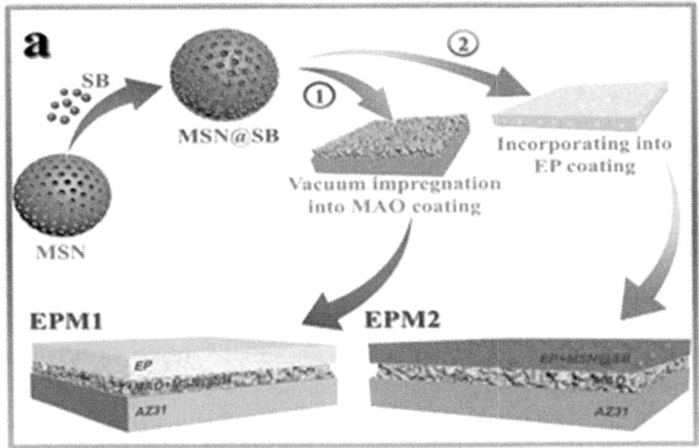

FIGURE 13.21 The preparation of smart EPM composite coatings.

the coating was able to sense temperature changes repeatably and stably in the temperature range of –20°C to 110°C. Hence the self-sensing capacity was proved by its ability to detect and track its own damage with varying levels of impact.

Ai-meng Zhang et al., developed a smart micro-arc oxidation (MAO)/epoxy (EP) resin composite coating on AZ31 magnesium alloy. Then nanocontainers of mesoporous silica containing an anticorrosion inhibitor of sodium benzoate were incorporated into the MAO as microcapsules in the outermost layer of the epoxy [41, 42]. A study was carried out to study the effect of positioning the nanocontainers in the layers of the protective coating. It was found that the superior resistance to corrosion of this particular hybrid was due to the mechanism provided by the nanocontainers, that is, by reducing the admittance value, which reduced the distance between the substrate and the coating and also by densifying the coating, thereby reducing the defects and hence achieving self-healing (Figure 13.21). The coating was able to resist corrosion effectively up to 90 days after the immersion. Hence the nanocontainers should be placed near the substrate for effective protective performance of corrosion [43].

L. Vertuccio et al. also confirmed that it is feasible to manufacture smart coatings with self-diagnostic ability from epoxy-based coatings incorporated with carbon nanotubes, which they applied on carbon fiber–reinforced plastics in aeronautics industries. The smart coating is able to provide real-time health of structures for effective monitoring. Hence it was very highly sensitive to any damage to the structure (Figure 13.22). Additionally, the coating worked efficiently in the normal working temperatures of aircraft; hence the coating acted as both an anticorrosion protector and a sensor for self-response [44].

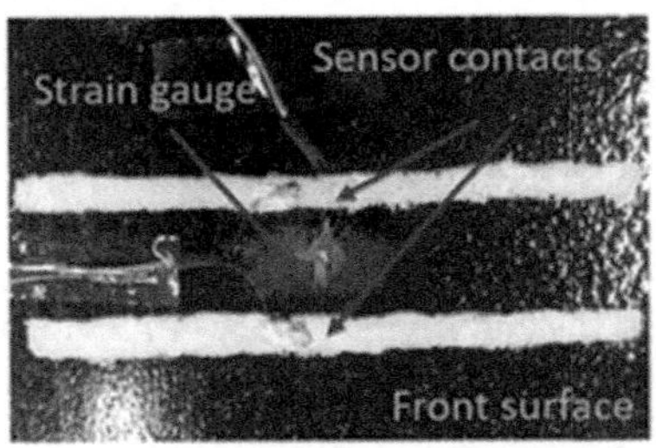

FIGURE 13.22 A self-responsive coating based on epoxy and carbon nanotubes.

13.5 CONCLUSION AND OUTLOOK

Over the years, researchers and scholars have explored the area of corrosion control and prevention using different techniques and methodologies. Corrosion has been mitigated and controlled by the use of self-healing, smart, nanocomposite, de-icing, and anti-icing plus superhydrophobic coatings aided by the various available techniques of application of such protective surfaces. Furthermore, the realization and deeper study of nanomaterials has made it easy to engineer protective surfaces for corrosion control, such as titanium nanoparticles, tin oxides, and graphene oxides, that possess superior electrical, physical, mechanical, optical, and thermal properties. However, a few areas, especially in the aerospace industry, still need improvements in terms of coatings for corrosion protection, especially in zones that involve high temperatures, such as engines and propeller blades, which continuously suffer high-temperature corrosion. There are also other areas where there is a need to improve the bonding strength and prolong the service life of the coatings by improving the deposition techniques for coatings, inventing and innovating new coatings, and venturing more deeply into smart coatings and systems that could aid in reducing maintenance costs and time.

REFERENCES

1. R. Asmatulu, Nanocoatings for corrosion protection of aerospace alloys, in Viswanathan S. Saji and Ronald Cook (eds.), *Corrosion Protection and Control Using Nanomaterials* (pp. 357–374). Woodhead Publishing, 2012, doi: 10.1533/9780857095800.2.357.
2. J. R. Davis, *Corrosion: Understanding the basics.* ASM International, 2000.
3. J. Lince, "Coatings for Aerospace Applications," International Conference on Metallurgical Coatings and Thin Films San Diego, CA, 26 April, 2017. doi: 10.13140/RG.2.2.21638.16964.
4. D. Bémer, R. Régnier, I. Subra, B. Sutter, M. T. Lecler, and Y. Morele, "Ultrafine particles emitted by flame and electric arc guns for thermal spraying of metals," *Annals of Occupational Hygiene*, vol. 54, no. 6, pp. 607–614, 2010, doi: 10.1093/annhyg/meq052.
5. X. Chen, K. Chong, T. B. Abbott, N. Birbilis, and M. A. Easton, Biocompatible strontium-phosphate and manganese-phosphate conversion coatings for magnesium and its alloys, in T. S. N. Sankara Narayanan and Il-Song Park and Min-Ho Lee (eds.), *Surface Modification of Magnesium and its Alloys for Biomedical Applications* (pp. 407–432). Woodhead Publishing, 2015. doi: 10.1016/B978-1-78242-078-1.00015-3.
6. C. Krishnamoorthy and R. Chidambaram, Chapter 17 - Nanostructured thin films and nanocoatings, in Ahmed Barhoum and Abdel Salam Hamdy Makhlouf (eds.), *Emerging Applications of Nanoparticles and Architecture Nanostructures* (pp. 533–552). Elsevier, 2018. doi: 10.1016/B978-0-323-51254-1.00017-8.
7. B. Aleksandra, K. Cholewa-kowalska, M. Gajewska, B. Grysakowski, and T. Moskalewicz, "Surface & Coatings Technology Electrophoretic deposition, microstructure and selected properties of nanocrystalline SnO_2/Sr enriched bioactive glass/chitosan composite coatings on titanium," *Surface and Coatings Technology*, vol. 450, pp. 129004, 2022.
8. N. Rana, V. N. Shukla, R. Jayaganthan, and S. Prakash, "Degradation studies of micro and nanocrystalline NiCrAlY coatings for high temperature corrosion protection," *Procedia Engineering*, vol. 75, pp. 118–122, 2014, doi: 10.1016/j.proeng.2013.11.026.
9. U. Mamudu, M. Redza, J. Hernandez, and R. Chong, "Synthesis and characterisation of sulfated-nanocrystalline cellulose in epoxy coatings for corrosion protection of mild steel from sodium chloride solution," *Carbohydrate Polymer Technologies and Applications*, vol. 5, 2023.
10. U. Mamudu, L. A. Omeiza, M. R. Hussin, Y. Subramanian, A. K. Azad, M. S. Alnarabiji, E. E. Ebenso, and R. C. Lim, "Recycled eggshell waste in zinc-rich epoxy coating for corrosion protection of mild steel in a controlled elevated temperature saline environment," *Progress in Organic Coatings*, vol. 186, 2024, doi: 10.1016/j.porgcoat.2023.108025.
11. L. Palmolahti *et al.*, " Pinhole-resistant nanocrystalline rutile TiO_2 photoelectrode coatings," *Acta Materialia*, vol. 239, p. 118257, 2022.
12. Z. B. Zhang, X. B. Liang, Y. X. Chen, and B. S. Xu, "Abrasion resistance of Al-Ni-Mm-Fe amorphous and nanocrystalline composite coating on the surface of AZ91 magnesium alloy," Physics Procedia, vol. 50, pp. 156–162, 2013, doi: 10.1016/j.phpro.2013.11.026.
13. K. M. Hyie, N. A. Resali, W. N. R. Abdullah, and W. T. Chong, "Synthesis and characterization of nanocrystalline pure cobalt coating: Effect of pH," Procedia Engineering, vol. 41, pp. 1627–1633, 2012, doi: 10.1016/j.proeng.2012.07.360.

14. K. M. Hyie, N. A. Resali, W. N. R. Abdullah, and W. T. Chong, "Synthesis and characterization of nanocrystalline pure cobalt coating: Effect of pH," *Procedia Engineering*, vol. 41, pp. 1627–1633, 2012, doi: 10.1016/j.proeng.2012.07.360.
15. Y. J. Jang *et al.* "Tribological properties of multilayer tetrahedral amorphous carbon coatings deposited by filtered cathodic vacuum arc deposition," *Friction, vol.* 9, 1292–1302, 2021, doi: 10.1007/s40544-020-0476-y.
16. J. Kim, J. Kim, D. Hyun, Y. Jang, and N. Umehara, "Improved wear imbalance with multilayered nanocomposite nanocrystalline Cu and tetrahedral amorphous carbon coating," *Ceramics International*, vol. 47, pp. 25664–25673, 2021.
17. S. Habib, A. Qureshi, R. A. Shakoor, R. Kahraman, N. H. Al-Qahtani, and E. M. Ahmed, "Corrosion inhibition performance of polyolefin smart self-healing composite coatings modified with ZnO@β-Cyclodextrin hybrid particles," *Journal of Materials Research and Technology*, vol. 21, pp. 3371–3385, 2022, doi: 10.1016/j.jmrt.2022.10.148.
18. J. Wang, S. Wu, L. Ma, B. Zhao, H. Xu, and X. Ding, "Corrosion resistant coating with passive protection and self-healing property based on Fe_3O_4 -MBT nanoparticles," *Corrosion Communications*, vol. 7, pp. 1–11, 2022.
19. Y. Ma, D. Jiang, Y. Yang, L. Ma, J. Zhou, G. Huang, and C. Lin, "Synthesis of magnetic targeted delivery microcapsules and its anti-corrosion self-healing behavior in epoxy resin coatings," *Corrosion Communications*, vol. 10, pp. 27–37, 2023, doi: 10.1016/j.corcom.2022.08.004.
20. Z. Jin, Z. Zhao, T. Zhao, H. Liu, and H. Liu, "One-step preparation of inhibitor-loaded nanocontainers and their application in self-healing coatings," *Corrosion Communications*, vol. 2, pp. 63–71, 2021.
21. Y. Yin, H. Zhao, M. Prabhakar, and M. Rohwerder, "Organic composite coatings containing mesoporous silica particles: Degradation of the SiO_2 leading to self-healing of the delaminated interface," *Corrosion Science*, vol. 200, 2022.
22. Y. Ma, J. Zhang, G. Zhu, X. Gong, and M. Wu, "Robust photothermal self-healing superhydrophobic coating based on carbon nanosphere/carbon nanotube composite," *Materials & Design*, vol. 221, 2022, p. 110897.
23. Y. Huang, T. Liu, L. Ma, J. Wang, D. Zhang, and X. Li, " Saline-responsive triple-action self-healing coating for intelligent corrosion control," *Materials & Design*, vol. 214, 2022, p. 110381.
24. E. R. Ghomi *et al.* "Synthesis and characterization of TiO_2/acrylic acid-co-2-acrylamido-2-methyl propane sulfonic acid nanogel composite and investigation its self-healing performance in the epoxy coatings," *Colloid and Polymer Science*, vol. 298, pp. 213–223, 2020, doi: 10.1007/s00396-019-04597-0.
25. E. Rezvani, S. Nouri, M. Sadegh, and M. Dinari, "Synthesis of TiO 2 nanogel composite for highly efficient self-healing epoxy coating," *Journal of Advanced Research*, vol. 43, pp. 137–146, 2023.
26. B. Zhang, H. Fan, W. Xu, and J. Duan, "Thermally triggered self-healing epoxy coating towards sustained anti-corrosion," *Journal of Materials Research and Technology*, pp. 4–9, 2022.
27. A. Arellano, "Are you Legal?," in Pensoneau-Conway, S. L., Adams, T. E., Bolen, D. M. (eds.), *Doing Autoethnography* (pp. 197–203). SensePublishers, 2017, doi: 10.1007/978-94-6351-158-2_20.
28. E. Vazirinasab, R. Jafari, and G. Momen, "Application of superhydrophobic coatings as a corrosion barrier: A review," *Surface and Coatings Technology*, vol. 341, pp. 40–56, 2018, doi: 10.1016/j.surfcoat.2017.11.053.
29. A. Cherubin, J. Guerra, E. Barrado, C. García-Serrada, and F. J. Pulido, "Addition of amines to molasses and lees as corrosion inhibitors in sustainable de-icing materials," *Sustainable Chemistry and Pharmacy*, vol. 29, 2022, doi: 10.1016/j.scp.2022.100789.
30. V. G. Grishaev *et al.*, "Anti-icing fluids interaction with surfaces: Ice protection and wettability change," *International Communications in Heat and Mass Transfer*, vol. 129, 2021, doi: 10.1016/j.icheatmasstransfer.2021.105698.
31. L. Vertuccio, F. Foglia, R. Pantani, M. D. Romero-Sánchez, B. Calderón, and L. Guadagno, "Carbon nanotubes and expanded graphite based bulk nanocomposites for de-icing applications," *Composites Part B: Engineering*, vol. 207, 2021, doi: 10.1016/j.compositesb.2020.108583.
32. M. A. Khan *et al.*, "Fabrication of Ag nanoparticles on a Cu-substrate with excellent superhydrophobicity, anti-corrosion, and photocatalytic activity," *Alexandria Engineering Journal*, vol. 61, no. 8, pp. 6507–6521, 2022, doi: 10.1016/j.aej.2021.12.010.
33. M. Wu, J. Wang, S. Ling, R. Wheatley, and X. Hou, " Microporous metallic scaffolds supported liquid infused icephobic construction," *Journal of Colloid and Interface Science*, vol. 634, pp. 369–378, 2023.
34. X. F. Sánchez-Romate, R. Gutiérrez, A. Cortés, A. Jiménez-Suárez, and S. G. Prolongo, "Multifunctional coatings based on GNP/epoxy systems: Strain sensing mechanisms and Joule's heating capabilities for de-icing applications," *Progress in Organic Coatings*, vol. 167, 2022, doi: 10.1016/j.porgcoat.2022.106829.

35. T. Khaleque and S. Goel, "Repurposing superhydrophobic surfaces into icephobic surfaces," *Materials Today: Proceedings*, vol. 64, pp. 1526–1532, 2022, doi: 10.1016/j.matpr.2022.05.585.
36. A. Laroche, D. Bottone, S. Seeger, and E. Bonaccurso, "Silicone nanofilaments grown on aircraft alloys for low ice adhesion," *Surface and Coatings Technology*, vol. 410, 2021, doi: 10.1016/j.surfcoat.2021.126971.
37. S. Maharjan *et al.*, "Self-cleaning hydrophobic nanocoating on glass: A scalable manufacturing process," *Materials Chemistry and Physics*, vol. 239, p. 122000, 2019, doi: 10.1016/j.matchemphys.2019.122000.
38. M. Balordi, F. Pini, and G. Santucci de Magistris, "Superhydrophobic ice-phobic zinc surfaces," *Surfaces and Interfaces*, vol. 30, 2022, doi: 10.1016/j.surfin.2022.101855.
39. J. O. Carneiro, V. Teixeira, S. Azevedo, and M. Maltez-da costa, "Smart self-cleaning coatings for corrosion protection," in Abdel Salam Hamdy Makhlouf (ed.), *Handbook of Smart Coatings for Materials Protection* (pp. 489–509). Woodhead Publishing, 2014, doi: 10.1533/9780857096883.3.489.
40. Q. Meng, G. Guo, X. Qin, Y. Zhang, X. Wang, and L. Zhang, " Smart multifunctional elastomeric nanocomposite materials containing graphene nanoplatelets," *Smart Materials in Manufacturing*, vol. 1, no. June 2022, 2023.
41. B. Alderete, F. Mücklich, and S. Suarez, "Characterization and electrical analysis of carbon-based solid lubricant coatings," *Carbon Trends*, vol. 7, 2022.
42. J. Zhang, C. Wang, and L. Zhang, "Deployment of SMP Miura-ori sheet and its application : Aerodynamic drag and RCS reduction," *Chinese Journal of Aeronautics*, vol. 35, pp. 121–131, 2022.
43. A. Zhang, C. Liu, P. Sui, C. Sun, and L. Cui, "Corrosion resistance and mechanisms of smart micro-arc oxidation/epoxy resin coatings on AZ31 Mg alloy: Strategic positioning of nanocontainers," *Journal of Magnesium and Alloys*, vol. 12, 4562–4574, 2023, doi: 10.1016/j.jma.2022.12.013.
44. L. Vertuccio *et al.*, "Smart coatings of epoxy based CNTs designed to meet practical expectations in aeronautics," *Composites Part B: Engineering*, vol. 147, pp. 42–46, 2018.

14 Multifunctional Coatings for Biomedical Applications

Oktay Yigit, Mehmet Topuz, and Burak Dikici

14.1 INTRODUCTION

The development of multifunctional coatings for biomedical applications is a promising area of research in the field of biomaterials. Multifunctional coatings are designed to provide a range of properties, including biocompatibility, antibacterial activity, antifouling, and drug delivery. These coatings can be applied to various biomedical surfaces, such as implants, medical devices, and surgical instruments, to improve their performance and reduce the risk of infection and complications [1–3].

One of the primary benefits of multifunctional coatings is their ability to reduce infection rates, which is a significant concern in the medical field. According to the Centers for Disease Control and Prevention (CDC), healthcare-associated infections (HAIs) affect millions of patients worldwide and result in substantial health care costs. Multifunctional coatings can help prevent the growth of bacteria and other microorganisms on the surface of biomedical devices, thereby reducing the risk of infection [4].

The development of multifunctional coatings for biomedical applications involves the use of various materials and methods. Some of the materials used in these coatings include polymers, ceramics, metals, and composites. These materials are selected based on their properties, such as biocompatibility, durability, and drug-loading capacity [5]. The methods used to synthesize these coatings include physical vapor deposition, chemical vapor deposition, sol-gel, and electrospinning [1, 6]. Several studies have been conducted on the development of multifunctional coatings for biomedical applications up to now. The chapter provides an overview of the current state of multifunctional coatings.

14.2 TYPES OF MULTIFUNCTIONAL COATINGS

Biomedical implants play a vital role in addressing limb loss and tissue replacement in the human body. The success of these implants depends on the interaction between human physiology and the surface condition of engineered biomaterials [7]. However, metallic biomaterials and their alloys, despite meeting mechanical requirements, have weak interactions with tissues and suffer from problems like poor tissue attachment, corrosion, and reduced lifetime [8].

For example, corrosion is a problem that causes loosening and toxicity in metallic implants, as well as shortening the expected long life in patients of different age groups, significantly reducing implant success [9]. Besides, problems such as low osteoconductivity, decrease in wear resistance, and formation of wear residues can be seen as a result of corrosion [10, 11]. These issues stem from the surfaces of metallic implants, which are crucial in biological applications. Therefore, implant surfaces need to be carefully designed to address these problems and provide multiple advantages [12].

To overcome these challenges, multifunctional coatings provide several advantages, such as reduced inflammation, antibacterial action, enhanced biocompatibility, improved mechanical properties, superior corrosion resistance, and the ability to deliver drugs or growth factors. This modification is necessary to achieve good bone formability and desired biological interactions. Each feature has individual characteristics and is evaluated under its main headings for various multifunctional coatings developed for biological applications. These titles are listed next.

DOI: 10.1201/9781032635347-14

14.2.1 Antifouling Coatings

Biomedical antifouling coatings are designed to prevent or reduce the deposition of biological materials such as proteins, cells, and bacteria on medical implant material surfaces. These coatings, designed to be resistant to biological fluids, prevent biofilm formation on the surfaces and reduce the risk of infection [13]. Polymers such as polyethylene glycol (PEG) and polytetrafluoroethylene (PTFE) are antifouling coatings' preferred structures [3, 14]. With antifouling coatings, biofilm formation can be prevented by reducing the ability of bacteria to adhere to surfaces. In addition, antifouling coatings can improve the performance of medical devices by reducing friction between the implant and biological tissues. These coatings increase the performance of many medical devices, such as surgical instruments, heart valves, hip prostheses, and dental implants, and reduce the risk of infection. One of the essential features of biomedical antifouling coatings is that these coatings are biocompatible. Figure 14.1 shows the mechanism of antifouling coatings. Antifouling surfaces work by combining two elements into one system that can either prevent or release bacterial adhesion (hydrophilic polymers, zwitterionic, antiadhesive, topography, bioinspired surfaces, etc.) and eliminate attached bacteria (antibiotics, peptides, nitric oxide, ammonium salts, light, etc.) [13].

14.2.2 Biocompatible Coatings

The most important difficulty in biological applications is resolved by biocompatible coatings, which are made to produce a good interface between the implant material and biological tissues. There are two types of biocompatible materials; one group is materials that are originally biocompatible and ready to use, and the other needs surface modifications to exhibit biocompatibility [15]. The body's defense system responds naturally when it comes into contact with tissues. In this case, the tissues in the area are broken down, and fever and infection situations occur. To eliminate this situation, coatings similar to the natural body structure have been developed that will not react negatively to the body [16]. Thus, successful implantation can be performed with a high bioactivity between the tissue and the implant. The materials used for biocompatible coatings vary according to the place of use, and while coatings such as hydroxyapatite (HA), tricalcium phosphate (TCP), and zirconium oxide are used in implants that interact with bone tissues, polymers such as PEG and hydrogels are often preferred in other applications such as stents and pacemakers [17–19]. The biocompatibility of these materials has been examined in extended studies, and it has been shown that they can improve the interaction between implant materials and biological tissues.

14.2.3 Antibacterial Coatings

One of the biggest problems experienced after surgical implantation of metallic materials used in biomedical applications is the risk of infection in equipment or tissues that come from the environment or are already contaminated during the surgical procedure. To prevent bacteria formation,

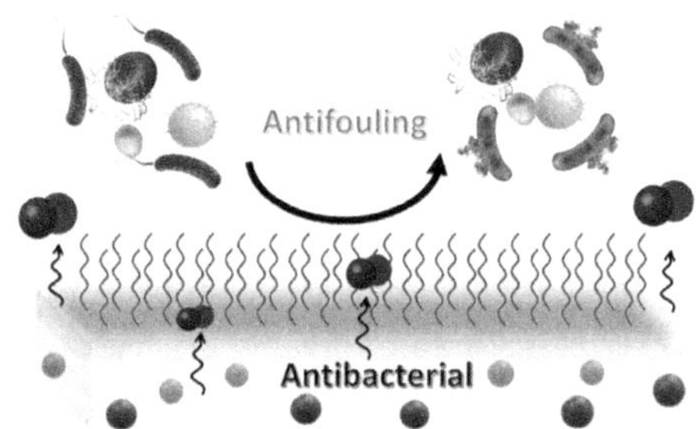

FIGURE 14.1 The mechanism of antifouling coatings [13].

adhesion, and reproduction on medical devices, implants, and other medical equipment, coatings are designed to reduce the risk of infection [20]. If a material with antibacterial action shows toxic properties, it will damage natural tissues and negatively affect biocompatibility [21]. Silver is a frequently used material due to its antibacterial properties. Silver is known to kill bacteria by piercing the cell wall. Silver's biocompatibility is also very high, making it an ideal coating material for skin contact medical devices [22]. Similarly, titanium is frequently used for medical implants due to its high biocompatibility and durability [1]. When it is used as an antibacterial coating, the growth of bacteria on the titanium surface is significantly reduced. Polymers are also frequently used for antibacterial coatings. Polymer coatings prevent the growth and reproduction of bacteria on the implant surface, thus reducing the use of antibiotics in the later stages, like other antibacterial additives. In addition, polymer coatings are highly biocompatible and well suited for eliminating skin and tissue sensitization conditions. Poly(hydroxy alkanoate) (PHA), polyvinylpyrrolidone (PVP), polyethylene glycol (PEG), and polycaprolactone (PCL) are frequently preferred polymer antibacterial materials [14, 23, 24]. However, some concerns remain regarding the use of antibacterial coatings. One of the concerns is that antibacterial coatings can cause bacteria to develop resistance to them. Therefore, more research is needed on the long-term effects of antibacterial coatings. Figure 14.2 shows the difference between antibacterial coatings and antifouling coatings. The initial adhesion of bacteria can be prevented or reduced by a passive surface. However, exposure to the physiological environment can cause a material's chemistry to change dramatically, which can cause antifouling material chemistry. Bacteria eventually manage to penetrate the modified surface, colonize, and result in a biofilm. Active surfaces with contact-based killing are harmed by protein and bacterial debris from deceased bacteria. Until the active agent's supply is exhausted, pathogens are still being eliminated by the release of active agents from these biomaterials. However, in prolonged use, both single-mechanism active and passive surfaces eventually result in the production of biofilms [13].

14.2.4 Drug-Delivery Coatings

The design of drug-delivery coatings gained momentum with the development of drug-delivery systems [25, 26]. While the traditional approach is to administer drugs to the whole body to affect a particular area, in recent years there has been more emphasis on providing direct drug delivery to the diseased area or particular need area with drug-delivery systems. Drug-delivery coatings can be applied to various medical devices, including stents, catheters, and orthopedic implants. The main benefit of drug-release coatings is the ability to provide targeted drug delivery and dosing, which can increase treatment efficacy and minimize side effects [27]. Polymer matrix coatings, lipid coatings, poly(lactic acid) coatings, hydrogel coatings, black carbon coatings, silver coatings, and

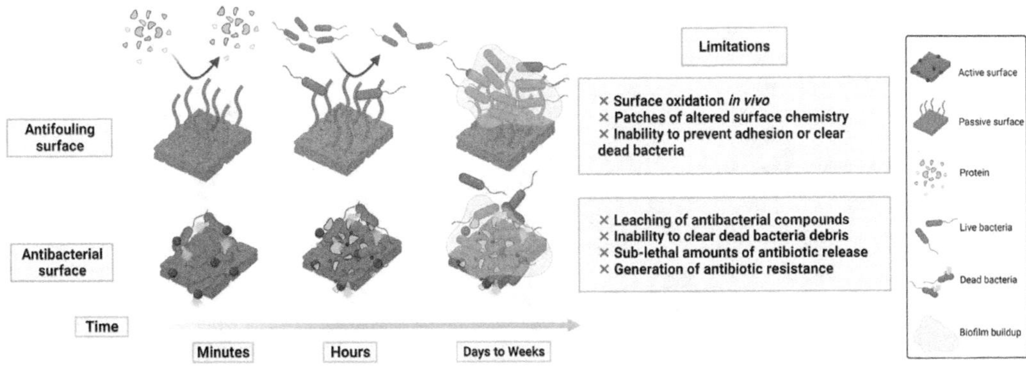

FIGURE 14.2 Progression of biofilm formation and proliferation on a medical device surface and failure of a singular approach to fully prevent infection on biomaterial surfaces [13].

magnetic nanoparticle coatings are the most widely used drug carrier coating types [27]. However, these coatings' bioavailability, toxicity, and effectiveness must be carefully investigated.

14.2.5 Hydrophilic Coatings

Hydrophilic means that a material can quickly absorb and retain water in its structure. For this reason, hydrophilic coatings are designs in which the surface energy of a material is increased, and its surfaces are made more attractive to water molecules. These coatings can be applied to various surfaces, including metals, ceramics, and polymers. Different purposes can be created for making hydrophilic coatings. These can be created for purposes such as reducing friction and improving the lubrication behavior of medical devices [28]. The most commonly used materials for hydrophilic coatings are polymers such as PEG and polyvinylpyrrolidone (PVP) [26, 29]. These polymer structures can be applied to medical devices to improve coating performance in coatings and to reduce the risks experienced in implant materials. Hydrophilic coatings can also be used in drug carrier coatings to provide better absorption and distribution of drugs to the coating surface and increase their bioavailability. Figure 14.3 shows hydrogels and drug-delivery systems. Smart biomaterials called stimuli-responsive hydrogels can release drugs in response to external triggers like pH, temperature, electrical and magnetic fields, light, and biomolecule concentration [27].

14.2.6 Anti-inflammatory Coatings

The primary purpose of designing anti-inflammatory coatings is to reduce the immune response that occurs when a foreign body enters the body [28]. The main purpose of these coatings is to minimize inflammation and improve the long-term performance of the implant material. Polymers such as PEG and hyaluronic acid are the most commonly used materials for anti-inflammatory coatings [19, 30]. These materials can reduce the immune response by blocking specific signaling pathways or releasing anti-inflammatory drugs. Anti-inflammatory coatings can also be used with biocompatible coatings to provide a comprehensive solution in medical applications. Biomedical anti-inflammatory coatings are being developed to reduce or help prevent inflammation [31]. These coatings exhibit similar behavior to drug-delivery applications, mainly because they act to reduce the risk of inflammation in the tissues around medical implants. Examples of biomedical

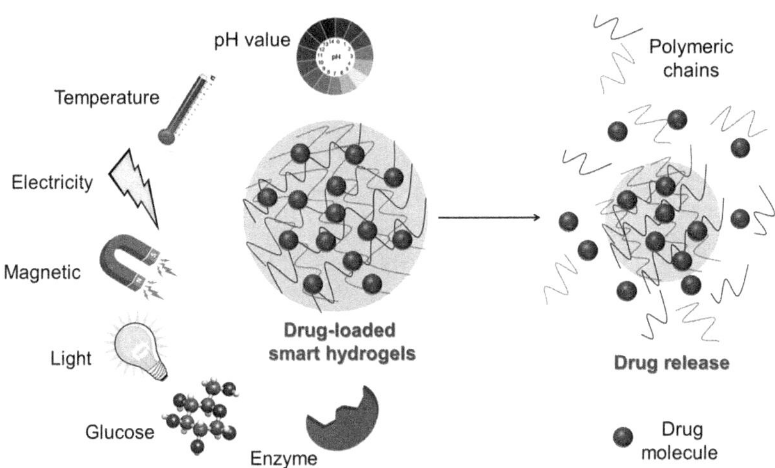

FIGURE 14.3 Various external stimuli, including pH, temperature, electricity, magnetics, light, and biomolecules (including glucose and enzyme), are controlling the drug release from a smart hydrogel [27].

anti-inflammatory coatings, hydrogel coatings, steroid coatings, nonsteroidal anti-inflammatory coatings, and coatings containing drug supplements can be listed [31].

14.2.7 Self-Healing Coatings

Self-healing coatings are unique structures designed to repair damage to a material's surface. The main benefit of self-healing coatings in biomedical applications is their ability to prolong the life of implant materials and reduce the need for replacement without further surgical intervention [32]. Polymers such as polyurethane and polydopamine are the most common materials used for self-healing coatings. These polymer structures are stimulated and responsive to pH or temperature changes within the desired range to initiate healing. Self-healing coatings can also be combined with other multifunctional coating types, such as antifouling and antibacterial coatings in medical implant materials [33]. Self-healing coatings can repair deformations such as scratches, holes, or damage on surfaces. In addition, these coatings increase the strength and functionality of the surface, making it ideal for long-term use [34]. Self-healing coatings generally consist of two main components: microcapsules and polymer matrix. Microcapsules are small capsules that contain unique ingredients, such as chemicals or monomers, used to repair surface damage. The polymer matrix is a unique matrix that allows these microcapsules to open in cracked or damaged areas and to perform the repair process. Some self-healing materials include various materials such as polymers, epoxy, glass, ceramics, and metal. For example, polymeric self-healing materials usually repair damage caused by high temperatures, while glass and ceramic materials repair scratches and holes in the surface. Metal self-healing materials are developed using special alloys to repair the deformations on the surface [33].

14.3 MATERIALS USED IN MULTIFUNCTIONAL COATINGS

Surface treatments in implant materials proceed with two different approaches. The first of these approaches is to cover the metallic surfaces with an organic or inorganic layer without any change in the implant material [35]. The second approach is to subject the substrate material to a chemical surface modification, changing the surface properties in the server—in other words, creating modified layers. With this process, the properties of the metallic implant are incorporated into the newly formed surface layers. Thus, a strong mechanical bond between the surface and the substrate is achieved with the coating layers with higher mechanical properties [36]. This approach, where both methods are used together, is used in modern biomedical studies, such as biocompatibility, corrosion resistance, antimicrobial behavior, drug delivery, and mechanical properties [19, 25, 37, 38]. In this section, we will talk about various ceramic, metal, and polymer-based materials used for biomedical coatings. The types and combinations of the materials define the multifunctional properties of the surfaces to be obtained.

14.3.1 Polymer-Based Materials

Polymer-based coatings are frequently used in functional coatings due to their excellent biocompatibility, ease of synthesis, and elastic properties. They can be functionalized with amino or carboxylic functional groups to increase biocompatibility and facilitate cell attachment. In addition, with the help of polymer-based coatings, drugs or growth hormones can be delivered to the relevant tissues. Examples of polymer-based coatings include polyethylene glycol (PEG) [19], chitosan [39], hyaluronic acid [40], polyether ether ketone (PEEK) [41], poly(hydroxy alkanoate) (PHA) [23], polyvinylpyrrolidone (PVP) [24], polyurethane (PU) [29], and polycaprolactone (PCL) [42]. Apart from these polymers, many applications and new polymers are synthesized. For example, polyethylene glycol (PEG) is a frequently used polymer due to its hydrophilic structure and high biocompatibility. It has been shown that coatings produced with this polymer, which can be functionalized with different functional groups such as carboxylic or amino groups, reduce inflammation and improve the mechanical properties of biomedical implants [19]. Chitosan, also one of the most used polymers, is a natural polysaccharide obtained from chitin and has been shown to have antibacterial properties. In addition, in many studies, its biodegradable and antibacterial properties are the main reason for its use as additives or functional coatings forming the main matrix [43]. Changing the chitosan structure's polymerization mechanisms can also change its mechanical and thermal properties. The most well-known polymer coating materials used in the biomedical field and their effects are summarized in Table 14.1.

In Table 14.1, PEEK is especially preferred as a bone substitute in clamps used in orthopedic and dental implants and removable dental prostheses [44]. With their superior mechanical properties, PEEK-coated substrates can increase the tribological properties of light alloys. Most sliding and bearing implant materials are coated with PEEK because of their properties and thermal stability [45]. On Ti-13Nb-13Zr titanium alloy, PEEK coatings (70–90 μm thickness) electrophoretically deposited showed exceptional wear resistance that was 200 times higher than the uncoated alloy [46]. This demonstrates its success in creating multifunctional coatings. Finally, polyurethane (PU) is a versatile polymer that can be engineered to have several properties, including biocompatibility, elasticity, and self-healing abilities. It reduces the risk of infection of PU-based coatings and increases the life of the coating and, thus, the performance of the implant material [29].

14.3.1.1 Polymeric Gels

Collagen, one of the main components of bone and a natural hydrophilic polymer that can be easily prepared in an aqueous solution, is a very suitable candidate for tissue studies and has several desirable properties for multifunctional coatings [49]. These polymeric gels are called hydrogels because of their high water-holding ability [50]. Hydrogels, which can swell when in contact with

TABLE 14.1
The Most Well-Known Polymer Coating Materials and Their Effects

Polymer	Effect	Ref.
Hyaluronic acid (HA)	It has antifouling properties that prevent infection, is preferred as an adhesion inhibitor and is used to reduce inflammation and fibrosis around implants.	[40]
Polyhydroxyalkanoates (PHA)	It has high biocompatibility and has an environmentally friendly structure. It has antibacterial properties, and this polymer produced by natural microorganisms is bacteria-inhibiting.	[23]
Polyvinylpyrrolidone (PVP)	It is a water-soluble polymer that is frequently used in many pharmaceutical applications. Its antibacterial properties make it preferred in products used for injuries. It can be used by adding to the coatings or by direct coating to eliminate the risk of infection of the implants.	[29]
Poly (N-vinylpyrrolidone) (PNVP)	It is a water-soluble antibacterial polymer with properties similar to PVP.	[47]
Polycaprolactone (PCL)	It is a highly biocompatible polymer used in many medical applications. Providing antibacterial properties allows the coatings to be used in cases with a risk of infection.	[42]
Poly ether ether ketone (PEEK)	It is a thermoplastic polymer that combines excellent hardness, superior chemical and physical properties, and toughness.	[48]
Polyurethane (PU)	It has several properties such as biocompatibility, elasticity, and self-healing abilities.	[29]
Chitosan	It can be used for many biomedical applications such as antibacterial and biocompatible coatings.	[39]
Hyaluronic acid	Its most common usage is for anti-inflammatory coatings.	[40]

water, dry out in an anhydrous environment [51]. Thanks to their hydrophilic behavior, it is possible to use water-soluble drugs, different hormones, and even other organic structures by incorporating them into hydrogels [52]. Furthermore, surface coatings designed with hydrogels become activatable structures with external stimuli such as pH [53], the introduction or presence of certain chemicals, the effect of temperature, or the presence of target molecules in the environment. Furthermore, a hydrogel has a structure that promotes cell attachment in coatings. If we list the natural hydrogels used primarily in bone tissue engineering (BTE), polysaccharides (e.g., cellulose) and polypeptide structures (e.g., alginate) can be mentioned.

Synthetic polymeric gels, which have the advantages of natural polymeric gels, are highly suitable candidates for molecular changes that facilitate appropriate designs—that is, mechanical properties and adjustment of biophysical and biochemical cues. They have a soft structure, high absorbency, and porous and biocompatible structures, and therefore they are the soft biomaterials preferred in multifunctional coating designs. However, hydrogels have disadvantages as well as superior properties; generally, poor mechanical strength limits their use on their own and requires a longer recovery and waiting time when used for bone regeneration. New technologies such as 3D printing for the manufacture of hydrogels and the manufacture of hydrogel-based components for their deposition on surfaces have recently begun to emerge and open up many possibilities for overcoming certain limitations and making new designs [54].

14.3.2 Ceramic-Based Materials

Ceramic materials are inorganic, nonmetallic, oxide, nitride, or carbide compounds of metallic materials. In their structural properties they are brittle, very hard, and resistant to compression loads but weak with regard to shear and tensile loads. On the other hand, they have a very durable structure compared to other metallic or polymeric materials in acidic or caustic environments. Examples of those used in biomedical applications include bioactive glasses (BG), aluminum oxide (Al_2O_3), zirconium oxide (ZrO_2), hydroxyapatite (HAp), tricalcium phosphate (TCP), titanium dioxide (TiO_2), silicon dioxide (SiO_2), strontium apatite (SrAp), and other calcium- and silica-based ceramic materials [55–58].

The most important property that makes ceramic materials excellent candidates for biomedical applications is their excellent biocompatibility. When ceramics are evaluated with their reactivity in the human body, they are classified into three categories: (i) bioactive ceramics, (ii) bioinert ceramics, and (iii) bioresorbable ceramics [59] (Figure 14.4). Bioactive ceramics are designed to increase the interaction of the implanted biomaterial surfaces with the surrounding tissues and show a higher level of reactivity. In this context, the ceramic-type HA structure is most commonly used and synthesized closest to the natural bone structure. In addition, other forms of apatite such as biphasic calcium phosphate (BCP) or strontium apatite (SrAp) are used frequently as bioactive coating materials [16].

Bioinert ceramics, on the other hand, are designed so that they do not show any reactivity with the surrounding tissues, in contrast to bioactive ceramics. Implant surfaces can form a physical bond with tissues, but they do not have an activity to increase the development of tissues [60].

The last type, bioabsorbable ceramics, shows a low reactivity with the tissues they come into contact with. In the later stages, they are gradually absorbed into the tissues from the implant surface and eventually replace the bone tissue, increasing the tissue–implant interaction. The most common uses of bioabsorbable ceramics are orthopedics and dental applications and highly biocompatible structures that can be designed for drug-delivery applications [61].

Ceramic structures show higher resistance to compression loads, have a low-toxicity chemical character and high corrosion resistance, and accelerate complex tissue formation after implantation. The frequently used hydroxyapatite structure is a synthetic ceramic with the same chemical structure designed closest to natural bone and dental hard tissues. The same Ca/P ratio (~1.67) as bone makes it a unique substitute [62]. These excellent properties of ceramics make them one of

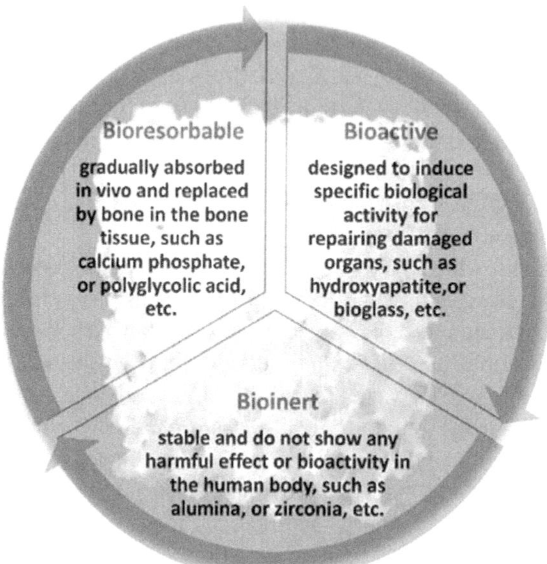

FIGURE 14.4 Bioceramic classification categories.

the indispensable components of multifunctional coatings. However, there are some limitations to ceramics. Ceramics are fragile in implant applications that are exposed to high and sudden load changes. Suppose ceramic surfaces are of high density and have higher mechanical properties than bone. In that case, the load cannot be transmitted through the bone, which leads to damage and deterioration of the bone tissue [63].

The use of ceramic materials as composite materials in multifunctional coatings provides superior properties compared to their use alone in ceramic coatings. Besides, the weight-strength ratio can be improved with composite structures in multifunctional coatings, and ceramic composite coatings with antibacterial, antifouling, anti-inflammatory, and drug-delivery properties can be produced with various designs [64].

By the way, new types of composites prepared with various compositions of bioactive and bioinert ceramics can provide both high bioactivity and mechanical properties. The most common examples are composites with Al_2O_3, ZrO_2, or TiO_2 doped into HAp, which exhibits good osseointegration with bone and good bioactivity [65]. The most preferred of these additives is TiO_2. TiO_2-based coatings both improve the antibacterial properties of coatings and increase their mechanical properties. For this reason, they are often preferred in multifunctional coatings. For example, TiO_2-based coatings are used in drug-delivery applications [66] and in orthopedic and dental implant coatings [67]. In addition, they exhibit high catalytic activity under photo and chemical corrosion conditions and provide long-term stability with their antibacterial properties [68]. When the TiO_2 structure is in the coating, it supports the formation of apatite or calcium phosphate on the surface of the coating. Thus, it is a successful coating element for reconstruction and bone replacement applications.

14.3.2.1 Transition Metal Nitride Materials

Transition metal nitrides and carbides are used against high wear, delamination, and corrosion. Recent studies have focused on the usability of metal nitrides in orthopedic implants. In addition, they form a diffusion barrier layer that prevents toxic ion release from occurring on implant metal surfaces [69]. TiN-based coatings have high scratch resistance as well as high hardness and low coefficient of friction. Therefore, TiN makes a suitable candidate for coating various metals used in arthroplasty. The formation of TiN-based coatings in orthopedic applications provides better biological performance than other nitrides [70]. TiAlN, another version of TiN, is another biocompatible coating type that is a second alternative in biomedical coatings despite its aluminum (Al) content [71]. In addition, transition metal carbonitrides (TiCN, ZrCN) provide long life in implant materials against wear resistance in biological environments [72]. Quaternary carbonitride coatings (TiAlCN, TiCrCN, TiNbCN, etc.), which are their quaternary compounds, show higher corrosion resistance and mechanical and tribological success compared to triple carbonitride coatings [72, 73]. In addition, carbon-based carbonitride coatings show successful biocompatibility properties [74]. The distinctive properties of such coatings are expressed in thermal stability, a decrease in the coefficient of friction, and, accordingly, the improvement of high wear resistance and mechanical properties (hardness, Young's modulus). Ali et al. [75] investigated the corrosion and surface properties of β-type Ti alloys coated with TiN, ZrN, and CrN thin layers using the PVD technique under in vitro conditions. The results revealed that all nitride-based coatings exhibited significantly higher adhesion resistance compared to the expected adhesion strength for biomaterial coatings. Furthermore, the coated alloys have greater corrosion resistance than the untreated metals. The study also showed that coatings of TiN, ZrN, and CrN, TNTZ might be used as biomaterials with advantageous surface characteristics and antibacterial activity.

14.3.2.2 Calcium Phosphate–Based Materials

Calcium phosphate (CaP) compounds are in the ceramic-based materials group, and the inorganic composition of bone is composed of Ca and P compounds similarly. For this reason, they are the most preferred structures in implant coatings. There are different CaP compounds according to their Ca/P ratio. Some of them are monocalcium phosphate (MCP) (Ca/P: 0.50), hydroxyapatite Ca/P: 1.67), tricalcium phosphate (Ca/P: 1.50), or tetracalcium phosphate (Ca/P: 2.0). In particular, the chemical composition of hydroxyapatite (HAp, $Ca_{10}(PO_4)_6(OH)_2$) is very similar to the chemical composition of complex tissues such as bones and teeth. Therefore, all properties, such as bioactivity, crystallinity, stability, and ion release, vary according to the CaP type [76].

Synthetic HAp shows excellent biocompatibility, osteoconductivity, osteoinductivity, and bioactivity and is the first choice over all other implant materials [77]. When HAp coatings are used in the body, they release calcium and phosphate ions into the tissues. Thus, it improves bone regeneration by facilitating the activation of osteoclast and osteoblast cells [78]. The use of HAp in multifunctional coatings not only regenerates bones but also increases osteoconductivity for bone growth, and encapsulating growth factors increase bone mineralization. In orthopedic applications, HAp coatings form a thin bond layer with the natural bone, providing bone apposition. These thin connective tissue bonding properties have led HAp ceramic coatings to be accepted as bioactive-based coatings. Thanks to the studies aimed at improving the strength of HAp coatings, high-quality HAp coatings have been obtained, and these can be designed as multifunctional coatings.

Up to now, many techniques such as hydrothermal, sol-gel, or PEO have been used successfully to synthesize CaP-based coatings [77, 79, 80]. Dense coatings using CaP constructs reduce the risk of coating delamination during in vivo testing in human body fluids. In addition, the roughness of the coating surface is one of the most critical factors affecting dissolution behavior, bone apposition, and growth. Porous surfaces improve the cell attachment or formation of extracellular matrix in biomedical coatings. Still, the porous layer to be obtained should be on the coating surface because the formation of macro-sized pores at the coating/substrate interface may cause poor coating adhesion [81].

Sankar et al. [82] compared the HAp coatings produced by the electrophoretic deposition (EPD) method with the HAp coatings produced by the pulsed laser deposition (PLD) method. EPD coated surfaces' porous structure causes the corrosion resistance of HAp coatings to be lower than PLD coatings. To prevent this, different additives are added to the HAp coating. Corrosion resistance can be enhanced by adding antimicrobial additives such as graphene. Yigit et al. [62, 83] used PEO and hydrothermal methods in their studies to produce graphene nanosheets (GNS) and graphene oxide (GO) doped HAp coatings and investigated their corrosion behavior. Graphene layers acted as a barrier layer in the interior of the coating, which prevented the transmission of corrosion liquid from the pores to the substrate. Similarly, different studies have shown that adding graphene and its derivatives to HAp significantly improved corrosion resistance and increased HAp crystallization. While the obtained coatings were in a more compact structure, the surfaces were porous and formed to support osteoconductivity [40, 84]. By the way, even if all necessary precautions are taken, bacterial films may form on implants and surfaces to be used during surgery due to surgical equipment or cross-contamination. Infection in the implant causes damage and inflammation in the surrounding tissue. Adding antibacterial agents into the coating is frequently done to prevent this problem in ceramic materials and multifunctional coatings. For this purpose, Zn can be added into HAp composites, and enhanced antimicrobial activity against *C. albicans* fungal cells and *S. aureus* bacteria is observed in Zn-doped HAp coatings [85]. The results showed that biocompatibility, stability, and crystallization properties of the implant materials can be improved by adding various ions into HAp coatings.

14.3.2.3 Bioactive Glass Materials

The structure, also called bioglass (compound 45S5), developed commercially by Hench, created a new revolution in the medical field as a bioactive material [86]. Bioglasses with amorphous or semicrystalline structures exhibit superior biological properties, but their use alone in coatings produced for biomedical purposes is limited by insufficient mechanical properties. To eliminate this deficiency in their mechanical properties, bioactive glasses are modified with various metal oxides such as TiO_2, Al_2O_3, and ZrO_2 or with various derivatives of graphene (graphene oxide, GNS, and reduced graphene oxide) with various additives [87]. The purpose of preparing these composites is to increase the mechanical properties and corrosion resistance without reducing the superior bioactivity of the bioglass [88]. Like other ceramics in the design of bioactive glasses as biomedical materials, bioactive mesoporous particles can be designed as 3D designs, thin films, coatings of various thicknesses, or particles ranging from nano to micron sizes [89].

14.3.3 Metal-Based Materials

The use of ceramic materials in biomedical applications is preferred due to their properties, such as excellent biocompatibility and bioactivity. However, their brittleness and low strength make them less suitable than metallic materials for some applications. For this reason, metals and their alloys are generally used as load-bearing elements in biomedical implant designs. However, although metals show high success in terms of strength and durability, there is a risk of releasing various ions and their residues due to physiological interaction with body fluids and other biological environments. In addition, their long-term corrosion triggers the release of harmful ions into the tissues. It also negatively affects strength and load-carrying ability. There is a risk of ion release in most alloys [90]. Excessive release of ions into the blood carries a high risk of accumulation in organs such as the spleen and kidneys, lungs, and liver, which may damage the organs. Therefore, organ failure may occur due to cytotoxicity and long-term accumulation. Metallic materials are not naturally received by the human body and are seen as foreign elements, and as a result, inadequate attachment of the implant slows or inhibits tissue growth, which can result in pain, fever, and infection at the implant site [91]. In this context, metallic implants have a higher risk of infection than ceramic implant materials, and the patient's recovery time is relatively slow. However, the superior properties of metallic materials also open up areas for use in multifunctional coatings. Certain conditions are sought when biomedical metallic materials are preferred as the primary implant material. Chemically inert platinum and gold show superior corrosion resistance but are expensive materials. Instead of these, Ti and its alloys with high corrosion resistance are used, and Mg and its alloys are preferred in recent studies due to their biodegradable properties [16, 57]. Metallic materials, which have unique properties in multifunctional coatings, both add functionality to coatings and improve the mechanical properties of coatings. Especially recent studies have focused on the use of silver (Ag), gold (Au), copper (Cu), zirconium (Zr), titanium (Ti), and magnesium (Mg) based metals in surface coatings due to their different characteristics [92–95].

14.3.4 Carbon-based Coatings

Carbon-based materials are considered bioinert materials in biomedical fields. Carbon-based materials in coatings are generally used in applications such as increasing corrosion resistance and improving wear and friction properties [96]. In addition, the surfaces in carbon coatings show hydrophobic behavior. Therefore, minimal protein adhesion is achieved, and a superior increase in excellent biocompatibility is observed. Many different coatings or carbon-based additives are used. Nanocrystalline diamond (NCD), pyrolytic carbon (PyC), and diamond-like carbon (DLC) are the most commonly used coating types [74]. Biomedically used PyC coatings are used in heart valves due to their resistance to platelet adhesion and their biocompatible behavior [97]. There are also orthopedic applications of PyC coatings [98]. PyC-coated orthopedic implants are used in the arthroplasty of proximal interphalangeal joints and small joints such as the wrist [99]. Nanocrystalline diamond and DLC coatings are also successful coating types for medical implants. Coatings obtained by the CVD process can be produced as nanometer-scale grains, and NCD coatings have very low surface roughness. They are hydrophobic and exhibit excellent biocompatibility in the body [100]. For this reason, they are often preferred in cardiovascular applications. NCD coatings can also be used as hard antibacterial coatings, and they show outstanding success in applications with a risk of infection, especially in cardiovascular applications in direct contact with blood. Successfully the bacterial adhesion was prevented on the coating surface, and thus an interaction was formed to reduce bacterial colonization.

Other additives are graphene, graphene oxide, reduced graphene oxide, and carbon nanotubes, which are newly developed and used in different forms in many coatings. Graphene and carbon nanotube derivatives are generally used in implant applications to increase the mechanical and corrosion properties of coatings, improve the surface properties of coatings, and protect biological properties [62, 83]. For example, graphene and its derivatives, used in many studies using different methods, can be doped with various biomedical alloys, and hydrophilic surfaces can be obtained. In this way, cell adhesion capabilities can be improved. In addition, thanks to its layered structure, graphene acts as a barrier to prevent the transmission of corrosion fluids in the body to the substrate surfaces, thus increasing corrosion resistance.

14.4 METHODS OF COATING APPLICATION

Today, biomaterials, which are frequently used in biomedical applications as a result of the increasing population, are applied with various coating methods and different bioactive components, and biomaterial surfaces can be obtained that provide fast and stable adhesion between the implant and tissue and form a strong bond with the tissue. Moreover, while these coatings provide for regular and rapid tissue growth between the implant and the tissue, they also prevent the release of ions from the implant surface to the surrounding tissues, and as a result, the osseointegration process takes place in a much better way [8]. With these coatings, antibacterial properties, abrasion resistance, and other mechanical properties (adhesion resistance, etc.) can be improved by increasing the biocompatibility of the biomaterial [101]. Since each coating method has its own advantages and disadvantages, the specific characteristics of the coating method to be selected should be well known. Therefore, it is very important to optimize the coating methods for each biomaterial candidate that is considered to be coated, and to exhibit homogeneous, increased interfacial strength and biocompatible behavior. The most demanding coating methods used to produce multifunctional coatings are briefly summarized at following subheadings.

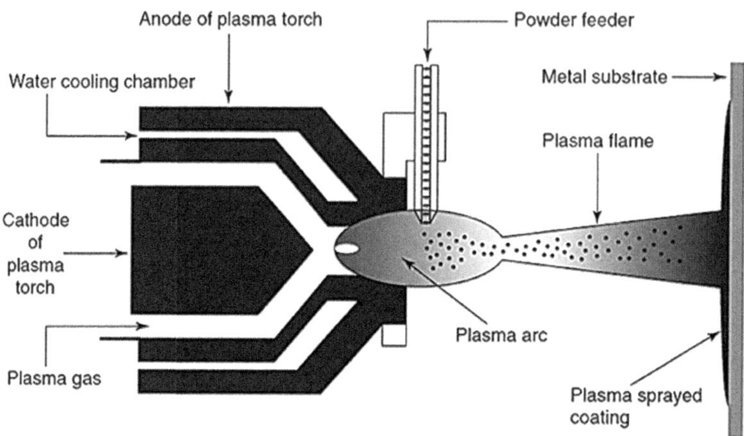

FIGURE 14.5 Schematic illustration of the plasma spray method [105].

14.4.1 Plasma Spray Coating

This coating method is the most common thermal spray coating method used to make metals more resistant to oxidation, abrasion, corrosion, and heat by coating them with various powders [102, 103]. The system in which this coating process is carried out consists of power unit, gas feeding unit, powder feeding unit, and spray gun. In this method, a DC electric arc is formed between the nozzle and the electrode. With inert gas (usually argon) or gaseous inert gas mixtures, its power is sent to the arc zone and heated by an electric arc. The energy required for the coating is provided from the plasma. Therefore, the high processing temperature of the plasma spray method permits working with ceramics or composites with high melting points [104]. The powders that melt in the high temperatures solidify by hitting the surface to be coated rapidly. The layer solidifies quickly, and the process ends with rapid cooling. A schematic working principle of the plasma spray method is presented in Figure 14.5. Table 14.2 presents the advantages and disadvantages of plasma spray coatings.

The quality of coatings made with this coating process depends on three factors: the nature of the powder fed, the substrate material, and the spray parameters. These interactions depend on the shape, size, density of the powder, whether the plasma flame is laminar or turbulent, its chemical composition, enthalpy, velocity of the plasma flame, and the way and speed of sending the powder

TABLE 14.2
Advantages and Disadvantages of Plasma Spray Coating Method [106–109]

Advantages	Disadvantages
• It can produce coatings from materials with high melting temperatures. • Coatings and substrate materials can be combined independently of each other. • It is possible to create pores in the desired ratio by changing the spraying parameters. • Smooth surfaces can be achieved without need for grinding or polishing. • Functional layered coatings can be produced. • Materials of different sizes can be coated. • High adhesion strength and production rate can be achieved.	• Evaporation or trajectory of the coating powders from the spray gun until they reach the substrate. • For substrates that will be affected by high temperatures, the heat input should be adjusted or a suitable cooling process should be applied. • There is a high probability of formation of undesirable phases in coatings due to high temperature. • They are high-cost systems.

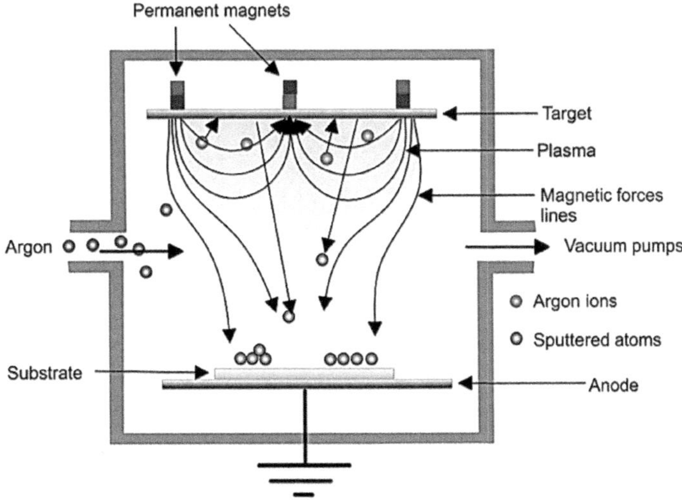

FIGURE 14.6 Working principle of magnetron sputtering process [102].

into the plasma flame [110]. However, when the powder encounters high temperatures, it can deteriorate and transform into different phases [109].

14.4.2 Magnetron Sputtering Coating

Magnetron sputtering is one of the most widely used methods for thin film coating of biomaterials. In this method, the target biomaterial to be coated is placed on the holder consisting of water-cooled magnets or electromagnets. The central axis of the substrate (biomaterial) to be coated forms one pole of the magnet, while the opposite pole is formed in the shape of a ring by magnets placed on the edges of the substrate. In the magnetic field formed by the arrangement of the magnets in this way, the secondary electrons emitted from the cathode surface by ion bombardment are kept on the substrate, causing the ionization and the plasma to be intense. Thus, thin films are produced on the target biomaterial [111]. Figure 14.6 presents a schematic representation of the magnetron sputtering method. Table 14.3 presents the advantages and disadvantages of magnetron sputtering coating systems.

Since the years when this method was first used, it has become possible to obtain coatings with the desired properties, e.g., high-entropy alloys, thanks to various developments made up to the present time [112]. In most cases, films deposited by magnetron sputtering is superior to other coating processes and perform the similar function as thicker films. As a matter of fact, considering

TABLE 14.3
Advantages and Disadvantages of Magnetron Sputtering Coating Method [102, 114]

Advantages	Disadvantages
• Simple and applicable to many materials	• Slow coating speed
• Coatings can be produced stoichiometrically	• Low ionization effect in plasma
• Low surface temperatures	• Irregular plasma concentration
• Enhanced adhesion, crystallographic, and homogeneous structures can be achieved	• High substrate temperature

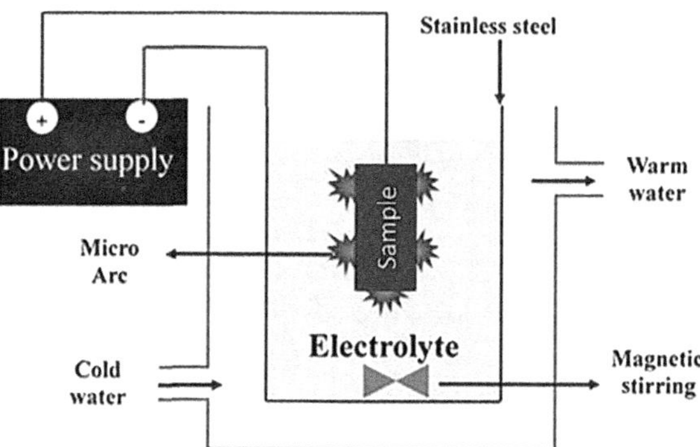

FIGURE 14.7 Schematic illustration of plasma electrolytic oxidation (PEO) coating [116].

the increased adhesion strength of the thinner film, it reveals a very successful situation in terms of the method. As a result, the magnetron sputtering method has an important effect in applications including hard, wear-resistant, low friction coefficient, corrosion-resistant, decorative and special optical or electrical properties [113, 114].

14.4.3 Plasma Electrolytic Oxidation (PEO)

The plasma electrolytic oxidation (PEO) or micro-arc oxidation method is one of the ceramic coating methods for metal, and its use has been increasing in recent years [115]. Metals such as Mg, Al, Zr, Ti, Nb, and their alloys are suitable for coating with the PEO method in that the stable oxide film they form on their surface creates resistance as a dielectric barrier layer and thus allows anodization. The oxides formed by these metals are stable under open circuit conditions and are difficult to corrode. If the potential of the metal electrode is increased in the positive direction by applying current to the metal, almost all of the current passing through the circuit is consumed by the growth of the oxide film on the surface. Depending on the chemical activity of the electrolytic solution, an oxide film layer or surface dissolution can be observed on the metal surface. Briefly, coating with PEO is an electrochemical process that occurs at the electrode–electrolyte interface, regulated by high-voltage sparks. The PEO process is both an environmentally friendly and safe coating method because it is made with a nontoxic, dilute, and basic electrolytic solution. Figure 14.7 shows the working principle of the PEO method. Table 14.4 presents both advantages and disadvantages of the PEO coating method.

TABLE 14.4
Advantages and Disadvantages of PEO Coating Method [102, 115]

Advantages	Disadvantages
• Sample preparation is simple and easy.	• It has high energy requirements.
• It is sufficient that the metal surfaces are free from dirt, oil, and oxide layers before coating.	• It is a somewhat dangerous coating method due to the use of high voltage.
• The surface preparation process is environmentally friendly and harmless to health as it is done without the use of any chemicals.	• A high-capacity cooler is also required to cool the heated solution due to arcs on the sample surface during coating.
• It provides an ideal nucleation center for the growth of living tissues on the surface due to the pores on the surface after coating.	• The adhesion strength of the coatings is relatively low compared to other methods.

There are some parameters that determine the quality and properties of the coatings produced with the PEO method. These parameters are generally electrical values such as process time, distance between anode and cathode, composition of substrate, composition of electrolyte, temperature of solution, current, and voltage [115, 117]. With changes to these parameters, many properties such as the morphology, thickness, surface roughness, hardness, and wear and corrosion resistance of the coating can be developed.

14.4.4 Sol-Gel Coating

Sol-gel (solution-gelation) is the process of coating surfaces using inorganic materials using a synthetic route. Metal, glass, ceramic, polymer, textile, paper, wood, and many other materials can be coated with the sol-gel coating method. Sol is the presence of solid particles in a liquid or solution in colloidal suspension. Gel, on the other hand, is the precipitate formed by the precipitation of these colloidal particles, and it contains a large amount of water [118]. This process consists of different steps; hydrolysis of the precursors, condensation of the sol to obtain gel, aging to increase its strength, predrying, and finally sintering. The precursor is a solution that triggers chemical reactions on the surface of the material. The solvents to be used in the mixture should be well dispersed and react in the mixture. During the coating process, the precursor is applied to the surface of the material, and after the coating process, the material is subjected to the sintering process [119]. As a result of the evaporation of the solvents in this mixture during the sintering process, the desired coating can be made on the material [120–122]. Coatings can be applied to the surface in the desired thickness and distribution after gelation (Figure 14.8). The sol-gel coating method allows the chemical and physical properties of coating materials to be controlled. The composition and temperature of the coating material can be controlled during the coating process. The sol-gel coating method allows the surfaces of materials to be made biocompatible. This feature provides a great advantage in biomedical applications. The advantages and disadvantages of the coatings obtained by the sol-gel process are summarized in Table 14.5.

In order to obtain sol-gel coatings that exhibit the required properties on the desired material, the surface must be cleaned properly, the coating medium must be clean, and the solution must be filtered. In sol-gel coatings, it is important that there are no foreign substances in the environment

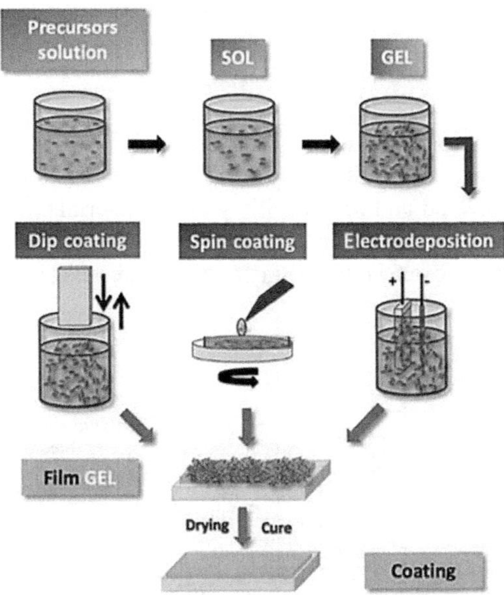

FIGURE 14.8 Schematic illustration of the sol-gel method [120].

TABLE 14.5
Advantages and Disadvantages of Sol-Gel Coating Method [124]

Advantages	Disadvantages
• Thin and thick coatings can be achieved. • It can cover materials with complex shapes. • It can be produced in high purity. • It has high homogeneity. • The coating composition can be easily changed by modifications of the precursors. • It is carried out at low temperatures. In this respect, it is a simpler, more economical and effective method than reactions that require high energy. • Since it is made at low temperature, no other phases are encountered in the coatings.	• Expensive precursor compounds • Low wear resistance • High permeability • Difficulty of porosity control • Observation of shrinkage crack formation during drying • Long processing time

and the atmosphere and that the application is carried out in a clean area. Sudden temperature changes should be avoided, and problems such as cracking and undesired rapid gelation should be also avoided [119, 123].

14.4.5 Hydrothermal Synthesis

The hydrothermal synthesis coating method is used in the production of advanced materials, especially electronic devices, ceramics, magnetic information storage devices, and biomedical materials [125]. Hydrothermal treatment can be defined as the dissolution of any heterogeneous reaction in a liquid solvent under high temperature and pressure in a closed environment and the dissolution and recrystallization of relatively insoluble materials under normal conditions. Hydrothermal synthesis in water or an aqueous solution is carried out under pressure at temperatures higher than its boiling point (Figure 14.9). In hydrothermal synthesis, high-temperature and pressure devices called autoclaves are used for material synthesis and coating [62, 77]. The advantages and disadvantages of the hydrothermal synthesis coating method are presented in Table 14.6.

14.4.6 Electrophoretic Deposition

The electrophoretic deposition (EPD) method is an electrochemical method that has become widespread recently in material processing. Recently, EPD has been used for the processing of functional and composite ceramics, layered and functionally graded materials, thin films, high-performance ceramic and composite coatings, and biomaterials, as well as for the deposition of nanoparticles and carbon nanotubes to produce advanced nanoconstructive materials [111, 127, 128].

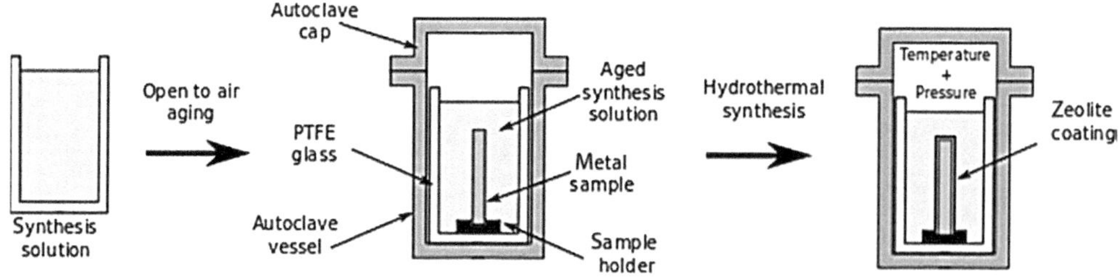

FIGURE 14.9 Procedures of hydrothermal coating process [126].

TABLE 14.6
Advantages and Disadvantages of Hydrothermal Synthesis Method [16, 77]

Advantages	Disadvantages
• Capability of coatings with high purity and homogeneity • Limited particle size distribution • Low sintering temperature • Wide selection of chemical compositions • One-step and fast coatings can be obtained • Low energy requirement	• Inability to observe the crystalline growth process • Requires strict security measures • High agglomeration rate • The need for expensive technical equipment such as autoclaves • High rate of chemical contamination due to autoclave corrosion

Electrophoretic deposition coating is a two–stage colloidal process that involves shaping the material from a stable suspension to which a DC electric field is applied. In the first step, an electric field is applied between the two electrodes and the charged particles are suspended in a suitable liquid and move toward the oppositely charged electrode. In the second step, the particles accumulate on the deposition electrode and produce a relatively compact and homogeneous film (deposition) (Figure 14.10) [102, 129]. Parameters such as the voltage, the conductivity of the substrate, and the powder size of the particle affect the efficiency of this coating type (see Table 14.7 for advantages and disadvantages of the method) [130].

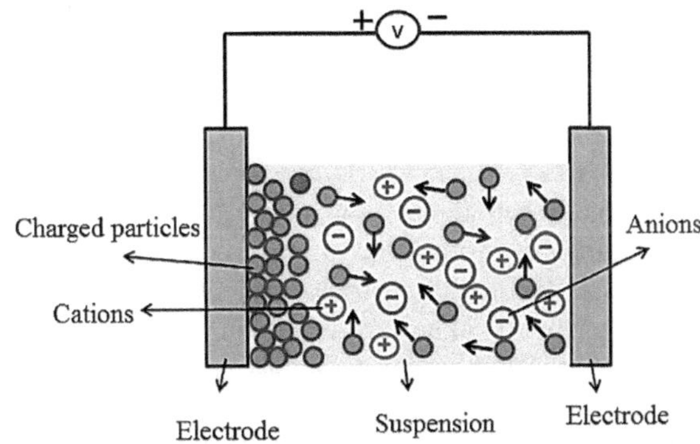

FIGURE 14.10 Schematic presentation of the electrophoretic deposition coating process [131].

TABLE 14.7
Advantages and Disadvantages of Electrophoretic Deposition Coating Method [102, 129, 132]

Advantages	Disadvantages
• Fast and low cost because it can be repeated in a short time • High microstructural homogeneity • Suitable for coating 3D complex and porous structures • Suitable for production with simple apparatus, for example, there is no need for a vacuum system • Environmentally friendly • Allowing the production of coatings of desired thickness (from a few nanometers to several millimeters)	• Sintering required at high temperatures, which makes it difficult to make crack-free coatings • Poor adhesion strength • Particle dissolution during the sintering process

After deposition of film on a substrate, a sintering process is needed to increase the density of the coating and eliminate porosity. The thickness of the coatings varies from 1 μm to 500 μm [131]. Two types of coating are formed, depending on the nature of the electrode on which the coating is formed. If the particles are positively charged, the coating is formed at the cathode and is called cathodic EPD coating. On the other hand, if particles are negatively charged, the coating is formed at the anode, which is called anodic EPD coating. After the coating is carried out, the substrate should be sintered at high temperatures. Smaller particle sizes can be achieved with EPD compared to other coating methods.

14.5 POTENTIAL DRAWBACKS IN THE PRODUCTION OF MULTIFUNCTIONAL COATINGS

Multifunctional coatings are used in biomedical applications to increase the biocompatibility of biomaterials, to reduce the risk of infection, to change the biological activity, and to provide features such as preventing interactions in the biological environment. In addition, multifunctional coatings can increase the durability of biomaterials by changing their mechanical properties and changing their biodegradability behavior. The biocompatibility of biomaterials used in the biomedical field is important in terms of the ability of the material to create an appropriate host response during its use in the body [2].

Multifunctional coatings used in biomedical applications are important to improve the performance, safety, and durability of implants and medical devices. These coatings increase the performance and longevity of the biomaterial by increasing abrasion resistance and improving other mechanical properties. In addition, some multifunctional coatings can also control biological reactions, particularly at the site of insertion of implants, by increasing cell adhesion or inhibiting bacterial growth [133]. Therefore, multifunctional coatings have an important role in reducing the complications that may occur during the use of medical devices. Multifunctional coatings can also increase the use of medical devices in many different applications. For example, multifunctional coatings based on nanotechnology can be used for the treatment of many diseases [134]. Overall, biomaterial coatings can be shown to improve the safety, efficacy, and durability of medical devices and implants, thereby improving the quality of life for patients.

The outcomes and responses that biomaterials experience in the biological setting highlight the complexity of material choice and implant design. In actuality, all interactions between implants and the tissues around them are dynamic ones. Water, dissolved ions, different biomolecules, and cells surround the implant surface shortly after the implantation procedure [135]. The reactions occurring at the biomaterial and tissue interface lead to time-dependent changes in the surface properties of both the implanted biomaterial and the surrounding tissues.

There are parameters that affect the properties of biocompatible coatings, films, and layers. These parameters are thickness (which influences the adhesion and fixation of the optimum layer), crystallinity (affects dissolution and biological behavior), phase and chemical purity, and porosity [79, 136].

The effects of biocompatible coatings on the biomaterial can be listed as the inclusion of the organism in metabolic processes, the biocompatibility of both surfaces or the entire material, the integration of the bioactive implant with the bone tissue at the molecular level, or the displacement of the degradable implant with healthy surrounding tissue [137].

The initial interaction that occurs when a biomaterial is introduced in a biological fluid medium is that water molecules quickly reach the surface and cover it. The characteristics of the surface affect how water molecules interact with it. The proteins and molecules that will stick to it after the hydration layer has formed are also determined by this characteristic. The extracellular matrix proteins cover the surface of the biomaterial during the second stage, which begins shortly after implantation and lasts for hours. In the third stage, cells are brought to the surface by proteins that have been adsorbed, and this is how the cell–surface interaction starts. After implantation, this period may begin immediately and persist for several days. Cell adhesion, migration, and differentiation occur over time [138–141]. Negative responses such as coagulation, fibrous capsule formation, and implant damage may occur [142]. Therefore, while developing a biomaterial, attention should be paid to reducing such effects and providing implant integration with controlled and rapid healing.

During the use of the implant in the body, toxic effects or allergies may occur due to its interaction with body fluid, proteins, enzymes, and various cells, and as a result, damage to the implant is inevitable [143]. In addition, factors such as chemical composition (alloying elements and their amount, segregations, foreign materials), microstructure (grain size and arrangement), surface properties (coated and uncoated), concentration of chemical compounds (pH, chlorine amount, etc.), temperature, and thermo-mechanical processing history (dislocation density, amount of thermal or residual stress, point errors, deformation rate) are the factors that greatly affect metallic biomaterials' corrosion behavior [117]. To explain more clearly, if corrosion occurs in the body, ion release occurs from the metal to the surrounding tissues [144]. The driving force in this mechanism is the potential difference between metal and body fluid. However, the oxide film formed on the metal surface or the resistance in the solution reduces the corrosion rate. Due to the biodegradability of metallic materials, these oxide films formed on the surface of metallic implants cause infection and stress accumulation in the orthopedic implant and cause the implant to lose its properties [145].

Another important issue is the slow diagnosis of implant-related infections, which makes it difficult to treat established infections with long-term, regular antibiotic therapy. As a result, complicated additional surgical procedures, including the removal and reimplantation of prosthetics, may frequently be needed to treat postoperative infections [111].

Following implantation, bacteria and tissue cells compete to attach to the biomaterial surface. The result of this event will determine the implant surface's future. The implant is less prone to bacterial infections if the tissue adheres to the surface, protecting it. If not, bacteria are everywhere around the implant surface, which triggers an infection quickly [146]. Since they can result in the loss of bone tissue, implant failure, and even mortality, implant-related infections are a serious issue for both patients and medical professionals [147]. As a result, the most dangerous and bothersome consequence in surgical procedures involving implantation is this surgical site infection.

Biofilm formation also causes infection in surrounding tissues. In order to completely eliminate the problem of implant-induced infection, it is necessary to prevent the formation of bacteria [146, 148]. Antibacterial surface coatings have been developed to prevent bacteria from adhering to the implant surface. Coatings are divided into active and passive coatings, depending on the release of antibacterial agents. Active coatings prevent the formation of implant-tissue infection by releasing antibacterial agents. These coatings can only prevent short-term postoperative infections. Passive coatings maintain their antibacterial properties for a long time without causing any unwanted side effects in the body [149]. Both active and passive coatings are illustrated in Figure 14.11 [150].

The current treatment approach is to administer high concentrations of antibiotics to the patient in order to prevent implant-associated infections. Even though tissue or biofilms that grow on implants can occasionally be penetrated by strong dosages of antibiotics, this can have hazardous adverse effects [151]. For these reasons, experiments on controlled and local antimicrobial release systems are being conducted to possibly replace conventional therapy. Silver nanoparticles and salts are used to protect medical devices from infection, including burn ointments, wound dressings and catheters, vascular grafts, and endotracheal tubes [152–154]. However, coatings functionalized with antibacterial bioactive polymers, such as chitosan, also inhibit bacterial adherence and growth. These polymers are frequently chosen as coating materials in biomedical applications because they give the implant surface antibacterial characteristics [155].

Drug-delivery carriers produced in nano sizes can cross various barriers such as blood-brain barriers, bronchioles in the respiratory system, and connections in the skin, and enable the delivery of drugs to the desired tissue [156]. Small molecules can be modified to bind proteins, peptides, and nucleic acids to target tissue types. Their surface properties can be modified for recognition by the immune system [133].

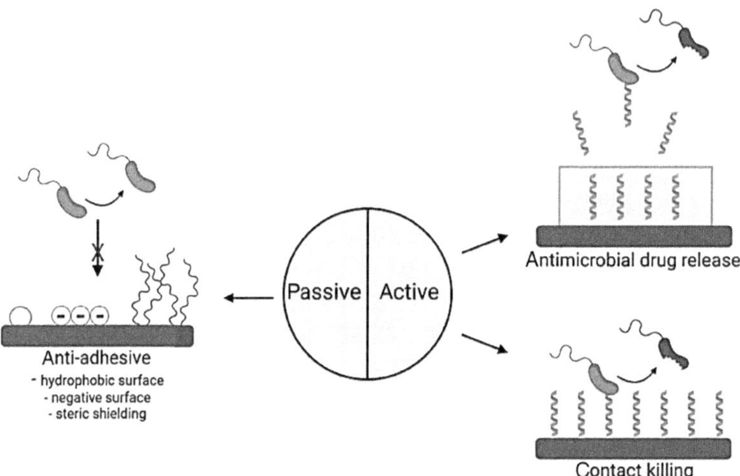

FIGURE 14.11 Active and passive coating strategy with scope of antibacterial coatings [150].

14.6 CHALLENGES AND FUTURE DIRECTIONS

Multifunctional coatings have shown great potential as biomaterials for various biomedical applications due to their ability to perform multiple functions such as antibacterial activity, drug delivery, and biocompatibility [157]. However, the development of these coatings still faces several challenges. In this chapter, we have discussed some of the challenges that researchers face when developing multifunctional coatings for biomedical applications and present some potential future directions for research in this area [102].

One of the major challenges in the development of multifunctional coatings is achieving a balance between different functions. For instance, coatings that have excellent antibacterial activity may not be biocompatible. Additionally, coatings that are good at drug delivery may not have good mechanical properties, which is crucial for implantable biomedical devices [5]. Therefore, researchers need to find a way to optimize the coatings' properties to achieve a balance between different functions [133]. Another challenge is the lack of standardized testing methods for multifunctional coatings. Currently, there is no consensus on the testing methods for evaluating the performance of multifunctional coatings [158]. This makes it difficult to compare the results of different studies and to develop a standard for the characterization of these coatings.

One potential future direction for research in this area is the use of nanotechnology. Nanoparticles have unique physicochemical properties that make them attractive for biomedical applications. For example, nanoparticles can be used to enhance the mechanical properties of coatings or to provide sustained drug release [159]. Moreover, the use of nanotechnology can also help overcome the challenges related to achieving a balance between different functions by allowing for the precise control of the coatings' properties [62, 77]. Another potential future direction is the use of biomimicry. Nature has provided us with a wealth of inspiration for the development of new materials. Researchers can learn from the structures and functions of natural materials to design and develop multifunctional coatings that mimic the properties of natural tissues. For instance, the use of biomimicry can lead to the development of coatings that are not only biocompatible but also possess the ability to promote tissue regeneration [160].

Multifunctional coatings have shown great potential as biomaterials for various biomedical applications [133]. However, the development of these coatings still faces several challenges. Researchers need to find a way to achieve a balance between different functions while optimizing the coatings' properties. Additionally, there is a need for standardized testing methods for multifunctional coatings. Finally, the use of nanotechnology and biomimicry holds great promise for the future development of multifunctional coatings as biomaterials [161, 162].

14.7 CONCLUSION

In conclusion, the development of multifunctional coatings for biomedical applications is an essential area of research that can improve the performance of medical devices and reduce the risk of infection. The selection of materials and methods used to create these coatings is critical in achieving their desired properties. However, further research in this area is necessary to optimize the design and efficacy of these coatings for various biomedical applications.

REFERENCES

1. O. Yigit, N. Ozdemir, B. Dikici and M. Kaseem, *Surface properties of graphene functionalized TiO2/ nHA hybrid coatings made on Ti6Al7Nb alloys via plasma electrolytic oxidation (PEO)*, Molecules. 26 (2021), p. 3903.
2. V. Vishwakarma, G.S. Kaliaraj and K.K. Amirtharaj Mosas, *Multifunctional coatings on implant materials—A systematic review of the current scenario*, Coatings. 13 (2022), p. 69.
3. K.K.A. Mosas, A.R. Chandrasekar, A. Dasan, A. Pakseresht and D. Galusek, *Recent advancements in materials and coatings for biomedical implants*, Gels. 8 (2022), pp. 1–35.
4. V. Puro, N. Coppola, A. Frasca, I. Gentile, F. Luzzaro, A. Peghetti et al., *Pillars for prevention and control of healthcare-associated infections: An Italian expert opinion statement*, Antimicrob. Resist. Infect. Control. 11 (2022), p. 87.
5. D. Aggarwal, V. Kumar and S. Sharma, *Drug-loaded biomaterials for orthopedic applications: A review*, J. Control. Release. 344 (2022), pp. 113–133.
6. B. Dikici, M. Niinomi, M. Topuz, S.G. Koc and M. Nakai, *Synthesis of biphasic calcium phosphate (BCP) coatings on β–type titanium alloys reinforced with rutile-TiO2 compounds: Adhesion resistance and in-vitro corrosion*, J. Sol-Gel Sci. Technol. 87 (2018), pp. 713–724.
7. G. Tang, Z. Liu, Y. Liu, J. Yu, X. Wang, Z. Tan and X. Ye. Recent *t*rends in the *d*evelopment of *b*one *r*egenerative *b*iomaterials, Front. Cell Dev. Biol. 9 (2021), p. 665813. doi: 10.3389/fcell.2021.665813
8. A. Goharian, *Fundamentals in Loosening and Osseointegration of Orthopedic Implants*, in *Osseointegration of Orthopaedic Implants*, Elsevier Academic Press, 2019.
9. Pierre Voisine, Rishi Puri, Philippe Pibarot, Josep Rodés-Cabau, Aortic Bioprosthetic Valve Durability: Incidence, Mechanisms, Predictors, and Management of Surgical and Transcatheter Valve Degeneration, Journal of the American College of Cardiology, Volume 70, Issue 8, 2017, Pages 1013–1028, https://doi.org/10.1016/j.jacc.2017.07.715.
10. N. Shayesteh Moghaddam, M. Taheri Andani, A. Amerinatanzi, C. Haberland, S. Huff, M. Miller et al., *Metals for bone implants: Safety, design, and efficacy*, Biomanufacturing Rev. 1 (2016), pp. 1–10.
11. O. Bazaka, K. Bazaka, P. Kingshott, R.J. Crawford and E.P. Ivanova, *Metallic implants for biomedical applications*, in Christopher Spicer (eds.), The Chemistry of Inorganic Biomaterials, The Royal Society of Chemistry, 2021.
12. W. Liu, S. Liu and L. Wang, *Surface modification of biomedical titanium alloy: Micromorphology, microstructure evolution and biomedical applications*, Coatings. 9 (2019), p. 249.
13. M.K. Chug and E.J. Brisbois, *Recent developments in multifunctional antimicrobial surfaces and applications toward advanced nitric oxide-based biomaterials*, ACS Mater. Au. 2 (2022), pp. 525–551.
14. S. Jiang and Z. Cao, *Ultralow-fouling, functionalizable, and hydrolyzable zwitterionic materials and their derivatives for biological applications*, Adv. Mater. 22 (2010), pp. 920–932.
15. Y. Oshida and Y. Guven, *Biocompatible coatings for metallic biomaterials*, in Cuie Wen (eds.), *Surface Coating and Modification of Metallic Biomaterials*, Elsevier, 2015.
16. O. Yigit, *Structural, chemical and osteogenic properties of GNS reinforced fluorine-doped strontiumapatite coatings on AZ31 Mg alloys for potential biomedical applications*, Surf. Coatings Technol. 451 (2022), pp. 1–13.
17. R.A. Surmenev, M.A. Surmeneva and A.A. Ivanova, *Significance of calcium phosphate coatings for the enhancement of new bone osteogenesis - A review*, Acta Biomater. 10 (2014), pp. 557–579.
18. H. Sampatirao, S. Amruthaluru, P. Chennampalli, R.K. Lingamaneni and R. Nagumothu, *Fabrication of ceramic coatings on the biodegradable ZM21 magnesium alloy by PEO coupled EPD followed by laser texturing process*, J. Magnes. Alloy. 9 (2021), pp. 910–926.

19. I.C. Pereira, A.S. Duarte, A.S. Neto and J.M.F. Ferreira, *Chitosan and polyethylene glycol based membranes with antibacterial properties for tissue regeneration*, Mater. Sci. Eng. C. 10 (2019), 1–20.
20. J. Andrade-Del Olmo, L. Ruiz-Rubio, L. Pérez-Alvarez, V. Sáez-Martínez and J. Luis Vilas-Vilela, Antibacterial *c*oatings for *i*mproving the *p*erformance of *b*iomaterials, Coatings. 10, no. 2 (2020), p. 139. doi: 10.3390/coatings10020139
21. M. Cloutier, D. Mantovani and F. Rosei, Antibacterial *c*oatings: *C*hallenges, *p*erspectives, and *o*pportunities, *t*rends in *b*iotechnology, 33, no. 11 (2015), pp. 637–652. doi: 10.1016/j.tibtech.2015.09.002
22. X. Zhang, H. Wu, Z. Geng, X. Huang, R. Hang, Y. Ma et al., *Microstructure and cytotoxicity evaluation of duplex-treated silver-containing antibacterial TiO_2 coatings*, Mater. Sci. Eng. C. 45 (2014), pp. 402–410.
23. K. P. Luef, F. Stelzer and F. Wiesbrock, Poly(hydroxy alkanoate)s in *m*edical *a*pplications, Chem. Biochem. Eng. Q. 29, no. 2 (2015), pp. 287–297.
24. S. Rehmat, N.B. Rizvi, S.U. Khan, A. Ghaffar, A. Islam, R.U. Khan et al., *Novel stimuli-responsive pectin-PVP-functionalized clay based smart hydrogels for drug delivery and controlled release application*, Front. Mater. 9 (2022), pp. 1–15.
25. R. Mohan Raj, P. Priya and V. Raj, *Gentamicin-loaded ceramic-biopolymer dual layer coatings on the Ti with improved bioactive and corrosion resistance properties for orthopedic applications*, J. Mech. Behav. Biomed. Mater. 82 (2018), pp. 299–309.
26. L. Shi, J. Zhang, M. Zhao, S. Tang, X. Cheng, W. Zhang, W. Li, X. Liu, H. Peng and Q. Wang, *Effects of polyethylene glycol on the surface of nanoparticles for targeted drug delivery*, Nanoscale. 13 (2021), pp. 10748–10764.
27. A. Bordbar-Khiabani and M. Gasik, Smart *h*ydrogels for *a*dvanced *d*rug *d*elivery *s*ystems, Int. J. Mol. Sci. 23 (2022), p. 3665. doi: 10.3390/ijms23073665
28. Y. Wang, Q. Li, T. Shao, W. Miao, C. You and Z. Wang, *Fabrication of anti-fouling and anti-bacterial hydrophilic coating through enzymatically-synthesized cellooligomers*, Appl. Surf. Sci. 600, no. 1–12 (2022), p. 154133.
29. A. Kuźmińska, B.A. Butruk-Raszeja, A. Stefanowska and T. Ciach, *Polyvinylpyrrolidone (PVP) hydrogel coating for cylindrical polyurethane scaffolds*, Colloids Surfaces B Biointerfaces. 192, no. 1–6 (2020), p. 111066.
30. H. Tan, C.R. Chu, K.A. Payne and K.G. Marra, *Injectable in situ forming biodegradable chitosan-hyaluronic acid based hydrogels for cartilage tissue engineering*, Biomaterials. 30 (2009), pp. 2499–2506.
31. G. Tan, J. Xu, W.M. Chirume, J. Zhang, H. Zhangm and X. Hu, Antibacterial and *a*nti-*i*nflammatory *c*oating *m*aterials for *o*rthopedic *i*mplants: A *r*eview, Coatings. 11 (2021), p. 1401. doi: 10.3390/coatings11111401
32. D. Zhang, F. Peng and X. Liu, Protection of magnesium alloys: From physical barrier coating to smart self-healing coating, J. Alloys Compd. 853 (2021), pp. 157010. doi: 10.1016/j.jallcom.2020.157010
33. A.E. Hughes, P. Johnston and T.J. Simons, *Self-healing coatings*, in Guoqiang Li and Xiaming Feng (eds.), Recent Advances in Smart Self-Healing Polymers and Composites, *Second* Edition, Elsevier, 2022.
34. G. Cui, C. Zhang, A. Wang, X. Zhou, X. Xing, J. Liu, Z. Li, Q. Chen and Q. Lu, Research progress on self-healing polymer/graphene anticorrosion coatings, Prog. Org. Coat. 155 (2021), p. 106231. doi:10.1016/j.porgcoat.2021.106231
35. P. Sikder, Y. Ren and S.B. Bhaduri, *Synthesis and evaluation of protective poly(lactic acid) and fluorine-doped hydroxyapatite-based composite coatings on AZ31 magnesium alloy*, J. Mater. Res. 34 (2019), pp. 3766–3776.
36. A. Bekmurzayeva, W.J. Duncanson, H. S. Azevedo and D. Kanayeva, Surface modification of stainless steel for biomedical applications: Revisiting a century-old material, Mater. Sci. Eng. C. 93 (2018), pp. 1073–1089. doi: 10.1016/j.msec.2018.08.049
37. Z. Geng, R. Wang, X. Zhuo, Z. Li, Y. Huang, L. Ma et al., *Incorporation of silver and strontium in hydroxyapatite coating on titanium surface for enhanced antibacterial and biological properties*, Mater. Sci. Eng. C. 71 (2017), pp. 852–861.
38. A.J. López, A. Ureña and J. Rams, *Wear resistant coatings: Silica sol-gel reinforced with carbon nanotubes*, Thin Solid Films. 519 (2011), pp. 7904–7910.
39. Hartatiek, P. Dwiasih, Yudyanto, N. Hidayat, R. Kurniawan and Masruroh, Sonochemical Synthesis of Nano-Hydroxyapatite/Chitosan Biomaterial Composite from Shellfish and Their Characterizations, in IOP Conference Series: Materials Science and Engineering, 2019.
40. M. Li, Q. Liu, Z. Jia, X. Xu, Y. Shi, Y. Cheng et al., *Electrophoretic deposition and electrochemical behavior of novel graphene oxide-hyaluronic acid-hydroxyapatite nanocomposite coatings*, Appl. Surf. Sci. 284 (2013), pp. 804–810.

41. H. Mahjoubi, E. Buck, P. Manimunda, R. Farivar, R. Chromik, M. Murshed et al., *Surface phosphonation enhances hydroxyapatite coating adhesion on polyetheretherketone and its osseointegration potential*, Acta Biomater. 47 (2017), pp. 149–158.
42. Z. Ansari, M. Kalantar, M. Kharaziha, L. Ambrosio and M.G. Raucci, *Polycaprolactone/fluoride substituted-hydroxyapatite (PCL/FHA) nanocomposite coatings prepared by in-situ sol-gel process for dental implant applications*, Prog. Org. Coatings. 147 (2020), p. 105873.
43. S. Islam, M.A.R. Bhuiyan and M.N. Islam. Chitin and Chitosan: Structure, *p*roperties and *a*pplications in *b*iomedical *e*ngineering. J. Polym. Environ. 25 (2017), pp. 854–866. doi: 10.1007/s10924-016-0865-5
44. D. Almasi, N. Iqbal, M. Sadeghi, I. Sudin, M.R. Abdul Kadir, T. Kamarul, Preparation *m*ethods for *i*mproving PEEK's *b*ioactivity for *o*rthopedic and *d*ental *a*pplication: A *r*eview, Int. J. Biomater. 2016 (2016), p. 8202653. doi: 10.1155/2016/8202653. Epub 2016 Apr 4. PMID: 27127513; PMCID: PMC4834406.
45. L. Lin, Y. Zhao, C. Hua and A.K. Schlarb, *Effects of the velocity sequences on the friction and wear performance of PEEK-based materials*, Tribol. Lett. 68 (2021), pp. 1–11.
46. A. Sak, T. Moskalewicz, S. Zimowski, Ł Cieniek, B. Dubiel, A. Radziszewska et al., *Influence of polyetheretherketone coatings on the Ti-13Nb-13Zr titanium alloy's bio-tribological properties and corrosion resistance*, Mater. Sci. Eng. C. 63 (2016), pp. 52–61.
47. X. Liu, Y. Xu, Z. Wu and H. Chen. Poly(N-vinylpyrrolidone)-modified surfaces for biomedical applications, Macromol Biosci. 13, no. 2 (2013 Feb), pp. 147–154. doi: 10.1002/mabi.201200269. Epub 2012 Dec 4. PMID: 23212975.
48. K. Friedrich, Z. Zhang and A.K. Schlarb, *Effects of various fillers on the sliding wear of polymer composites*, Compos. Sci. Technol. 65 (2005), pp. 2329–2343.
49. G.L. Koons, M. Diba and A.G., *Mikos, Materials design for bone-tissue engineering*, Nat. Rev. Mater. 5 (2020), pp. 584–603. doi: 10.1038/s41578-020-0204-2
50. A.K.S. Chandel, B. Nutan, I.H. Raval and S.K. Jewrajka, *Self-assembly of partially alkylated dextran-graft -poly[(2-dimethylamino)ethyl methacrylate] copolymer facilitating Hydrophobic/Hydrophilic drug delivery and improving conetwork hydrogel properties*, Biomacromolecules. 19, no. 4 (2018), pp. 1142–1153.
51. B. Nutan, A.K.S. Chandel and S.K. Jewrajka, *Liquid prepolymer-based in situ formation of degradable poly(ethylene glycol)-linked-poly(caprolactone)-linked-poly(2-dimethylaminoethyl)methacrylate amphiphilic conetwork gels showing polarity driven gelation and bioadhesion*, ACS Appl. Bio Mater. 1 (2018), pp. 1606–1619.
52. L. Tallet, V. Gribova, L. Ploux, N.E. Vrana and P. Lavalle, *New smart antimicrobial hydrogels, nanomaterials, and coatings: Earlier action, more specific, better dosing?*, Adv. Healthc. Mater. 10 (2021), pp. 1–16.
53. S. Paroha, A.K.S. Chandel and R.D. Dubey, Nanosystems for drug delivery of coenzyme Q10, Environ. Chem. Lett. 16 (2018), pp. 71–77. doi: 10.1007/s10311-017-0664-9
54. H.K. Chang, D.H. Yang, M.Y. Ha, H.J. Kim, C.H. Kim, S.H. Kim et al., *3D printing of cell-laden visible light curable glycol chitosan bioink for bone tissue engineering*, Carbohydr. Polym. 287, no. (1–14) (2022), p. 119328.
55. B. Aksakal, M. Gavgali and B. Dikici, *The effect of coating thickness on corrosion resistance of hydroxyapatite coated Ti6Al4V and 316L SS implants*, J. Mater. Eng. Perform. 19 (2010), pp. 894–899.
56. Y. Say, B. Aksakal and B. Dikici, *Effect of hydroxyapatite/SiO2 hybride coatings on surface morphology and corrosion resistance of REX-734 alloy*, Ceram. Int. 42 (2016), pp. 10151–10158.
57. O. Yigit, T. Gurgenc, B. Dikici, M. Kaseem, C. Boehlert and E. Arslan, *Surface modification of pure Mg for enhanced biocompatibility and controlled biodegradation: A study on graphene oxide (GO)/ Strontium apatite (SrAp) biocomposite coatings*, Coatings. 13 (2023), p. 890.
58. S. Sonmez, B. Aksakal and B. Dikici, *Influence of hydroxyapatite coating thickness and powder particle size on corrosion performance of MA8M magnesium alloy*, J. Alloys Compd. 596 (2014), pp. 125–131.
59. T. Yamamuro, *Bioceramics*, in *Biomechanics and Biomaterials in Orthopedics, Second Edition*, 2016. https://link.springer.com/book/10.1007/978-1-84882-664-9
60. C. Piconi and A.A. Porporati, *Bioinert ceramics: Zirconia and alumina*, in *Handbook of Bioceramics and Biocomposites*, Springer Nature, 2016.
61. M.A. McNally, J.Y. Ferguson, M. Scarborough, A. Ramsden, D.A. Stubbs and B.L. Atkins, *Mid- to long-term results of single-stage surgery for patients with chronic osteomyelitis using a bioabsorbable gentamicin-loaded ceramic carrier*, Bone Jt. J. 104-B (2022), pp. 1095–1100.
62. O. Yigit, B. Dikici, T.C. Senocak and N. Ozdemir, *One-step synthesis of nano-hydroxyapatite/graphene nanosheet hybrid coatings on Ti6Al4V alloys by hydrothermal method and their in-vitro corrosion responses*, Surf. Coatings Technol. 394 (2020), p. 125858.

63. X. Gao, Y. Zhao, M. Wang, Z. Liu and C. Liu, *Parametric design of hip implant with gradient porous structure*, Front. Bioeng. Biotechnol. 10 (2022), pp. 1–15.
64. L. Fiorillo, M. Cicciù, T.F. Tozum, M. Saccucci, C. Orlando, G.L. Romano, C. D'Amico and G. Cervino, Endosseous *d*ental *i*mplant *m*aterials and *c*linical *o*utcomes of *d*ifferent *a*lloys: A *s*ystematic *r*eview, Materials (Basel). 15, no. 5 (2022 Mar 7), p. 1979. doi: 10.3390/ma15051979. PMID: 35269211; PMCID: PMC8911578.
65. M. Li, S. Komasa, S. Hontsu, Y. Hashimoto and J. Okazaki, *Structural characterization and osseointegrative properties of pulsed laser-deposited fluorinated hydroxyapatite films on nano-zirconia for implant applications*, Int. J. Mol. Sci. 23 (2022), p. 2416.
66. C. Xie, P. Li, Y. Liu, F. Luo and X. Xiao, *Preparation of TiO2 nanotubes/mesoporous calcium silicate composites with controllable drug release*, Mater. Sci. Eng. C. 67 (2016), pp. 433–439.
67. V. Kumaravel, K.M. Nair, S. Mathew, J. Bartlett, J.E. Kennedy, H.G. Manning et al., *Antimicrobial TiO2 nanocomposite coatings for surfaces, dental and orthopaedic implants*, Chem. Eng. J. 416, no. 1–19 (2021), p. 129071.
68. Y.Q. Cao, T.Q. Zi, X.R. Zhao, C. Liu, Q. Ren, J. Bin Fang et al., *Enhanced visible light photocatalytic activity of Fe2O3 modified* TiO_2 *prepared by atomic layer deposition*, Sci. Rep. 10 (2020), pp. 1–10.
69. S.J. Gobbi, *Orthopedic implants: Coating with TiN*, Biomed. J. Sci. Tech. Res. 16 (2019), pp. 11740–11742.
70. R.P. van Hove, I.N. Sierevelt, B.J. van Royen and P.A. Nolte, Titanium-*n*itride *c*oating of *o*rthopaedic *i*mplants: A *r*eview of the *l*iterature, Biomed Res Int. 2015 (2015), p. 485975. doi: 10.1155/2015/485975. Epub 2015 Oct 25. PMID: 26583113; PMCID: PMC4637053.
71. B. Subramanian, C.V. Muraleedharan, R. Ananthakumar and M. Jayachandran, A comparative study of titanium nitride (TiN), titanium oxy nitride (TiON) and titanium aluminum nitride (TiAlN), as surface coatings for bio implants, Surf. Coat. Technol. 205, no. 21–22 (2011), pp. 5014–5020. doi: 10.1016/j.surfcoat.2011.05.004
72. M. Dinu, I. Pana, P. Scripca, I.G. Sandu, C. Vitelaru and A. Vladescu, *Improvement of CoCr alloy characteristics by Ti-based carbonitride coatings used in orthopedic applications*, Coatings. 10 (2020), p. 495.
73. Anwar Ul-Hamid, *Microstructure, properties and applications of Zr-carbide, Zr-nitride and Zr-carbonitride coatings: a review,* Mater. Adv., 1 (2020), pp. 1012–1037.
74. D.K. Rajak, A. Kumar, A. Behera and P.L. Menezes, Diamond-*l*ike *c*arbon (DLC) *c*oatings: Classification, *p*roperties, and *a*pplications, Appl. Sci. 11 (2021), p. 4445. doi: 10.3390/app11104445
75. L.A. Ali, B. Dikici, N. Aslan, Y. Yilmazer, A. Sen, H. Yilmazer et al., *In-vitro corrosion and surface properties of PVD-coated β-type TNTZ alloys for potential usage as biomaterials: Investigating the hardness, adhesion, and antibacterial properties of TiN, ZrN, and CrN film*, Surf. Coatings Technol. 466 (2023), p. 129624.
76. M. Valletregi, *Calcium phosphates as substitution of bone tissues*, Prog. Solid State Chem, 32 (2004), pp. 1–31.
77. O. Yigit, B. Dikici and N. Ozdemir, *Hydrothermal synthesis of nanocrystalline hydroxyapatite–graphene nanosheet on Ti-6Al-7Nb: Mechanical and in vitro corrosion performance*, J. Mater. Sci. Mater. Med. 32 (2021), pp. 1–14.
78. J. Jeong, J.H. Kim, J.H. Shim et al., Bioactive calcium phosphate materials and applications in bone regeneration, Biomater Res. 23, no. 4 (2019). doi: 10.1186/s40824-018-0149-3
79. B. Dikici, M. Niinomi, M. Topuz, Y. Say, B. Aksakal, H. Yilmazer et al., *Synthesis and characterization of Hydroxyapatite/*TiO_2 *coatings on the β-type titanium alloys with different sintering parameters using sol-gel method*, Prot. Met. Phys. Chem. Surfaces. 54 (2018), pp. 457–462.
80. O. Yigit, B. Dikici, N. Ozdemir and E. Arslan, *Plasma electrolytic oxidation of Ti-6Al-4V alloys in nHA/GNS containing electrolytes for biomedical applications: The combined effect of the deposition frequency and GNS weight percentage*, Surf. Coatings Technol. 415 (2021), p. 127139.
81. R.B. Heimann and H.D. Lehmann, *Deposition, Structure, Properties and Biological Function of Plasma-Sprayed Bioceramic Coatings*, in Robert B. Heimann and Hans D. Lehmann (eds.), Bioceramic Coatings for Medical Implants, Wiley Online Library, 2015.
82. M. Sankar, S. Suwas, S. Balasubramanian and G. Manivasagam, *Comparison of electrochemical behavior of hydroxyapatite coated onto WE43 Mg alloy by electrophoretic and pulsed laser deposition*, Surf. Coatings Technol. 309 (2017), pp. 840–848.
83. O. Yigit, B. Dikici, M. Kaseem, M. Nakai and M. Niinomi, *Facile formation with HA/Sr–GO-based composite coatings via green hydrothermal treatment on β-type TiNbTaZr alloys: Morphological and electrochemical insights*, J. Mater. Res. 37 (2022), pp. 2512–2524.
84. Y. Yan, X. Zhang, H. Mao, Y. Huang, Q. Ding and X. Pang, *Hydroxyapatite/gelatin functionalized graphene oxide composite coatings deposited on* TiO_2 *nanotube by electrochemical deposition for biomedical applications*, Appl. Surf. Sci. 329 (2015), pp. 76–82.

85. S. Mallakpour and M. Okhovat, *Hydroxyapatite mineralization of chitosan-tragacanth blend/ZnO/ Ag nanocomposite films with enhanced antibacterial activity*, Int. J. Biol. Macromol. 175 (2021), pp. 330–340.
86. R. Sergi, D. Bellucci and V. Cannillo, A *r*eview of *b*ioactive glass/*n*atural *p*olymer *c*omposites: State of the *a*rt, Materials (Basel). 13, no. 23 (2020 Dec 6), p. 5560. doi: 10.3390/ma13235560. PMID: 33291305; PMCID: PMC7730917.
87. R. Pachaiappan, S. Rajendran, P.L. Show, K. Manavalan and M. Naushad, *Metal/metal oxide nanocomposites for bactericidal effect: A review*, Chemosphere. 272 (2021), p. 128607.
88. H.G. Augustin, K. Braun, I. Telemenakis, U. Modlich, W. Kuhn, Y. Feng et al., *Angiogenesis imaging methodology: AIM for clinical trials: The role of imaging in clinical trials of anti-angiogenesis therapy in oncology*, Complexity (2016).
89. M. Cannio, D. Bellucci, J.A. Roether, D.N. Boccaccini and V. Cannillo, Bioactive *g*lass *a*pplications: A *l*iterature *r*eview of *h*uman *c*linical *t*rials, Materials (Basel). 14, no. 18 (2021 Sep 20), p. 5440. doi: 10.3390/ma14185440. PMID: 34576662; PMCID: PMC8470635
90. D. Cadosch, E. Chan, O.P. Gautschi and L. Filgueira, Metal is not inert: *R*ole of metal ions released by biocorrosion in aseptic loosening--*C*urrent concepts, J. Biomed. Mater. Res. A. 91, no. 4 (2009 Dec 15), pp. 1252–1262. doi: 10.1002/jbm.a.32625. PMID: 19839047
91. S. Gupta, H. Gupta and A. Tandan, *Technical complications of implant-causes and management: A comprehensive review*, Natl. J. Maxillofac. Surg. 6 (2015), pp. 3–8.
92. W. Zhu, Z. Zhang, B. Gu, J. Sun and L. Zhu, *Biological activity and antibacterial property of nano-structured TiO_2 coating incorporated with Cu prepared by micro-arc oxidation*, J. Mater. Sci. Technol. 29 (2013), pp. 237–244.
93. A. Dubnika and V. Zalite, *Preparation and characterization kof porous Ag doped hydroxyapatite bio-ceramic scaffolds*, Ceram. Int. 40 (2014), pp. 9923–9930.
94. N.M. Zain, R. Hussain and M.R.A. Kadir, *Surface modification of yttria stabilized zirconia via polydopamine inspired coating for hydroxyapatite biomineralization*, Appl. Surf. Sci. 322 (2014), pp. 169–176.
95. Y. Bai, I.S. Park, S.J. Lee, T.S. Bae, W. Duncan, M. Swain et al., *One-step approach for hydroxyapatite-incorporated TiO2 coating on titanium via a combined technique of micro-arc oxidation and electrophoretic deposition*, Appl. Surf. Sci. 257 (2011), pp. 7010–7018.
96. L. Rathmann, T. Rusche, H. Hasselbruch, A. Mehner and T. Radel, *Friction and wear characterization of LIPSS and TiN/DLC variants*, Appl. Surf. Sci. 584 (2022), p. 152654.
97. R.B. More, A.D. Haubold and J.C. Bokros, *Pyrolytic carbon for long-term medical implants*, in Buddy D. Ratner, Allan S. Hoffman, and Jack E. Lemons (eds.), *Biomaterials Science: An Introduction to Materials, Third* Edition, Elsevier, 2013.
98. M. Ross, D. Williams, G. Couzens and J. Klawitter, *Pyrocarbon for joint replacement*, in Peter Revell (eds.), Joint Replacement Technology, Elsevier, 2021.
99. J.M. Adkinson and K.C. Chung, *Advances in small joint arthroplasty of the hand*, Plast. Reconstr. Surg. 134 (2014), pp. 1260–1268.
100. P.A. Nistor and P.W. May, *Diamond thin films: Giving biomedical applications a new shine*, J. R. Soc. Interface. 14 (2017), p. 20170382.
101. T.Y. Liao, A. Biesiekierski, C.C. Berndt, P.C. King, E.P. Ivanova, H. Thissen et al., *Multifunctional cold spray coatings for biological and biomedical applications: A review*, Prog. Surf. Sci. 97 (2022), p. 100654.
102. M.P. Nikolova and M.D. Apostolova, *Advances in multifunctional bioactive coatings for metallic bone implants*, Materials (Basel). 16 (2023), pp. 1–53.
103. B. Dikici and M. Topuz, *Production of annealed cold-sprayed 316L stainless steel coatings for biomedical applications and their in-vitro corrosion response*, Prot. Met. Phys. Chem. Surfaces. 54 (2018), pp. 333–339.
104. R. Palanivelu and A.R. Kumar, *Scratch and wear behaviour of plasma sprayed nano ceramics bilayer Al_2O_3 -13 wt%TiO_2/hydroxyapatite coated on medical grade titanium substrates in SBF environment*, Appl. Surf. Sci. 315 (2014), pp. 372–379.
105. M. Wang, *Composite coatings for implants and tissue engineering scaffolds*, in Luigi Ambrosio (eds.), *Biomedical Composites*, Woodhead Publishing Limited, 2009, pp. 127–177.
106. I. Ratha, P. Datta, V. K. Balla, S. K. Nandi and B. Kundu, *Effect of doping in hydroxyapatite as coating material on biomedical implants by plasma spraying method: A review*, Ceram. Int. 47, no. 4 (2021), pp. 4426–4445. doi: 10.1016/j.ceramint.2020.10.112
107. J. Henao, C. Poblano-Salas, M. Monsalve, J. Corona-Castuera and O. Barceinas-Sanchez, *Bio-active glass coatings manufactured by thermal spray: A status report*, J. Mater. Res. Technol. 8 (2019), pp. 4965–4984.

108. G. Prashar and H. Vasudev, *Thermal sprayed composite coatings for biomedical implants: A brief review*, J. Therm. Spray Eng. 2 (2020), pp. 50–55.
109. R.B. Heimann, *Plasma-sprayed hydroxylapatite-based coatings: Chemical, mechanical, microstructural, and biomedical properties*, J. Therm. Spray Technol. 25 (2016), pp. 827–850.
110. T.J. Levingstone, M. Ardhaoui, K. Benyounis, L. Looney and J.T. Stokes, *Plasma sprayed hydroxyapatite coatings: Understanding process relationships using design of experiment analysis*, Surf. Coatings Technol. 283 (2015), pp. 29–36.
111. S. Roy, *Functionally graded coatings on biomaterials: A critical review*, Mater. Today Chem. 18 (2020), p. 100375.
112. A. Motallebzadeh, *Evaluation of mechanical properties and in vitro biocompatibility of TiZrTaNbHf refractory high-entropy alloy film as an alternative coating for TiO_2 layer on NiTi alloy*, Surf. Coatings Technol. 448 (2022), p. 128918.
113. B. Gui, H. Zhou, J. Zheng, X. Liu, X. Feng, Y. Zhang et al., *Microstructure and properties of TiAlCrN ceramic coatings deposited by hybrid HiPIMS/DC magnetron co-sputtering*, Ceram. Int. 47 (2021), pp. 8175–8183.
114. M.A. Surmeneva, T.M. Mukhametkaliyev, H. Khakbaz, R.A. Surmenev and M. Bobby Kannan, *Ultrathin film coating of hydroxyapatite (HA) on a magnesium-calcium alloy using RF magnetron sputtering for bioimplant applications*, Mater. Lett. 152 (2015), pp. 280–282.
115. F. Songur, B. Dikici, M. Niinomi and E. Arslan, *The plasma electrolytic oxidation (PEO) coatings to enhance in-vitro corrosion resistance of Ti–29Nb–13Ta–4.6Zr alloys: The combined effect of duty cycle and the deposition frequency*, Surf. Coatings Technol. 374 (2019), pp. 345–354.
116. S. Sridhar, A. Viswanathan, K. Venkateswarlu, N. Rameshbabu and N.L. Parthasarathi, *Enhanced visible light photocatalytic activity of P-block elements (C, N and F) doped porous TiO_2 coatings on Cp-Ti by micro-arc oxidation*, J. Porous Mater. 22 (2015), pp. 545–557.
117. S. Mahmud, M. Rahman, M. Kamruzzaman, H. Khatun, M.O. Ali and M.M. Haque, *Recent developments in hydroxyapatite coating on magnesium alloys for clinical applications*, Results Eng. 17 (2023), p. 101002.
118. S. Noreen, A. Maqbool, I. Maqbool, A. Shafique, M.M. Khan, Y. Junejo et al., *Multifunctional mesoporous silica-based nanocomposites: Synthesis and biomedical applications*, Mater. Chem. Phys. 285 (2022), p. 126132.
119. R. Ahmadi and A. Afshar, *In vitro study: Bond strength, electrochemical and biocompatibility evaluations of TiO2/Al2O3 reinforced hydroxyapatite sol–gel coatings on 316L SS*, Surf. Coatings Technol. 405 (2021), p. 126594.
120. R.B. Figueira, *Hybrid sol-gel coatings for corrosion mitigation: A critical review*, Polymers (Basel). 12 (2020), pp. 9–12.
121. M. Topuz, *Hydroxyapatite – Al2O3 reinforced poly– (lactic acid) hybrid coatings on magnesium: Characterization, mechanical and in-vitro bioactivity properties*, Surfaces and Interfaces. 37 (2023), p. 102724.
122. M. Topuz, *Effect of ZrO2 on morphological and adhesion properties of hydroxyapatite reinforced poly– (lactic) acid matrix hybrid coatings on Mg substrates*, Res Eng. Struct. Mater. 8 (2022), pp. 721–733.
123. M. Topuz, B. Dikici, S. Güngör Koç, H. Yılmazer, M. Niinomi and M. Nakai, *Zirkonya Takviyeli Hidroksiapatit (HA) Bazlı Biyoaktif Hibrid Kaplamaların Korozyon Duyarlılıkları*, e-Journal New World Sci. Acad. 12 (2017), pp. 66–77.
124. A.C. Pierre, Introduction to Sol-Gel Processing, Vol. 1, The Kluwer International Series in Sol-Gel Processing: Technology and Applications, Springer US, Boston, MA, 1998.
125. L. Ling, S. Cai, Y. Zuo, M. Tian, T. Meng, H. Tian et al., *Copper-doped zeolitic imidazolate frameworks-8/hydroxyapatite composite coating endows magnesium alloy with excellent corrosion resistance, antibacterial ability and biocompatibility*, Colloids Surfaces B Biointerfaces. 219 (2022), p. 112810.
126. L. Calabrese, *Anticorrosion behavior of zeolite coatings obtained by in situ crystallization: A critical review*, Materials (Basel). 12 (2018), pp. 1–25.
127. J. Alipal, S. Saidin, H.Z. Abdullah, M.I. Idris and T.C. Lee, *Physicochemical and cytotoxicity studies of a novel hydrogel nanoclay EPD coating on titanium made of chitosan/gelatin/halloysite for biomedical applications*, Mater. Chem. Phys. 290 (2022), p. 126543.
128. H.A.Y. Al-mashhdani, Corrosion and Corrosion Protection Studies of Carbon Steel alloy in Seawater using; Zirconia, Silicon Carbide and Alumina Nanoparticles, University of Baghdad, 2014.
129. I. Corni, M.P. Ryan and A.R. Boccaccini, *Electrophoretic deposition: From traditional ceramics to nanotechnology*, J. Eur. Ceram. Soc. 28 (2008), pp. 1353–1367.

130. C.C. Kee, K. Ng, B.C. Ang and S.C. Metselaar, *Synthesis, characterization and in-vitro biocompatibility of electrophoretic deposited europium-doped calcium silicate on titanium substrate*, J. Eur. Ceram. Soc. 43 (2023), pp. 1189–1204.
131. V.S. Yadav, M.R. Sankar and L.M. Pandey, *Coating of bioactive glass on magnesium alloys to improve its degradation behavior: Interfacial aspects*, J. Magnes. Alloy. 8 (2020), pp. 999–1015.
132. T. Moskalewicz, M. Warcaba, A. Łukaszczyk, M. Kot, A. Kopia, Z. Hadzhieva et al., *Electrophoretic deposition, microstructure and properties of multicomponent sodium alginate-based coatings incorporated with graphite oxide and hydroxyapatite on titanium biomaterial substrates*, Appl. Surf. Sci. 575, no. 1–17 (2022), p. 151688.
133. M. Pagel and A.G. Beck-Sickinger, *Multifunctional biomaterial coatings: Synthetic challenges and biological activity*, Biol. Chem. 398 (2017), pp. 3–22.
134. I.O. Silva, R. Ladchumananandasivam, J.H.O. Nascimento, K.K.O.S. Silva, F.R. Oliveira, A.P. Souto et al., *Multifunctional chitosan/gold nanoparticles coatings for biomedical textiles*, Nanomaterials. 9 (2019), pp. 1–21.
135. R. Jain and D. Kapoor, *The dynamic interface: A review*, J. Int. Soc. Prev. Community Dent. 5 (2015), pp. 354–358.
136. A. Ładniak, M. Jurak and A.E. Wiącek, *Physicochemical characteristics of chitosan-TiO_2 biomaterial. 2. Wettability and biocompatibility*, Colloids Surf. A: Physicochem. Eng. Asp. 630, no. 1–14 (2021), p. 127546.
137. C.-H. Huang and M. Yoshimura, *Biocompatible hydroxyapatite ceramic coating on titanium alloys by electrochemical methods via growing integration layers [GIL] strategy: A review*, Ceram. Int. 49 (2023), pp. 24532–24540.
138. Y. Zhao, A. Li, L. Jiang, Y. Gu and J. Liu, *Hybrid membrane-coated biomimetic nanoparticles (HM@BNPs): A multifunctional nanomaterial for biomedical applications*, Biomacromolecules. 22 (2021), pp. 3149–3167.
139. P. Li, X. Li, W. Cai, H. Chen, H. Chen and R. Wang et al., *Phospholipid-based multifunctional coating via layer-by-layer self-assembly for biomedical applications*, Mater. Sci. Eng. C. 116 (2020), pp. 111237.
140. X. Chen, J. Zhou, Y. Qian and L.Z. Zhao, *Antibacterial coatings on orthopedic implants*, Mater. Today Bio. 19 (2023), p. 100586.
141. Z. Zhu, Q. Gao, Z. Long, Q. Huo, Y. Ge, N. Vianney et al., *Polydopamine/poly(sulfobetaine methacrylate) co-deposition coatings triggered by CuSO4/H2O2 on implants for improved surface hemocompatibility and antibacterial activity*, Bioact. Mater. 6 (2021), pp. 2546–2556.
142. S. Hassan, A.Y. Nadeem, M. Ali, M.N. Ali, M.B.K. Niazi and A. Mahmood, *Graphite coatings for biomedical implants: A focus on anti-thrombosis and corrosion resistance properties*, Mater. Chem. Phys. 290 (2022), p. 126562.
143. P. Tong, Y. Sheng, R. Hou, M. Iqbal, L. Chen and J. Li, *Recent progress on coatings of biomedical magnesium alloy*, Smart Mater. Med. 3 (2022), pp. 104–116.
144. D. Zhang, F. Peng, J. Qiu, J. Tan, X. Zhang, S. Chen et al., *Regulating corrosion reactions to enhance the anti-corrosion and self-healing abilities of PEO coating on magnesium*, Corros. Sci. 192 (2021), p. 109840.
145. F. Cemin, L. Luís Artico, V. Piroli, J. Andrés Yunes, C. Alejandro Figueroa and F. Alvarez, *Superior in vitro biocompatibility in NbTaTiVZr(O) high-entropy metallic glass coatings for biomedical applications*, Appl. Surf. Sci. 596, no. 1–11 (2022), p. 153615.
146. X. Lan, H. Zhang, H. Qi, S. Liu, X. Zhang and L. Zhang, *Custom-design of triblock protein as versatile antibacterial and biocompatible coating*, Chem. Eng. J. 454 (2023), p. 140185.
147. M. Du, M. Peng, B. Mai, F. Hu, X. Zhang, Y. Chen et al., *A multifunctional hybrid inorganic-organic coating fabricated on magnesium alloy surface with antiplatelet adhesion and antibacterial activities*, Surf. Coatings Technol. 384 (2020), p. 125336.
148. J. Woo, Y. Na, W. Il Choi, S. Kim, J. Kim, J. Hong et al., *Functional ferrocene polymer multilayer coatings for implantable medical devices: Biocompatible, antifouling, and ROS-sensitive controlled release of therapeutic drugs*, Acta Biomater. 125 (2021), pp. 242–252.
149. S. Qin, K. Xu, B. Nie, F. Ji and H. Zhang, *Approaches based on passive and active antibacterial coating on titanium to achieve antibacterial activity*, J. Biomed. Mater. Res. - Part A. 106 (2018), pp. 2531–2539.
150. L. Stillger and D. Müller, *Peptide-coating combating antimicrobial contaminations: A review of covalent immobilization strategies for industrial applications*, J. Mater. Sci. 57 (2022), pp. 10863–10885.
151. A. Dehghanghadikolaei and B. Fotovvati, *Coating techniques for functional enhancement of metal implants for bone replacement: A review*, Materials (Basel). 12, no. 1–23 (2019), p. 1795.

152. L.C. Ardelean and L.-C. Rusu, *Advanced biomaterials, coatings, and techniques: Applications in medicine and dentistry*, Coatings. 12 (2022), p. 797.
153. B. Aksakal, Y. Say and Z.A. Sinirlioglu, *Effects of silver/selenium/chitosan doped hydroxyapatite coatings on REX-734 alloy: Morphology, antibacterial activity, and cell viability*, Mater. Today Commun. 33 (2022), p. 104246.
154. T.C. Senocak, K.V. Ezirmik and S. Cengiz, *The antibacterial properties and corrosion behavior of silver-doped niobium oxynitride coatings*, Mater. Today Commun. 32 (2022), p. 103975.
155. J. Andrade Del Olmo, L. Pérez-Álvarez, V. Sáez Martínez, S. Benito Cid, L. Ruiz-Rubio, R. Pérez González et al., *Multifunctional antibacterial chitosan-based hydrogel coatings on Ti6Al4V biomaterial for biomedical implant applications*, Int. J. Biol. Macromol. 231 (2023), p. 123328.
156. C. Ma, Y. Duan, C. Wu, E. Meng, P. Li, Z. Zhang et al., *Spatiotemporally co-delivery of triple therapeutic drugs via HA-coating nanosystems for enhanced immunotherapy*, Asian J. Pharm. Sci. 16 (2021), pp. 653–664.
157. A.J. Nathanael and T.H. Oh, *Biopolymer coatings for biomedical applications*, Polymers (Basel). 12 (2020), p. 3061.
158. A. Trentin, A. Pakseresht, A. Duran, Y. Castro and D. Galusek, *Electrochemical characterization of polymeric coatings for corrosion protection: A review of advances and perspectives*, Polymers (Basel). 14 (2022), p. 2306.
159. I. Khan, K. Saeed and I. Khan, *Nanoparticles: Properties, applications and toxicities*, Arab. J. Chem. 12 (2019), pp. 908–931.
160. J.C. Kwan, J. Dondani, J. Iyer, H.A. Muaddi, T.T. Nguyen and S.D. Tran, *Biomimicry and 3D-printing of mussel adhesive proteins for regeneration of the periodontium—A review*, Biomimetics. 8 (2023), p. 78.
161. P.J. Sreelekshmi, V. Devika, M.M. Sreejaya, S. Sadanandan, M.S. Mathew, A. Saritha et al., *Biomaterials and biomimetics*, in *Antiviral and Antimicrobial Smart Coatings*, Elsevier, 2023, pp. 23–69.
162. F. Ordikhani, F. Mohandes and A. Simchi, *Nanostructured coatings for biomaterials*, in Mehdi Razavi and Avnesh Thakor (eds.), *Nanobiomaterials Science, Development and Evaluation*, Elsevier, 2017, pp. 191–210.

15 Multifunctional Coatings for Household and Other Applications

Khadija El-Moustaqim, Jamal Mabrouki, and Driss Hmouni

15.1 INTRODUCTION

Existing demands and scarcity of freshwater resources, and the expected water stress caused by current climate change, underline the need for environmentally sustainable technologies to treat wastewater and recover resources at the same time. Water is much more than just a human need [1]. It represents an irreplaceable and essential part of the human life cycle. However, it can also be a source of serious illness [2–4]. The impact of disease on the world's population is automatically linked to water, and the transmission of disease through water pollution does not stop at the level of developing countries [5]. According to the literature, each year, several million newborns and children suffer from diseases resulting from polluted food and drinking water [6, 7]. It may be noted that the lack of balance between availability and demand is the subject of water scarcity, which is of course caused by the rapid growth of urban populations, the irregular distribution of water resources, periods of severe drought, and surface water and groundwater pollution [8, 9].

As a key factor in the prevention of water-related diseases, drinking water must be given special attention. Water destined for human use must not contain hazardous chemical substances or germs harmful to health. This is why all the actors in the water sector must constantly monitor the quality of the groundwater resource, which conditions the safety and reliability of the drinking water supply, and continue to recover the quality of surface water wherever it collects. Furthermore, the use of water from either open public wells or individual unprotected wells, coupled with inadequate sanitation facilities and ignorance of basic hygiene rules, favors the propagation of fecal-oral infections and may cause serious conditions like gastroenteritis, hepatitis, and typhoid. Given that the rapid increase in the human population produces a large amount of wastewater, contaminating rivers and reservoirs, freshwater must be treated as a scarce, precious resource. Not only climate change, but also ecosystem depletion, the inadequate deployment of sustainable assets, and the lack of ecologically sound management are inextricably linked to the reduction of watercourses through low groundwater flow. This reduced flow degrades reservoirs, adversely impacting water supply and accessibility [10]. The world today is facing many challenges, such as climate and global change; people have needs for energy and raw materials, and at the same time, they need more secure borders.

The utility of the Internet of Things (IoT) is known from the detection of microalgae species suitable for the extraction of biofuels during downstream processing. IoT is based on the detection, transfer, and processing of data by a process based on sensors and machine learning, which can be adapted and adjusted in real time when changes are detected. In the field of environmental chemical analysis, IoT is the foundation stone of a system founded on many connected intelligent sensors capable of measuring environmental parameters. That, combined with cloud computing, is able to provide data in a range that can fluctuate from microscopic to global [11, 12]. Therefore, smart, distributed sensing systems are of paramount importance in the IoT context and are relevant to wearable sensors for environmental monitoring. The current application of intelligent wireless sensors in real-world applications, however, faces major issues such as the consumption of energy, cost-effectiveness, and, naturally, the assurance of an adequate service life [13].

DOI: 10.1201/9781032635347-15

This chapter discusses a general study on how to make the most of the Internet of Things for wastewater monitoring systems to detect various types of particles in wastewater. In the second part of this chapter, we will approach a study in the form of a literature review on the evolution of different techniques used in IoT-based monitoring systems to detect various materials. All we discuss are their main properties and their potential.

15.2 IoT AND WASTEWATER

The complexity of environmental problems is growing as a result of extreme weather conditions, new contaminants, evolving landscapes, global climate change, and other factors. To characterize and model these problems, it is generally required to treat and store large volumes of data in real time. Developments in sensor and communication technologies, such as IoT, have enabled on-site monitoring to be coupled to remote capabilities for data processing [14, 15]. The importance of IoT from an industrial point of view becomes clear when devices connected to the global network of machines can interact with each other for customer support systems, intelligence applications for businesses, inventory systems for vendors based on requirements, and also to enable analytics operations to be performed for the enterprise [1, 11]. The technology sector has undergone a veritable evolution in the last few years. It has also become a must-have tool in our daily lives. One of the world's most recent technologies is IoT. The number of connected devices increases every day [16]. Such growth is having a positive impact on a number of sectors, including home security, smart grids, and so on. Using technology for water quality monitoring and management, the use of remote wireless sensors on IoT network installations enhances existing centralized systems and conventional processes, providing for shared intelligence on water quality systems to accommodate growing urban dynamics and disparate water availability networks [17]. City environmental infrastructures, such as water pollution, air pollution, solid waste, and transport, have not been designed for the significant population growth and waves of migration that often occur, and the dynamics of cities prevent the rebuilding of these infrastructures. However, urban infrastructures are expected to meet the needs of an ever-growing population, and it is therefore necessary to find intelligent solutions that do not entail high costs [18]. The term IoT describes an emerging wide-area network that allows any device to be connected to the Internet, exchanging data and working as a remote monitor (Figure 15.1) [19]. The technology for secure communication is designed to transmit data from a physical device to a smart device analytical tool via a wireless channel. It has been developed to enhance the protection and quality of wastewater treatment. Remote IoT relies upon connecting from a physical setting to an Internet connection for the purpose of monitoring and controlling wastewater [20, 21]. To monitor water in connected environments, IoT technologies are deployed with transducers. Monitoring water assets through IoT techniques can address the main barriers to

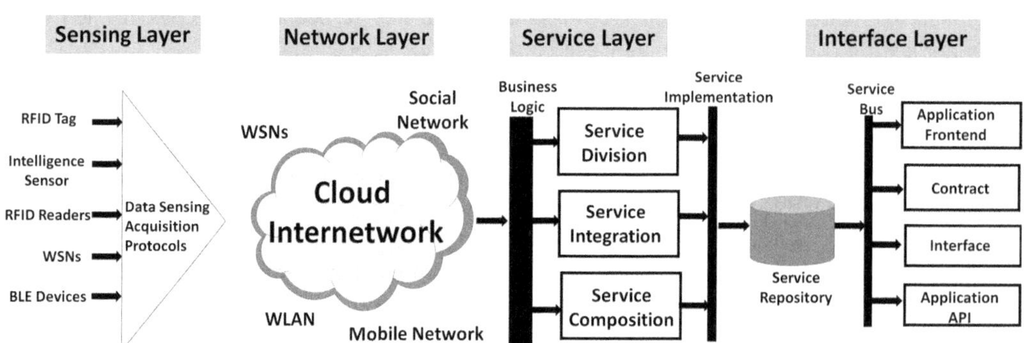

FIGURE 15.1 Architectural layers of IoT.

effective water management. Conventional monitoring methods can address water use, efficiency, quantitative detection, and treatment. Consequently, storage and condition indicators could be set up in sewage and wastewater treatment facilities. Furthermore, IoT provides effective control of wireless connections and remote water monitoring bodies, contaminants, data on water, ground use, sources, and surface runoff. The ability of IoT water management to deliver massive treatment capabilities and environmental, structural, and ecological attributes of various forms of media, is considerable [22, 23]. As such, it is a catalyst for creativity to incorporate detection processes for key decision-making, comprising sensor technology, remote sensing and more, and it can help control the commentary on the performance of water by deploying innovative technologies to safeguard people's health [12, 13, 24].

To better understand the architecture of IoT, one of its essential requirements is that the objects in the network must be connected to each other. IoT operation depends on the system architecture, which connects the physical and virtual worlds. When designing the IoT architecture, scalability, extensibility, and operability of devices should be considered. Objects can move, and they need to interact with others in real time [25]. Worldwide, protecting the environment against a wide range of hazards has rapidly become a major priority. Environmental parameters such as air quality and many others can be continuously monitored using the Internet of Things. Wastewater treatment systems are frequently used to eliminate and treat pollutants in effluent, allowing them to be applied directly to the water source with limited impact on the environment. A reliable wastewater treatment system is designed to address environmental changes directly. The key issue lies in monitoring waste disposal in treatment plants [26]. Multimedia sensor devices are also increasingly common as devices are widely rolled out, and new degrees of integration among multi-sensors and other devices are offering new opportunities. In addition, sensor technology, through emerging fields such as artificial intelligence and IoT, is now finding many applications in the design of future systems. Clearly, an even more autonomous and intelligent sensor future is sustainable and will offer new opportunities for operational effectiveness and cost reduction, as well as the creation of new capabilities to protect the planet's environment, human health, and general well-being in industry 4.0 (Figure 15.2).

Although the current IoT infrastructure can be applied to wastewater management, the existing technology is better adapted to consumer applications with a low-risk impact factor. But IoT technology

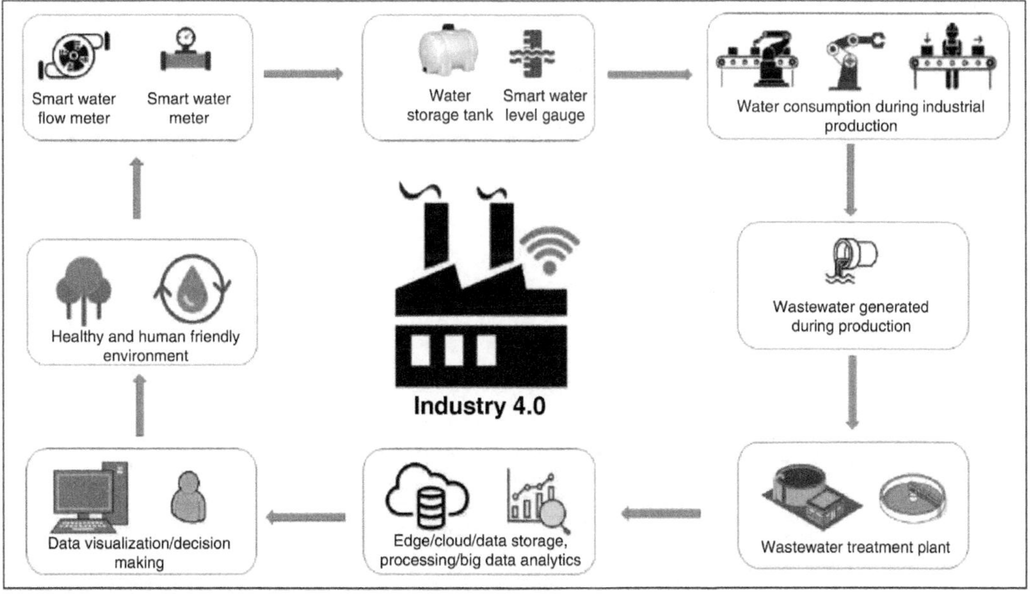

FIGURE 15.2 The Industry Access 4.0 approach to industrial process wastewater engineering [27].

could facilitate the evolution of a large number of transactions, drive better management of manufacturing waste, and help track waste flows with no risk of data falsification [28]. The concept of smart sensors is a new technology for the processing and capture of digital signals. The advantage of these systems over conventional sensors is that as well as the sensor hardware, the signal conditioning, which can be implemented via an electrical network to transform the analog signal of the sensor into a temporal, complex frequency or digital signal, is also integrated, along with a processing element such as a meter or a microcontroller. This means that a smart sensor can be linked to a digital or quasi-digital network bus, allowing faster, more effective information transfer. It also enables multiple sensors to be linked to a central network, and a central element assigned to each sensor to determine its identity [29]. Environmental sensors are essential to enable a more networked world. From providing information about our direct environment to helping combat climate change worldwide, sensors and sensor networks are transforming the way we experience the environment surrounding us. More and more intelligent sensors are being deployed in environmental, industrial, commercial, and personal spaces to enable new approaches to security, control, and surveillance. New developments are taking shape in a number of areas: sensor abilities are being rapidly extended, while precision and accuracy are being enhanced, and power consumption is reducing.

15.3 THE EFFECTS OF TOXIC CHEMICALS

Particular toxic chemicals emitted into the air, like benzene or vinyl chloride, can cause cancer, birth defects, long-term lung damage, and damage to the brain and nervous system. In addition, in some cases, inhalation of such chemicals can even lead to death. Some pollutants rise into the atmosphere, resulting in a thinning of the Earth's protective ozone layer. This phenomenon has led to modifications in the environment and a dramatic increase in skin cancer and age-related cataracts [30]. The development of our societies has led to the increased production of potentially toxic chemical molecules, spilled accidentally or intentionally into the environment. As early as the 1930s, a link between air pollution and respiratory pathologies was established, but it was not until the early 1950s that the first scientific reports on the chemical contamination of food appeared. The contamination of food revealed by frequent health scandals produces increased vigilance by populations and health authorities. The toxicity of many food contaminants on human cells and organs is now better and better documented [31]. Chemical contaminants' toxic effects on organisms is influenced by their bioavailability and durability, the capacity of organisms to collect and process contaminants, and the potential interplay of contaminants with ecological, metabolic, or specific functions. In recent years, an extensive range of toxicity test methods has been deployed to enable rapid and precise detection of the adverse effects of chemical contaminants (Figure 15.3) [32]. Such contaminants are also highly toxic because they are rarely biodegradable and are generally fragmented into huge quantities of macro, micro, and

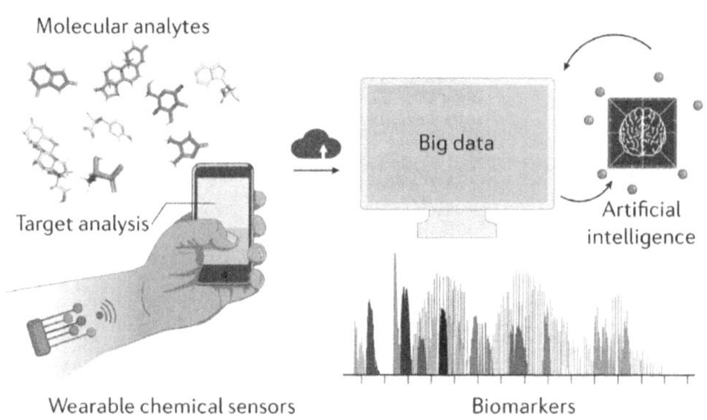

FIGURE 15.3 Chemical sensor array.

nanoparticles. By various different processes and natural circumstances, they become ubiquitous pollutants and a danger to the global environment [33].

15.4 THE EVOLUTION OF WATER BIOSENSING TECHNIQUES

An intelligent farm-scale piggery wastewater treatment system (SFS-PWT) was presented by Jung-Jeng Su et al. [34]. Conventional wastewater management processes for swine barns are primarily handled manually by skilled workers. With SFS-PWT, IoT technologies from a 1000-pig farm are applied to produce an advanced intelligent wastewater treatment process that is brought up to date by a fully automated, self-designed wastewater treatment system. Based on sensor data collected before and since the water performance data was measured and found the biochemical oxygen demand (BOD), chemical oxygen demand (COD), and suspended solids (SS) for the pig farm's wastewater, simultaneously.

L. Campanella et al. [35] developed a method for monitoring toxicity in estuarine waters. To achieve this, a suitable algal bioreceptor (the cyanobacterium *Spirulina subsalsa*) was coupled to a gas diffusion amplifying electrode. The sensor was used to monitor changes in photosynthetic O_2 and to detect modifications due to the toxic effects of pollutants detected in the surrounding environment. Using standardized natural water as a test medium, a series of four chemical species were tested at various concentrations, covering three major categories of contaminants (heavy metals, triazine herbicides, and carbamate insecticides). In each case, a toxic effect was detected, that is, a dose-dependent suppression of the photosynthetic activity was registered with excellent reproducibility. When toxic agents were present or absent, an automated system was developed to monitor a suitable bioreceptor's photosynthetic activity, monitoring the oxygen generated by means of a lifetime of at least 2 days. Replicable dose-dependent responses were detected in the presence of toxic agents. The suggested biosensor showed a high sensitivity from atrazine and intermediate sensitization to carbaryl, while toxic heavy metal detection was a long process. This might have been caused by a biological factor linked to the cell's adsorption surfaces and the specific sequestration, metabolization, and release pathways from the algal organisms involved.

Shumei Wang et al. suggested a system for online water quality monitoring (OWQMS) [36]. This is a key environmental component of IoT, comprising an onboard subsystem for monitoring water quality, and implementing a transmission subsystem for digital data and data processing elements, in order to operate an urban tourist stream. This included recycled municipal wastewater and freshwater from the surface of the existing Xinglin canal to maintain the river's water level, and the water system's landscape cycle to maintain the examined water quality. Subsequently, they demonstrated that no deterioration of the water had been found during this development period, further evidence that the OWQMS system had effectively managed the scenic river and maintained stable water quality in the landscape over the past four years. Processing performance yield results suggested that the campus treatment plant was effective in degrading the organic carbon. As regards ammoniacal nitrogen, through mixing directly with laboratory wastewater, the influent was maintained at a relatively low level with wastewater and a relatively low concentration of between 1.27 and 12.33 mg L^{-1}. This enabled the authors to confirm that the OWQMS system had been successful in managing the picturesque stream and maintaining the stable quality of the water. A high degree of self-cleansing of pollutants by the water body, particularly the degradation of nitrogen and phosphate, was noted, while deterioration of the water remained undetected throughout the period. Reclaimed wastewater must be treated on-site in the humid zone before being observed discharging into on-campus streams. Recovered wastewater must be processed at the treatment wetland before being observed entering the riverside on campus. The findings demonstrate that after treatment in the wetland, various parameters of water quality attained the required standard, suggesting recycling of water from the river is an appropriate means of stabilizing river quality in the years to come.

A wastewater treatment monitoring system (STMS) was proposed by Chunbo Xiu et al. [37]. An IoT-enabled STMS was introduced to enhance the process productivity of industrial commercial treatment of wastewater and experiences of incomplete operative guidance. The setup consists of a sensor component and an enabling environment. The detection layer captures plant variables from wastewater via a range of different sensors and device requirements. The STMS provides a holistic and highly reliable practical approach to the activity and servicing of wastewater treatment facilities.

The automated water quality monitoring (AWQM) system was reviewed by Rizqi Putri Nourma Budiarti et al. [38]. Here, AWQM monitored the quality of water sources, including rivers, lakes, and pools in relation to water consumption. Naturally, as a result of various water contaminations, water is increasingly becoming a key environmental resource. Thanks to the AWQM system, we can ensure the sustainability of water's functionality as a resource and control water quality. With an integrated IoT-based solution, AWQM can measure water levels through sensor-based environmental management systems.

Due to demographic increase, the urbanization of rural areas, and the extensive exploitation of marine resources for salt extraction, the exhaustion of water resources is becoming critical, leading to an increase in water demand. Varcha Lakchmikantha et al. [39] propose an intelligent system to monitor water quality. This system is composed of sensors that allow measurement of the values in real time, accompanied by an Arduino ATMEGA 328 software that can convert analog values into digital ones, with an LCD display indicating the sensor outputs. The connection was ensured by wireless access module between the equipment and associated software. They showed that after testing the three water samples and depending on their results, they were able to classify the water as potable or not. Hence they recommended the use of these sensors for detecting various quality parameters, utilizing enhanced wireless connectivity and IoT standards to create a more accurate monitoring system with the goal of securing an immediate response.

Marielle Thomas presented in her research thesis a new biological detection process. Its principle is based on the automatic characterization of electrical signals by a tropical fish. This work dealt with the influence of several physicochemical parameters (temperature, pH, conductivity, and dissolved oxygen) on the characteristics of the studied material concerning six polluting substances. The implementation of this new process allowed the detection in less than 35 minutes.

The development of intelligent detectors means that we will probably be able to better comprehend our surroundings and the changes taking place within them. Detection devices are improving rapidly toward more intelligent and independent modes of operation. However, the key to this effectiveness lies in the correct processing of sensory information. In order to provide real information, data must be gathered from multiple sources. Trend data is also sometimes more precious than one-off observations. These trends are all set to transform our world in completely unprecedented directions. Collectively, advances in cloud systems, sensor technology, wearable devices, and smartphones are all likely to provide greater insight into the environments around us in which we live, work, and play. We can easily envisage a world in which sensor networks send information about air pollution levels, weather conditions, radiation, and water quality in our immediate environment directly to our smartphones. Though some of these applications are city-based, there is an equally significant set of data feeds that will become increasingly available in nature and environmentally sensitive areas, such as forest fire monitoring, ozone level detection, melt rate, and ice and salinity detection. This data will not only enable us to monitor environments worldwide, it will also enable us to comprehend them in a completely different way.

15.5 CONCLUSION

The present state of research is focused on water pollution, which is considered a major threat to the world, affecting health, the economy, and altering biodiversity. Therefore, in this work, the cause-effects of water pollution are being discussed on the basis of a literature review of works that have adopted different water monitoring techniques, and which are IoT-based methods addressing water quality. While there have been numerous excellent intelligent water quality monitoring systems, the research field is still challenging. In this chapter, we highlight examples of recent work by scientists to improve the intelligence of water quality monitoring systems, as well as their efficiency in identifying water toxicity. Such systems feature low power consumption so that data monitoring, streaming, and reporting are forwarded to required position for more detailed treatment. Such systems have proved highly effective in detecting water quality. The latest trends suggest that "smart" sensors are being developed that can be easily deployed, not only in rivers and lakes but also in water supply systems and connected to different types of wastewater treatment plants. These sensors have the significant benefit of making the information they provide accessible to the general public via the Internet.

REFERENCES

1. N. Chamara, M. D. Islam, G. (Frank) Bai, Y. Shi, and Y. Ge, "Ag-IoT for crop and environment monitoring: Past, present, and future," *Agric. Syst.*, vol. 203, p. 103497, 2022, doi: 10.1016/j.agsy.2022.103497.
2. D. K. Lamprea Maldonado, "Caractérisation et origine des métaux traces, hydrocarbures aromatiques polycycliques et pesticides transportés par les retombées atmosphériques et les eaux de ruissellement dans les bassins versants séparatifs péri-urbains. Hydrologie. Ecole Centrale de Nantes (ECN)", 2009. Français. ⟨NNT:⟩. ⟨tel-00596847⟩.
3. D. Hilborn, "AÉRATION DU FUMIER LIQUIDE," p. 8 https://www.ontario.ca/fr/page/aeration-du-fumier-liquide#:~:text=A%C3%A9ration%20naturelle,5%20m%20(5%20pi%20).
4. Y. A. Idrissi, A. Alemad, S. Aboubaker, H. Daifi, K. Elkharrim, and D. Belghyti, "Caractérisation physico-chimique des eaux usées de la ville d'Azilal -Maroc- [physico-chemical characterization of wastewater from Azilal city -Morocco-]," *Afrique Science: Revue Internationale des Sciences et Technologie.*, AJOL vol. 11, n° 3, p. 12, 2015.
5. Z. Lei *et al.*, "Biochar enhances the biotransformation of organic micropollutants (OMPs) in an anaerobic membrane bioreactor treating sewage," *Water Res.*, vol. 223, p. 118974, 2022, doi: 10.1016/j.watres.2022.118974.
6. C. Chevrier, C. Petit, G. Limon, C. Monfort, G. Durand, S. Cordier, Biomarqueurs urinaires d'exposition aux pesticides des femmes enceintes de la cohorte Pélagie réalisée en Bretagne, France (2002–2006), BEH Hors-série / 16 juin 2009, p. 23–27.
7. X. Wan, M. Lei, and T. Chen, "Cost–benefit calculation of phytoremediation technology for heavy-metal-contaminated soil," *Sci. Total Environ.*, vol. 563–564, p. 796–802, 2016, doi: 10.1016/j.scitotenv.2015.12.080.
8. R. Hasan, "Bioremediation of swine wastewater and biofuel potential by using *Chlorella vulgaris*, *Chlamydomonas reinhardtii*, and *Chlamydomonas debaryana*," *J. Pet. Environ. Biotechnol.*, vol. 05, n° 03, 2014, doi: 10.4172/2157-7463.1000175.
9. H. Rajhi and A. Bardi, "Chapter 4 - Phytoremediation of endocrine disrupting pollutants in industrial wastewater," in *Current developments in biotechnology and bioengineering*, I. Haq, A. Kalamdhad, and A. Pandey, Ed., Elsevier, 2023, p. 55–84. doi: 10.1016/B978-0-323-91902-9.00002-X.
10. M. Achalhi, "Chronostratigraphie et sédimentologie des bassins néogènes de Boudinar et d'Arbaa Taourirt (Rif oriental, Maroc)," PhD Thesis, Université Mohammed Premier, Faculté des sciences Oujda (Maroc), 2016.
11. M. Amini and S. Chang, *A review of machine learning approaches for high dimensional process monitoring.* 2018. *Conference: Proceedings of the 2018 Industrial and Systems Engineering Research Conference At*: Orlando, FL.
12. A. Kamilaris, A. Kartakoullis, and F. X. Prenafeta-Boldú, "A review on the practice of big data analysis in agriculture," *Comput. Electron. Agric.*, vol. 143, p. 23–37, 2017, doi: 10.1016/j.compag.2017.09.037.
13. A. Yusuf *et al.*, "A review of emerging trends in membrane science and technology for sustainable water treatment," *J. Clean. Prod.*, vol. 266, p. 121867, 2020, doi: 10.1016/j.jclepro.2020.121867.
14. A. Pekar, J. Mocnej, W. K. G. Seah, and I. Zolotova, "Application domain-based overview of IoT network traffic characteristics," *ACM Comput. Surv.*, vol. 53, n° 4, p. 1–33, 2020, doi: 10.1145/3399669.
15. M. Durresi, A. Subashi, A. Durresi, L. Barolli, and K. Uchida, "Secure communication architecture for internet of things using smartphones and multi-access edge computing in environment monitoring," *J. Ambient Intell. Humaniz. Comput.*, vol. 10, n° 4, p. 1631–1640, 2019, doi: 10.1007/s12652-018-0759-6.
16. K. Samhat, "Contribution à l'optimisation de la production d'astaxanthine en photobioréacteur à partir de la microalgue Haematococcus pluvialis," Nantes Université; Université Libanaise, 2023.
17. S. A. Razzak, S. A. M. Ali, M. M. Hossain, and H. deLasa, "Biological CO_2 fixation with production of microalgae in wastewater–a review," *Renew. Sustain. Energy Rev.*, vol. 76, p. 379–390, 2017.

18. M. Y. Salman and H. Hasar, "Review on environmental aspects in smart city concept: Water, waste, air pollution and transportation smart applications using IoT techniques," *Sustain. Cities Soc.*, vol. 94, p. 104567, 2023, doi: 10.1016/j.scs.2023.104567.
19. J. Person, "Algues, filières du futur," *Livre Turquoise Adebiotech*, 2010.
20. M. K. Kagita, N. Thilakarathne, D. S. Rajput, and D. S. Lanka, "A detail study of security and privacy issues of internet of things." arXiv, 2020. doi: 10.48550/arXiv.2009.06341.
21. A. S. Rajawat, K. Barhanpurkar, R. N. Shaw, and A. Ghosh, "Chapter five - IoT in renewable energy generation for conservation of energy using artificial intelligence," in *Applications of AI and IOT in renewable energy*, R. N. Shaw, A. Ghosh, S. Mekhilef, and V. E. Balas, Eds., Academic Press, 2022, p. 89–105. doi: 10.1016/B978-0-323-91699-8.00005-X.
22. M. Azrour, J. Mabrouki, A. Guezzaz, and Y. Farhaoui, "New enhanced authentication protocol for internet of things," *Big Data Min. Anal.*, vol. 4, n° 1, p. 1–9, 2021, doi: 10.26599/BDMA.2020.9020010.
23. M. Azrour, J. Mabrouki, A. Guezzaz, and A. Kanwal, "Internet of things security: Challenges and key issues," *Secur. Commun. Netw.*, vol. 2021, p. 1–11, 2021, doi: 10.1155/2021/5533843.
24. F. Giannino, S. Esposito, M. Diano, S. Cuomo, and G. Toraldo, " A predictive decision support system (DSS) for a microalgae production plant based on internet of things paradigm," *Concurr. Comput. Pract. Exp.*, vol. 30, n° 15, p. e4476, 2018, doi: 10.1002/cpe.4476.
25. W. Zhang, F. Ma, M. Ren, and F. Yang, "Application with internet of things technology in the municipal industrial wastewater treatment based on membrane bioreactor process," *Appl. Water Sci.*, vol. 11, n° 3, p. 52, 2021, doi: 10.1007/s13201-021-01375-8.
26. P. M. Kumar and C. S. Hong, "Internet of things for secure surveillance for sewage wastewater treatment systems," *Environ. Res.*, vol. 203, p. 111899, 2022, doi: 10.1016/j.envres.2021.111899.
27. "Industry 4.0 Supported by Machine Learning | Bench Talk." https://www.mouser.com/blog/industry-40-supported-by-machine-learning (accessed August 19 2023).
28. J. Mabrouki, M. Azrour, D. Dhiba, Y. Farhaoui, and S. El Hajjaji, "IoT-based data logger for weather monitoring using Arduino-based wireless sensor networks with remote graphical application and alerts," *Big Data Min. Anal*, vol. 4, n° 1, p. 25–32, 2021.
29. V. Garrido-Momparler and M. Peris, "Smart sensors in environmental/water quality monitoring using IoT and cloud services," *Trends Environ. Anal. Chem.*, vol. 35, p. e00173, 2022, doi: 10.1016/j.teac.2022.e00173.
30. L. Cheng *et al.*, "Towards minimum-delay and energy-efficient flooding in low-duty-cycle wireless sensor networks," *Comput. Netw.*, vol. 134, p. 66–77, 2018, doi: 10.1016/j.comnet.2018.01.012.
31. S. Hakak, W. Z. Khan, G. A. Gilkar, N. Haider, M. Imran, and M. S. Alkatheiri, "Industrial wastewater management using blockchain technology: Architecture, requirements, and future directions," *IEEE Internet Things Mag.*, vol. 3, n° 2, p. 38–43, 2020, doi: 10.1109/IOTM.0001.1900092.
32. S. Comtet-Marre, P. Mosoni, and P. Peyret, "Effets des polluants environnementaux et alimentaires sur le microbiote intestinal," *Cah. Nutr. Diététique*, vol. 55, n° 5, p. 255–262, 2020 doi: 10.1016/j.cnd.2020.07.004.
33. J. McDowell Capuzzo, M. N. Moore, and J. Widdows, "Effects of toxic chemicals in the marine environment: Predictions of impacts from laboratory studies," *Aquat. Toxicol.*, vol. 11, n° 3–4, p. 303–311, 1988, doi: 10.1016/0166-445X(88)90080-X.
34. J.-J. Su, S.-T. Ding, and H.-C. Chung, "Establishing a smart farm-scale piggery wastewater treatment system with the internet of things (IoT) applications," *Water*, vol. 12, n° 6, p. 1654, 2020.
35. L. Campanella, F. Cubadda, M. P. Sammartino, and A. Saoncella, "An algal biosensor for the monitoring of water toxicity in estuarine environments," *Water Res.*, vol. 35, n° 1, p. 69–76, 2001, doi: 10.1016/S0043-1354(00)00223-2.
36. S. Wang, Z. Zhang, Z. Ye, X. Wang, X. Lin, and S. Chen, "Application of environmental internet of things on water quality management of urban scenic river," *Int. J. Sustain. Dev. World Ecol.*, vol. 20, n° 3, p. 216–222, 2013, doi: 10.1080/13504509.2013.785040.
37. C. Xiu and L. Dong, "Design of sewage treatment monitoring system based on internet of things," in *2019 Chinese Control and Decision Conference (CCDC)*, IEEE, 2019, p. 960–964.
38. R. P. N. Budiarti, A. Tjahjono, M. Hariadi, and M. H. Purnomo, "Development of IoT for automated water quality monitoring system," in *2019 International Conference on Computer Science, Information Technology, and Electrical Engineering (ICOMITEE)*, IEEE, 2019, p. 211–216.
39. V. Lakshmikantha, A. Hiriyannagowda, A. Manjunath, A. Patted, J. Basavaiah, and A. A. Anthony, "IoT based smart water quality monitoring system," *Glob. Transit. Proc*, vol. 2, n° 2, p. 181–186, 2021, doi: 10.1016/j.gltp.2021.08.062.

Index

U

V

W

X

Y

Z

For Product Safety Concerns and Information please contact our EU representative GPSR@taylorandfrancis.com Taylor & Francis Verlag GmbH, Kaufingerstraße 24, 80331 München, Germany

Batch number: 10392095

Printed by Printforce, the Netherlands